Inland Fishes and Fisheries of NAMIBIA

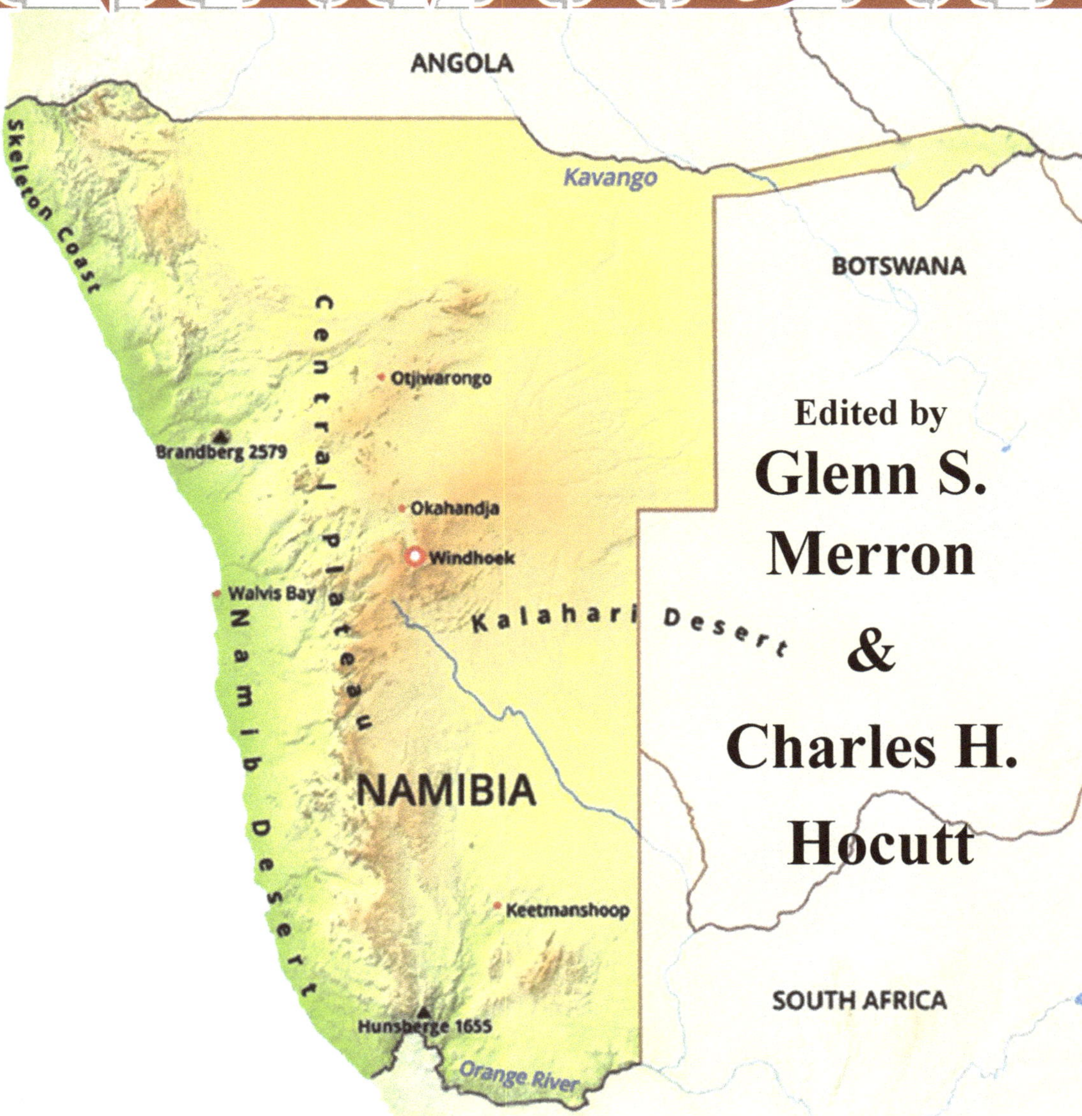

Edited by
Glenn S. Merron
&
Charles H. Hocutt

Splash Books

For information or additional copies, contact LeRue Press (LRP), 280 Greg Street, Suite 10, Reno, NV 89502

Aerial view of Namibia by Hellmut von Seydlitz

Cover design by LeRue Press, LLC and Adam Rice

www.Lrpnv.com

Library of Congress Number: 2022906914

ISBN: 978-1-938814-38-9

First Edition, 2023

Merron, Glenn S. and Charles H. Hocutt (Eds.). 2023. Inland Fishes and Fisheries of Namibia. LeRue Press, LLC, Splash Books, Reno NV: 367 pp.

DEDICATION

It is with great satisfaction that ***INLAND FISHES AND FISHERIES OF NAMIBIA*** is dedicated to Dr. Burger Wilhelm Oelofsen, former Director of Fisheries Resources, Ministry of Fisheries & Marine Resources (MFMR), Republic of Namibia. Burger has been an outstanding friend, mentor and colleague to so many of us who have contributed to this tome. His leadership in fisheries during the Republic's formative years as an independent nation cannot be overstated. He remains admired throughout government, academia, private industry and international fisheries sectors.

Burger, we all know him that way, was born on 9 December 1945 in Noupoort, South Africa, an 8th generation descendant from the first Oelofsen who arrived in Cape Town in 1685 from Norway. He is very proud of his dual heritage.

Burger received a Bachelor of Science degree from Stellenbosch University (SU) in 1968; Honour's Degree in Zoology in 1970, graduating Cum Laude from the University of Port Elizabeth (now the Nelson Mandela Metropolitan University); Master's Degree in Zoology in 1973 from SU, graduating Cum Laude; and Ph.D. in Palaeoecology in 1982 from SU. Thereafter, he was a Senior Lecturer at SU until 1987 when he joined the MFMR as the Head, Marine Research and later was promoted as Director of Research in 1996.

At the MFMR, Burger's duties were wide and varied having served the Ministry prior to Independence in 1990, then three terms under Founding President Samuel Nujoma (1990–2005) and two terms with President Hifikepunye Pohamba from 2005 to 2013 before retiring. He remains an advisor to both the Ministry and private fishing sector.

Burger's accomplishments are too numerous to mention. By example, he drafted the first Fresh Water Fisheries Act for the Republic; was Leader of the Namibian delegation that negotiated the South East Atlantic Fisheries Organisation (SEAFO) Treaty; initiated and successfully negotiated with donors to develop the tripartite Benguela Environment Fisheries Interaction and Training (BENEFIT) Programme between Angola, Namibia and South Africa that evolved into the Benguela Current Large Marine Ecosystem (BCLME) Programme; and successfully promoted the development of modern research facilities such as the Kamutjonga Inland Fisheries Institute (KIFI).

It is for these reasons…and so many others…that this book is dedicated to such a fine leader by example.

Burger with his beloved wife, Antonette (Tonie), who married in 1970 and together have five children.

FOREWORD

It is indeed a privilege and a great honour to be involved in this groundbreaking contribution to our knowledge of the freshwater fishes of Namibia, including important relevant management requisites and factors like dispersal, ecology, conservation and responsible utilisation.

The Government's mandate to manage our country's freshwater fish initially rested with Namibia's nature conservation authorities. During those early years there was a well-established management component for marine resources, but that function was later transferred to South Africa, at that time the mandate-holder in terms of the United Nations. Fortunately, during the pre-independence period, a sector to manage both marine resources and freshwater fisheries was established under the auspices of the South West Africa/Namibia Administration.

As Director of Fisheries during those years, I was directly involved in the process of establishing a management cadre. It grew from a component of three staff members to a fully-fledged department, which became the Ministry of Fisheries and Marine Resources (MFMR) after Namibia's Independence. Suddenly, management of the resources of Namibia's Exclusive Economic Zone became the responsibility of the new government and because of the economic importance thereof, the responsible utilisation and conservation of those resources were treated as a priority. Soon, during that initial period, management of the marine resources was so successful that an official of the World Bank considered it to be 'one of the best regulated fisheries of the world'.

Unfortunately, during those initial stages, attention to the freshwater fishes was secondary. However, this situation changed with the appointment of dedicated scientists like Dr. Ben van Zyl and Dr. Clinton Hay, supported by Dr. Burger Oelofsen.

Namibia is a vast country of more than 824,000 km^2, varying geographically from a real virgin desert to areas with relatively high rainfall. Namibia and large areas of southern Africa are gradually getting dryer and probably also warmer. There is ample evidence of past perennial and ephemeral lakes which now are dry pans. Some riverbeds end between sand dunes with no outflow either to other rivers or to the sea. There were even glaciers in Namibia. No wonder there are endemic relicts like *Clarias cavernicola* which consequently have extremely limited distributions. Such climatic changes, although it occurred over many thousands of years, present formidable challenges to our scientists to manage present-day species against the background of a changing environment.

A changing climatic environment presents challenges, but also does the circumstance that our country is home to an ever increasing population without jobs and with limited access to food. The impact on and demand for natural resources increase constantly. Consequently, the tasks of preserving and conserving those resources become more and more difficult. Without the cooperation of involved communities, exploitation control has become a difficult, if not an impossible, exercise. This in itself brings a new dimension to the formidable challenges facing the management cadre.

The authors of ***INLAND FISHES AND FISHERIES OF NAMIBIA*** are scientists with many years of experience in their fields of work and research. Some were my students when I was a Lecturer of Zoology at Stellenbosch University many years ago. Some were members of my staff here in Namibia while I was Director of Fisheries and later Executive Director (Permanent

Secretary) of the MFMR. There are no other scientists *cum* authors better equipped for the task of offering a high quality overview covering the freshwater fishes of Namibia and related issues linked to this subject.

This book is a groundbreaking, first of its kind for Namibia, a scientifically orientated record of the topic of freshwater fish in our country. However, at the same time, it has been written in such a way that it also offers a leisurely journey for the layman wishing to increase general knowledge about Namibia's freshwater fishes.

Dr. Jan Jurgens

Windhoek
June 2023

Namibia Nature Foundation
Committed to conservation

INLAND
ECOSYSTEMS

ACKNOWLEDGEMENTS

The editors are indebted to all those who preceded them to make this work possible. While unusual to do so, we especially acknowledge the contributors themselves: Ben van der Waal was (and is!) a true pioneer fishery scientist in Namibia, having worked in the country, Angola, Botswana, South Africa, Zambia and Zimbabwe since the 1970s. Ben van Zyl and Clinton Hay were 'freshwater fisheries' during the 1980–90s within the Ministry of Fisheries & Marine Resources before moving into different professional directions. Their void within the Ministry has been amply filled by Francois Jacobs and Renier Burger and their cadre of associates.

The former J.L.B. Smith Institute of Ichthyology, now the South African Institute for Aquatic Biodiversity (SAIAB), and the affiliated Department of Ichthyology & Fisheries Science, Rhodes University, cannot be disputed as regional centres of research, education and outreach. These institutions have been vital to our individual careers in Africa, notably influenced by Mike Bruton, Tom Hecht and Paul Skelton, and Jim Cambray of the Albany Museum.

Chapter 3 is a narrative based on published literature; the author trusts that his interpretations fairly represent the views of the authorities he has cited. Special acknowledgement is given to Mike de Wit, Denis Tweddle and Paul Skelton who reviewed various drafts of the chapter, and to Mandy Kambikambi and Wilbert Kadye for their discussion of the *Enteromius anoplus* complex. Additional input was received from Colin Reeves, Stephan Koblmuller, Graham Andrews and David Nash. Lastly, but not least, a very special 'thank you' is given to the family of Prof. Max Poll (1908–1991) who provided the author so many years ago with a copy of the treatise *Contribution à la faune ichthyologique de l'Angola*.

Numerous persons contributed to Chapter 4, particularly R. Bills, D. Tweddle and P. Skelton, all affiliated with SAIAB. Recognition is also extended to P. Bragança, M. Riddin, R. Tate and N. Jacobs for their photographs. Special gratitude is offered to Prof. Dr. Bernd Kramer of the Institut für Zoologie Universität Regensburg, Germany, who provided photographs of all type specimens of mormyrids that he has described.

Chapter 7 was influenced by Michael Ekandjo, Ministry of Fisheries and Marine Resources, Ongwediva, Namibia who took part in surveys of fish life during floods in the *iishana*, and Dr. Martin Hipondoka, Department of Geography, History and Environmental Studies, University of Namibia, who suggested possible fish migration over the divide between the Kunene and Cuvelai systems and participated in field work.

Thanks are due to Ilari Järvinen, National Museum of Finland for permission to reproduce a photo in Chapter 9 of a fishing harpoon from the Martti Rautanen's 1983 Ambo Collection, Helsinki.

Chapter 10 received valuable contributions and data from the Namibia Federation for Freshwater Anglers (NFFA), Nwanyi Angling Club and Crockango Angling Club. Information on species and angling techniques was provided by Max Pieper and Naas Jacobs for Largemouth Bass, Douglas Wagner for bream/tilapia and Frikkie Maartens for the yellowfishes.

We would also like to thank Lenore Halfide and Robert Klundt for technical support in compiling this book.

ABOUT THE EDITORS AND CONTRIBUTING AUTHORS

Glenn S. Merron has conducted extensive fisheries research in southern Africa including the Kavango River, Okavango Delta, Botswana and the Phongolo River and its floodplain in northern Zululand, South Africa. He is the founder of Inland Ecosystems Inc., a leading environmental consulting firm based in Reno, Nevada and can be contacted at gmerron@inlandecosystems.com.

Prof. Charles H. Hocutt has nearly 40 years fisheries experience in southern Africa. He was twice awarded Senior Fulbright Research Fellowships to Zimbabwe and Namibia, and was instrumental in the initiation of the Department of Ichthyology & Fisheries Science (DIFS), Rhodes University, South Africa. He is a Resident of Namibia, lives in Omaruru and can be contacted at hocutt@afol.com.na.

Renier Burger has been the Deputy Director at Kamutjonga Inland Fisheries Institute in the Ministry of Fisheries and Marine Resources since 2013. He previously was with the Ministry of Agriculture, Water and Rural Development through 2012 as (a) the chief agriculture advisor in the Oshikoto-Oshana and Erongo Regions of Namibia, (b) registrar for agro-chemicals, and (c) the national plant protection officer (NPPO) for the International Plant Protection Convention (IPPC) and can be reached at Reiner.Burger@mfmr.gov.na.

Dr. Clinton J. Hay is a Senior Lecturer in the Department of Environmental Science, University of Namibia, Windhoek and a Fisheries Technical Advisor for the Namibia Nature Foundation. Formerly, he was Head of the Hardap Fisheries Institute, Ministry of Fisheries & Marine Resources from 1996–2007 and can be reached at chay@unam.na.

Dr. Francois J. Jacobs is the Chief Fisheries Biologist at the Kamutjonga Inland Fisheries Institute (KIFI) in the Namibian Ministry of Fisheries and Marine Resources. His current research (a) contributes towards the understanding of natural and human-related impacts of freshwater fish communities and (b) includes aspects of sustainable utilisation, conservation and fisheries ecology. He can be contacted at Francois.Jacobs@mfmr.gov.na.

Dr. Kyle J. McHugh is a freshwater parasitologist with special interest in infectious diseases. He is currently employed by Pharmaceutical Product Development, Johannesburg, South Africa, involved in medically-important infections disease research and can be reached at sharky.mchugh@gmail.com.

Dr. Jay Richard Stauffer, Jr. has completed research in Malawi, Uganda, Namibia, Mozambique and the USA. He is a Distinguished Professor of Ichthyology at The Pennsylvania State University and Honourary Research Associate, South African Institute for Aquatic Biodiversity and can be contacted at vc5@psu.edu.

Ms. Sarah Tribelhorn is a former research assistant at the JLB Smith Institute of Ichthyology (now SAIAB) in Makhanda, South Africa. She is currently the Sciences Librarian at San Diego State University and can be reached at tribelhorn.sarah@gmail.com.

Dr. Ben van der Waal (right) is one of Namibia's pioneer fisheries scientists with nearly 50 years in-country experience. His extensive research has included aquaculture, biodiversity, ecology and distribution of the region's fishes. Prof. Charles Hocutt is seen on left. Dr. van der Waal can be contacted at bcwvis@gmail.com.

Dr. Ben van Zyl was the former (a) Head of the Hardap Fisheries Institute (HFI) responsible for all research on the Republic's freshwater fisheries and ecosystems (1988–1996) and (b) Deputy Director (1996–2008) of the National Marine & Information Centre (NatMIRC). He is currently affiliated with the Food and Agricultural Organization (FAO) as a specialist on the impacts of climate change on aquatic ecosystems and can be reached at ben@benguelaac.org.

Table of Contents

ACRONYMS

AGM - Annual general meeting

CAN - Canal and Olushandja Dam
CANAMCO - Oxfam Canada-Namibia Cooperation Programme
CBM/CBNRM - Community-based management/ Community-based natural resources management
CCRF - FAO Code of Conduct for Responsible Fisheries
CCU - Climate change unit
CFB/CFM - Cape Fold Belt/Cape Fold Mountains
CIPS - International Sport Fishing Confederation
CITES - Convention on International Trade in Endangered Species
COVID - Corona virus disease
CPUE - Catch per unit effort
CUVECOM - Cuvelai Watercourse Commission
CV - Cuvelai basin

DEA - Directorate of Environmental Affairs
DIFS - Department of Ichthyology & Fisheries Science, Rhodes University
DRC - Democratic Republic of the Congo

EAF - Ecosystem approach to fisheries
EbA - Ecosystem-based adaptation
EFF - Epalela Fish Farm
EGT - Etosha–Griqualand–Transvaal Axis
ENWC - Eastern National Water Carrier
EOD - Electric organ discharge
EUS - Epizootic Ulcerative Syndrome

FAO - Food and Agricultural Organization
FCR - Feed conversion ratio
FF - Fish farm
FFF - Fonteintjie Fish Farm
FIPSed - Fédération Internationale de la Pêche Sportive en eau Douce

FMP - Fisheries management plan
FR - Fish reserve
FWFFP - Freshwater Fish Farming Project

GRN - Government of the Republic of Namibia

HFI - Hardap Fisheries Institute
HIAC - Hardap Inland Aquaculture Centre

IAC - Inland aquaculture centre
IBI - Index of Biotic Integrity
IGFA - International Game Fish Association
IIS - *iishana*
IPPC - International Plant Protection Convention
IUCN - International Union for the Conservation of Nature and Natural Resources

Ka - Kavango = Cubango River
KAZA/KAZA TFCA - Kavango–Zambezi Transfrontier Conservation Area
Kb - Kuiseb River
KIAC - Katima Inland Aquaculture Centre
KIFI - Kamutjonga Inland Fisheries Institute
KL - Kwando (Cuando)-Linyanti River
KTWSP - Kunene Transboundary Water Supply Project
Ku - Kunene = Cunene River

LFF - Leonardville Fish Farm
LPIA - Late Palaeozoic Ice Age

MAWLR - Ministry of Agriculture, Water and Land Reform
MEFT - Ministry of the Environment, Forestry and Tourism
MFF - Mpungu Fish Farm
MFMR - Ministry of Fisheries & Marine Resources

NALAA - Namibia Artificial Lure Angling Association
NAMP - Namibia Aquaculture Master Plan
NamPol - Namibia Police
NatMIRC - National Marine & Information Centre
NBAA - Namibian Bass Angling Association

NCCC - National Climate Change Committee
NCE - Namibia Chamber of Environment
NDP - National Development Plan
NFBAA - Namibia Freshwater Bank Angling Association
NFFA - Namibia Federation for Freshwater Anglers
NGO - Non-government organisation
NLTBFA - Namibia Light Tackle Boat Freshwater Angling Association
NNF - Namibia Nature Foundation
NUST - Namibia University of Science and Technology
NW - National waterbody

OIAC - Onavivi Inland Aquaculture Centre
OIE - World Organisation for Animal Health
OKACOM - Permanent Okavango River Basin Water Commission
OKZ - Ovambo–Kalahari–Zimbabwe Axis
OM - Omadhiya Wetlands
OR - Orange-Fish River
ORASECOM - Orange-Senqu River Commission

PJTC - Permanent Joint Technical Commission – Kunene River

RAN - River Angling Namibia Association
RDC - Rural Development Centre
RIAC - Rundu Inland Aquaculture Centre
RSD - Red Spot Disease

SADC - Southern Africa Development Community
SAIAB - South African Institute for Aquatic Biology
SAN - Species Angling Namibia
SCP - Southern Coastal Plain
sGnRHa - Salmon gonadotropin releasing hormone
SOP - Standard Operating Procedure
SRTM - Shuttle Radar Topographic Mission
SWA - South-West Africa Anglers Union

TA - Traditional authority

UNAM - University of Namibia

UNEP - United Nations Environment Programme
UNFCCC - United Nations Framework Convention on Climate Change
UZC - Upper Zambezi-Chobe River
UZR - Upper Zambezi River

WCP - Western Coastal Plain
WWF - World Wildlife Fund

ZAMCOM - Zambezi Watercourse Commission

Inland Fishes and Fisheries of NAMIBIA

CHAPTER 1
INTRODUCTION

by
Glenn S. Merron and Charles H. Hocutt

I. The Republic of Namibia

The Republic of Namibia gained its independence on 21 March 1990 with its name derived from the Namib Desert which extends 1,572 km along the Southeast Atlantic Ocean. It is a vast country encompassing approximately 825,000 km^2 and situated along the Tropic of Capricorn. Its population base of 2.55 million persons makes it the 2nd least densely populated country in the World at 3.1 persons/km^2, second only to Mongolia. The Republic has often been referred to as the "Gem of Africa" since it has abundant natural resources including diamonds, gold and uranium in addition to its fisheries.[1]

Namibian fisheries prior to independence nominally fell under the Ministry of Agriculture that included departments of Nature Conservation, Forestry, Water Affairs and Veterinary Services with Fisheries managed by Nature Conservation. However, in reality the administration, management and research of fisheries, especially marine, prior to Namibian independence were led from Cape Town, South Africa. The Ministry of Agriculture was subdivided at independence and fisheries were briefly under the Ministry of Agriculture, Fisheries, Water and Rural Development until February 1991 when the Ministry of Fisheries and Marine Resources (MFMR) was created.[2] MFMR incorporated both marine and freshwater fisheries into its portfolio and was guided by its first administration, Honourable Minister Gert Hanekom, Permanent Secretary Carl-Hermann "Callie" Schlettwein and Director of Resource Management Dr. Jan Jurgens. The Ministry of Agriculture, Water and Land Reform (MAWLR) still retains sub-programmes on water and wetlands.

Hardap Dam was completed in 1963. The Hardap Fisheries Institute (HFI) was established in 1968 after which it was responsible for aquaculture activities and all research on Namibia's inland fisheries (Chapter 6), although the facility was 800 km away from the Kavango River with no *perennial* rivers in between and 1,200 km from the Upper Zambezi River. The *ephemeral*, but important Cuvelai system (Chapter 7) lies over 600 km air distance north of Hardap. The Hardap staff was eventually integrated into MFMR and it is now known as the Hardap Inland Aquaculture Centre (HIAC). The Kamutjonga Inland Fisheries Institute (KIFI) located on the Kavango River is now the main MFMR centre for inland fisheries and aquaculture research in Namibia; however, various regional offices have been established (Chapter 12). Also, the faculties of the University of Namibia (UNAM) and Namibia University of Science & Technology (NUST) provide tertiary training, each based in Windhoek with satellite campuses.

Namibia recognised that sustainable inland fisheries development (e.g., artisanal fisheries, aquaculture) had important roles in food security and improvement of socioeconomics for future generations.[3] A White Paper on "The Responsible Management of Inland Fisheries in Namibia" was published in 1995 and legislated in 1996 with an emphasis on protecting the floodplain fisheries and the establishment of ownership and protection of the *iishana* (see *oshana* in Glossary). The Aquaculture Policy of 2000 and Aquaculture Act of 2002 were passed, while simultaneously acknowledging the need to protect *indigenous* fish communities from importation of non-indigenous stocks (Chapter 12) and the introduction of diseases (Chapter 13). The Inland Fisheries Act was gazetted in 2003.

The inland waters and their fisheries are unique in several ways. Foremost, Namibia routinely experiences drought with only 1% of the land mass considered arable. Rainfall, when it occurs, is generally from December to April and distributed unevenly across the country with the northeast typically experiencing ~650 mm, the south <50 mm and much of the coast 0 mm. All northern rivers of Namibia have their headwaters in rain-rich Angola and Zambia with nearly 50% of Namibia's population living in the eight northern regions located on or near the rivers courses as they pass through the country (Fig. 1). The National poverty level was reported to be about 28% in 2018; extreme poverty levels ranged up to 53% in these same eight regions and were expected to increase 2.7% in 2021 influenced by the COVID-19 pandemic, based on World Bank estimates. The effect of climate change on Namibia's inland fisheries and subsistence of riparian communities is a key concern (Chapter 11).

Inland fisheries contribute significantly to household food and nutritional security in Africa, where some of the world's most marginalised communities are found.[4] The Kavango Region is considered the poorest area in Namibia with 94% of its populace living within 5 km of the river to relieve their poverty[5] with 91% of the households using fish as a food source.[6] These subsistence communities employ both traditional and modern gears to capture fish (Chapter 9). Marine fish and fisheries products from the Benguela Current fishery of the Southeast Atlantic Ocean are widely distributed to provide relatively inexpensive sources of protein to rural inland communities. Per capita fish consumption in the country increased from 4 to 13 kg over a 5-year period 2012–2017, although inland species remain preferred.

Aerial view of the vast river channels and floodplains of northeast Namibia (Photo: F. Jacobs).

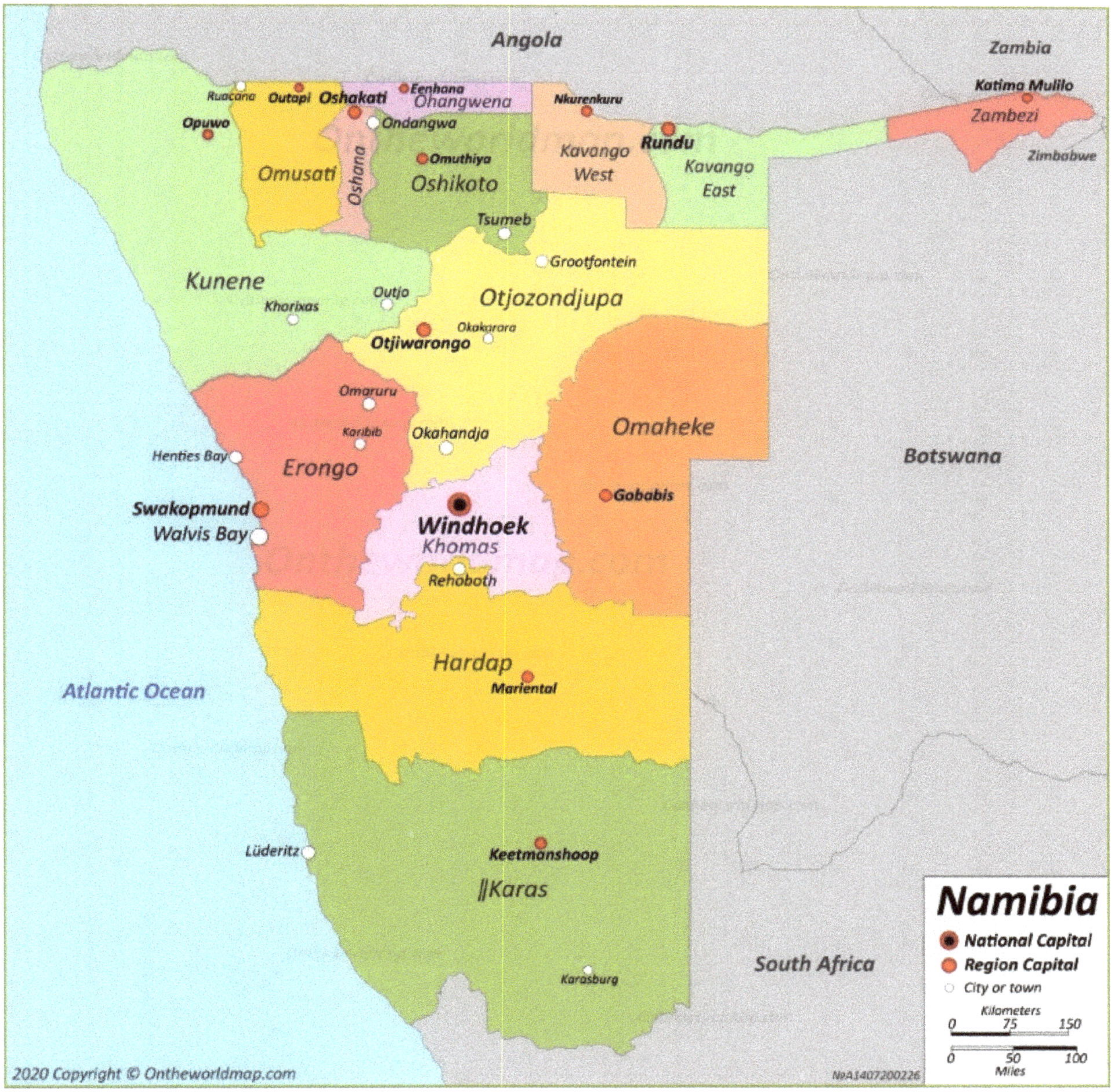

Figure 1. Map depicting 14 political regions of Namibia.[7]

II. The Waterbodies

A. Perennial Rivers

Namibia has five perennial rivers (Fig. 2), only two of which discharge into the Atlantic Ocean: the Orange River on the southern border with South Africa and the Kunene River in the northwest bordering Angola, separated by >1,500 km of coastline (Chapter 6). Both rivers have minor *estuaries* with no anadromous species and little influx of marine fishes upstream; it is for this reason that marine and estuarine species are not considered in this book.

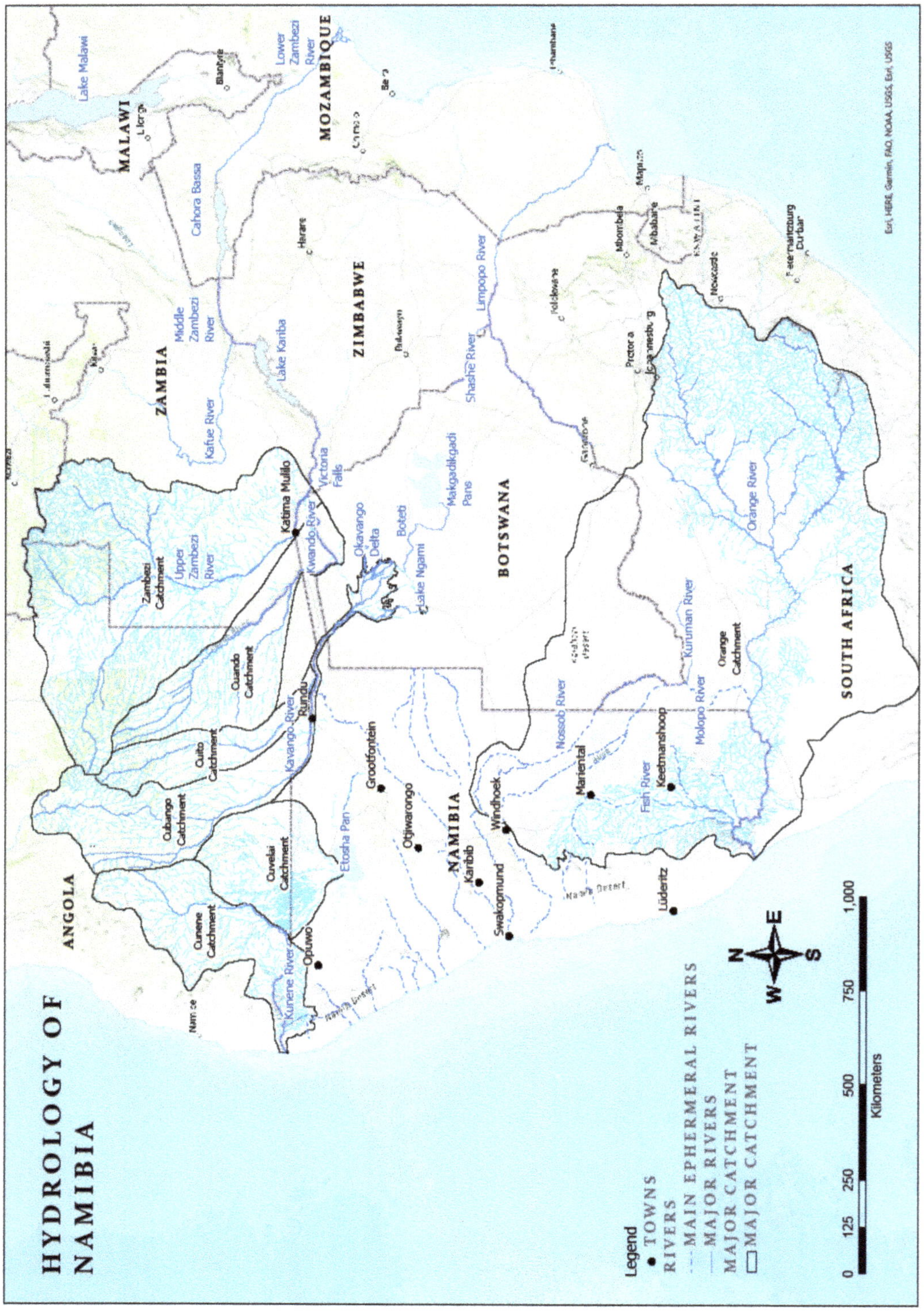

Figure 2. Namibian permanent basins, international boundaries and regional basins (map prepared by G. Jacobs).

The remaining perennial rivers (Kavango, Kwando-Linyanti and Upper Zambezi-Chobe) are all along the northern borders with Angola, Zambia and Botswana with each historically draining into the Indian Ocean on the eastern side of the continent (Chapter 3). The lower Kwando River and Linyanti Swamps are a maze of interconnected channels and wetlands; therefore they are considered a single basin in the following chapters. The Cubango of Angola and the Kavango of Namibia are the same river, the primary water source of the world renowned *endorheic* Okavango Delta, Botswana.

The eastern tip of Namibia's Zambezi Region (Fig. 1) is at the confluence of the Upper Zambezi and Chobe Rivers, and <2 km across the water from Zimbabwe. The Chobe is a virtual backwater of the Upper Zambezi, especially during the latter's flood stage; thus, they are considered as a single system here. The sharing of international borders along all the perennial rivers with neighbouring countries poses challenges to Namibia for management, monitoring and surveillance (Chapter 8).

B. Ephemeral Rivers

The interior of Namibia has no perennial rivers although several arise in the central highlands and cut through the Great Escarpment (Chapter 3) toward the Atlantic Ocean. These include the Kuiseb, Swakop, Omaruru, Ugab, Huab, Hoanib and Hoarusib, amongst others. These systems may flood during the rainy season leaving intermittent pools utilised by wildlife and livestock. They lack value, however, from an inland fisheries perspective (Chapter 6).

The seasonal Fish River is Namibia's primary tributary to the Orange River. It has been highly altered through water diversion, extraction and irrigation schemes with numerous agricultural activities and dams, similar to the Orange-Vaal-Senqu system in South Africa (Chapter 6). The Orange-Fish basin in Namibia has a *depauperate*, but distinct fish fauna.

The Cuvelai River system is unique in itself. It is a seasonal river basin with waters derived from several *disjunct* tributaries with headwaters in Angola that all terminate at the Etosha Pan, about 4,800 km^2 in size. The pan can be completely flooded or dry depending upon annual rainfall patterns (Fig. 3). This is an endorheic system with no surface outlet, akin to Botswana's Okavango Delta, where most of the water is lost to evaporation, transpiration and seepage into Kalahari sands instead of draining into the sea. Ironically, the vast flooding of the Cuvelai inundates the most populous area in Namibia and brings about a huge influx of fishes targeted by subsistence fishers. It is such an exceptional system that Chapter 7 is devoted to it.

Figure 3. Etosha Pan (left) during a dry year, 2022 (Photo: C. Wirth) and (right) inundated by the Cuvelai system, 2009 (Photo: C. Hocutt).

C. Natural Lakes

Lakes are usually thought of as permanent waterbodies visible to the eye with extensive shorelines and adjacent *littoral* zones, perhaps with scattered islands, coves and inlets, with a mouth discharging into a river. That is not the case in water-scarce Namibia.

Central Namibia, although lacking perennial rivers, has substantial and interesting underground natural water bodies in the arid Namib Desert and karst systems. The Aigamas Cave complex and Lake Guinas near Tsumeb each support endemic fishes considered threatened by water extraction and introduction of *non-native* species (Chapter 6).

Lake Liambezi in the Zambezi Region (Fig. 1), although ephemeral, is extremely important for food security and commercial fishing when filled. Its primary source is the Kwando-Linyanti system with its outlet via the Chobe River to the Upper Zambezi River. The Chobe River, however, can reverse its flow back into Lake Liambezi during extreme high floods in the Upper Zambezi (Chapter 6).

D. Reservoirs

Several man-made reservoirs occur in south-central Namibia for municipal water supply, irrigation and process water for mining. The most important of these are Hardap, Neckartal and Naute Dams, all in the Fish-Orange River basin; Von Bach and Swakoppoort Dams on the Swakop River; Oanob Dam on the endorheic Oanob River; and Omdel Dam on the Omaruru River (Chapter 6). They vary in importance to fisheries, e.g., angling tournaments are often held on Hardap and Von Bach, whereas the Omdel Dam can be completely dry for years.

III. Fish Diversity

Chapter 2 discusses how Namibia's inland fishes are identified, classified and named, whereas Chapter 3 considers their origins. There are 137 fish species known from Namibian waters, six of which have not been formally described (Chapter 4) and 12 which have been introduced (Chapter 5), principally through aquaculture activities (Chapter 12). However, populations of non-native fishes may occur in private or public waters as a consequence of aquarium or ornamental pond releases. It is expected that the number of indigenous species could increase through the recognition of new species and the influx from upstream waters of Angola and Zambia (Chapter 4). Conversely, the numbers could decrease through overharvesting, climate change, invasive species, water extraction and/or increased pollutant loading if conservation measures are not considered (Chapter 14).

Although the *diversity* of indigenous Namibian species is not high for African freshwater ecosystems, the diversity of families (17) and genera (45) is very high, especially compared to the African Great lakes which are dominated by a single family (Cichlidae) and a few genera.[8] Six families and 23 genera are represented in Namibia by a single species, while an additional six families have ≤ 5 species each. This high diversity above the species level is a consequence of the complex history of Namibia's rivers and the diversity of habitats in these systems.

There is a clear distinction in the composition of the fish communities in the northern perennial rivers versus the southern riverine habitats and state reservoirs. The northern Kunene, Cuvelai, Kavango, Kwando-Linyanti and Upper Zambezi-Chobe Rivers are characterised by a greater *richness* of fish species; all are linked historically to the *palaeo*-Upper Zambezi ichthyofauna. In contrast, the Orange-Fish River and other south-central ephemeral rivers have had a different geological and climatic history, this influencing their reduced fish fauna (Chapter 3). The Straightfin Barb, Sharptooth Catfish, Southern Mouthbrooder and Banded Tilapia are the only native species shared by all Namibian perennial rivers.

IV. Fish Ecology

A. Habitats

The distribution patterns and abundance of Namibia's inland fishes are clearly related to specific habitats requirements, such as amphiliid catfish (found in rocky habitats with fast-flowing water), anabantids (prefer slow-moving well vegetated backwaters) and Tigerfish (require well oxygenated waters). Others are generalists and adaptable to a variety of habitats (Banded Tilapia) or can survive extreme environmental conditions (Sharptooth Catfish).

Within a cross section of a waterbody, there is both vertical and horizontal zonation. Many species prefer to live near the *benthic* environment, such as catfishes, whereas the Tigerfish, a pursuit predator, prefers open-water *pelagic* habitats. The Southern African Pike, an ambush predator, waits concealed within aquatic vegetation for unsuspecting prey to pass. A few species, especially the topminnows and small barbs, live in permanent backwaters and lagoons near the water surface from which they feed. Some habitat specialists, such as the spiny eels live among dense reed, papyrus mats and rootstocks and along the edges of permanent channels, whereas the underside of papyrus mats and logs are foraged by suckermouths and squeakers, otherwise known as "upside down" catfish.

B. Feeding Strategies

A diverse fish community has representatives at several *trophic levels*. For instance, Tigerfish are apex predators eating other fishes (*piscivore*) and live prey (*carnivore*). At the bottom of the food chain are primary consumers (*herbivores*) with a diet primarily of algae or macrophytes, e.g., Redbreast Tilapia. In between, secondary consumers come in different forms, eating *periphyton* (or *aufwuchs*), diatoms and microorganisms. Moggel and other *Labeo* spp. are *detritivores* with long intestines to digest and absorb nutrients from decaying organic matter; *omnivores* such as the Sharptooth Catfish eat a wide variety of plant and animal foods; and many barbs and mormyrids are *insectivores* feeding on drifting insects in the water column or by pecking insects from vegetation, rocks and other submerged substrates. Under extreme environmental conditions, unique foraging habits have evolved, such as the Cave Catfish that is dependent on bat droppings. Moreover, it is normal for fish to pass through various trophic levels as they develop from larvae to adult, e.g., graduating from an insectivore to a piscivore.

The top predators differ between the perennial northern and southern rivers. Tigerfish, Southern African Pike and several largemouth bream are the top fish predators in the northern rivers, but absent from the Orange-Fish River drainage where Sharptooth Catfish and Vaal-Orange Largemouth Yellowfish are the apex fish carnivores. These top predators, especially, are targeted by the recreational fisheries (Chapter 11).

C. Reproductive Strategies

1. Non-guarders, guarders and bearers

All of Namibia's indigenous inland species lay eggs (*oviparous*), but three introduced forms (i.e., Guppy, Mosquitofish and Swordtail) are livebearers (*ovoviviparous*) with internal fertilisation and the eggs developing inside the female's body until they are born live. In ovoviviparous fishes, there is no direct connection of the embryos to the mother (i.e., placenta or similar structure), which occurs in some *viviparous* sharks. Caprivi Killifish are unique in that their eggs lay dormant for years in dried-up ponds during drought to hatch when the rains resume.

Thirty-two types of reproductive associations have been recognised amongst fishes worldwide, broadly classified within three breeding strategies: *non-guarders, guarders* and *bearers*.[9] Non-guarders release their eggs into the water or a *nest* where they are fertilised and left unprotected, either drifting in midwater, settling on the bottom or adhering to plants. Some non-guarders hide their eggs under sand, logs or leaves (brood hiders). Guarders protect the young in nests, which may be a hollow scooped-out depression in the sand or a foam nest. Bearers protect the young in a body cavity such as the mouth (external bearers) or oviduct (internal bearers). Males of mouthbrooding cichlids construct a *bower* in the substrate to attract females for mating and to fertilise eggs that are laid. Bowers provide no protection for the eggs or young, rather females (or in rare cases, the males) brood the eggs and larvae in their mouths.[10]

Non-guarders in Namibia's perennial waters comprise 81 species (65% of all native species) and include all the cyprinids (36), kneriids (1), citharines (4), alestids (4), catfishes (27), spiny eels (2), lampeyes (4) and labyrinth (3) fishes.

There are 21 native species (17%) of *guarders* in Namibia, including the snoutfishes (15); Southern African Pike (1) that build bubble nests on the surface of the water amongst vegetation; and five cichlids of the genera *Coptodon, Hemichromis* and *Tilapia* spp. that deposit eggs in a nest.

Bearers comprise 21 cichlids (17%) in the genera *Chetia, Oreochromis, Orthochromis, Pharyngochromis, Pseudocrenilabrus, Sargochromis*, *Serranochromis* and *Thoracochromis*. The males build *bowers* to attract females and where eggs are laid. The females immediately collect the fertilised eggs in their mouths where they are brooded until the complete absorption of the yolk sac by the developing larvae, thus being affording protection as they mature. In some cases, young will temporarily forage freely, but are orally gathered by the female if threatened.

Reproductive guilds have been reviewed relative to ecosystem stability.[9,10] The highest percentages of bearers and guarders were found in the ancient African Great Lakes, characterised by relatively stabile and predictable physicochemical regimes. The highest percentage of non-guarders was found in rivers and wetlands with less predictable regimes.

2. *r*- and K-selection

Reproduction strategies[11] known as *r*- and *K-selection* have been linked to ecosystem stability, environmental variability, population parameters and reproductive characteristics.[12] While these strategies have been refined by more-rigorous population models,[13] their basic tenants remain unchanged. *r*-selected inland species are more likely to occur in unstable environments, have a small body size, reach sexual maturity fast, have a high *fecundity* and short generation time with rapidly dispersed/highly adapted offspring. The Caprivi Killifish is an excellent example of an *r*-selected species that completes its life cycle in only a few months within a very unstable environment.

K-selected species require more stable conditions, usually grow larger at a slower rate, have a low fecundity with greater parental care and live longer with reproduction extended usually over a few years. K-selected species must use their resources more efficiently and are less adaptable to environmental stress. It goes without saying that there is a gradient between the extremes of *r*- and K-selection and a single species might exhibit characteristics of either strategy, dependent upon the environment in which it is living. For instance, Tigerfish exhibit high fecundity, a long life cycle and no parental care, while all the *mouthbrooders* have a true K-selected reproductive strategy. The Cave Catfish and Otjikoto Tilapia are examples where their populations must live in equilibrium with an environment that is limited in resources.

The parental investment in offspring and how much energy is invested in each individual egg or fry by the parents differ widely between breeding strategies. They are lowest in non-guarders that produce large numbers of small eggs with no parental care, i.e., *r*-strategists. Whereas, parental investment and body energy investment are highest in bearers (K-strategists), which produce small numbers of large eggs and mouthbrood. Those fishes which deposit eggs within a nest or on plants and guard the eggs and fry exhibit intermediate energy investment.

V. Flood Pulse Concept

The perennial Kavango, Kwando-Linyanti and Upper Zambezi-Chobe Rivers are *floodplain* rivers in Namibia (Fig. 4), some of which can spread over 10+ km either side of the main channel during peak flooding, giving rise to a diverse range of habitats, fish species and biomass, exploited by fishers (Chapter 6). The ephemeral Cuvelai basin has a massive floodplain fanning out over 130 km along the Namibia-Angolan border (Chapter 7). Floodplains are low-lying areas adjacent to the river covered by terrestrial grasses and shrubs that are subjected to flooding when rivers overflow their banks. The annual flooding regime is termed the *flood pulse*, which is the primary driver of ecosystem productivity in floodplain rivers.[14] The timing, frequency, magnitude and rates of rise and fall of flooding will be reflected in the ecological character of the associated biota.[15] The annual pulse of water connects the lagoons, floodplains and river channels into massive aquatic ecosystems that facilitate migrations and spawning of fishes found in the various aquatic habitats.

The dry season is as important as the wet season allowing nutrients to be deposited on the floodplains through vegetation production and terrestrial animal interactions. Riparian communities use the fertile floodplains for subsistence agriculture and to graze their livestock. Large mammals (e.g., hippo, buffalo and elephant) also graze on the floodplains, converting grasses into various stages of organic decomposition, enhancing the release of nutrients for assimilation in the following year's flood cycle. In effect, the terrestrial component of the floodplain contributes nutrients to the aquatic component and enhances productivity.[16]

The flood amplitude in a given year is the dominant influence on the nutrient cycle and hydrochemistry of the ecosystem, which in turn controls decomposition, primary and secondary productivity and the nutrient cycle. The previous year's flooding provides nutrients, minerals and organic matter to promote vegetative growth in the subsequent year(s). Lateral flooding sequentially stimulates nutrient uptake by primary producers; greatly expands niches for consumers; increases fish reproductive grounds; and provides refugia for juveniles. Water quality is enhanced during the flooding stage with higher levels of dissolved oxygen, reduced anoxic conditions, lower methane production, neutral pH and moderate water temperatures.[17] Large floods tend to have a homogenising effect on the nutrient regimes between the mainstream and floodplains due to increased circulation, as well as on the fish assemblages moving onto the floodplains.[18] Detritus- and nutrient-rich water return to the river during the flood recession.

Fish life cycles are synchronised with the ebb-and-flow of the flood pulse relative to movement activities, spawning and feeding behaviour. Fish movement is either (a) longitudinal where fish move upstream or downstream or (b) lateral where fish move onto or off the floodplains, dependent upon flood stage. Inundated floodplains provide enhanced food, shelter and spawning sites essential for survival and successful recruitment of juveniles, while fish play a vital role of converting resources at the bottom of the food chain such as detritus, algae, insects and plants into food for higher forms of life, including humans.

Fishes leave the floodplains with the retreat of floodwaters are exposed to high levels of natural and fishing mortality. Clariids, for instance, gorge themselves on migrating mormyrids and cyprinids, while subsistence and commercial fishers concentrate their gear along discharging channels. As waters recede to be confined between river banks, fishes are subjected to increasingly harsh conditions: mobile substrate, low productivity, minimal vegetation cover and excessive pressure from predators and fishers in an environment greatly reduced in area. Survivors of each species compete for food resources and refugia in order to initiate their life cycles in the next annual flood cycle. The abundant mormyrid downstream migration during June and July at Impalila Rapids is now popularly known as the 'Stonebasher Bash'.

Floodplain rivers and freshwater wetlands are amongst the most threatened ecosystems in the world. Any factor impacting floodplain river hydraulic functioning is certain to dually threaten (a) biodiversity and biomass production of the fish community[20] as well as (b) food security of subsistence communities dependent upon them. Such factors are not limited to those *en vogue*, such as population and climate change models, but also includes a plethora of concerns that policy makers should act upon and resolve, such as riparian zone management including deforestation, agriculture and irrigation schemes; impoundments; eutrophication and inorganic pollution; aquaculture of non-native species; and, especially, overfishing.

VI. SUMMARY

Inland fisheries have been undervalued globally in both developing and developed countries. Freshwater ecosystems and the fish they support, as well as the ecosystem services (Chapter 11) generated by inland fisheries, are largely ignored in sustainable development goals.[19] This oversight is particularly relevant given the role that inland fish and fisheries play in food security and supporting livelihoods in low-income countries such as Namibia.

It is the objective of ***INLAND FISHES AND FISHERIES OF NAMIBIA*** to describe the fascinating diversity and utilisation of the country's inland fisheries, including their diversity; life history habits; environmental forces that drive productivity; their regulation and management; their farming and value in the food security chain; and their conservation. Fishers and decision makers may use this information to assist in planning fisheries management protocol, while others may find this a valuable reference concerning Namibia's environmental heritage.

VII. REFERENCES

[1]Miller, D. 1990. Agency establishes assistance program in Namibia. USAID Front Lines December 1990: 7.
[2]Morcos, S., G. Hempel, T. Stromme and S. Kristmannsson. 1993. The Republic of Namibia: Development of marine sciences. UNESCO IOC/OPS/DEVELOPMENT No.3: 39 pp. + annex.
[3]Ministry of Fisheries and Marine Resources. 2017. Strategic plan 2017/18-2021/22. Windhoek: 32 pp.
[4]Chan, C.Y., N. Tran, S. Pethiyagoda, C. Crissman, T.B. Sulser and M.J. Phillips. 2019. Prospects and challenges of fish for food security in Africa. Global Food Security, Volume 20: 17–25.
[5]Tiyeho, D.C. 2020. Assessment of fishers' catches on the Kavango River floodplains, Namibia. MSc. Thesis, Univ. Namibia, Windhoek: 82 pp.
[6]Hay, C. J., T. Naesje, J. Breistein, K. Harsaker, J. Kolding, O.T. Sandlund and B. van Zyl. 2000. Fish population, gill net selectivity and artisanal fisheries in the Okavango River, Namibia. Recommendations for sustainable fishery. NINA-NIKU Project Report 10: 1–105.
[7]http://ontheworldmap.com/Namibia.
[8]Stauffer, J.R., Jr., M. Geerts, A.F. Konings, K.R. McKaye and K. Black. 2007. Cichlid fish diversity and speciation, pp. 213–225. In: T.R. Hodkinson and J.A.N. Parnell (Eds.). Reconstructing the tree of life: Taxonomy and systematics of species rich taxa. Systematics Association Special Series, Volume 72.
[9]Balon, E.K. 1975. Reproductive guilds of fishes: A proposal and definition. J. Fish. Res. Bd. Can. 32(6): 821–864.
[10]McKaye, K.R., J.R. Stauffer, Jr., G.F. Turner, A. Konings and T. Sato. 2001. Fishes, as well as birds, build bowers. J. Aquaricult. Aq. Sci. 9: 121–133.
[11]Bruton, M.N. and G.S. Merron. 1990. The proportion of different eco-ethological sections of reproductive guilds of fishes in some African inland waters. Environ. Biol. Fish. 28: 179–187.
[12]MacArthur, R.H. and E.O. Wilson. 1967. The theory of island biogeography. Monographs in population biology 1. Princeton University Press, Princeton, N.J.
[13]Reznick, D., M.J. Bryant and F. Bashey. 2002. *r*- and K-selection revisited: The role of population regulation in life-history evolution. Ecol. 83(6): 1509–1520.
[14]Junk, W.J., P.B. Bayley and R.E. Sparks. 1989. The flood pulse concept in river-floodplain systems, pp. 110–127. In: D.P. Dodge (Ed.). 1986. Proceedings of the International Large River Symposium (LARS), Honey Harbour, Ontario, September 1986. Can. Spec. Publ. Fish. Aquat. Sci. 106.
[15]King, A.J., P. Humphries and P.S. Lake. 2003. Fish recruitment on floodplains: The roles of patterns of flooding and life history characteristics. Can. J. Fish. Aq. Sci. 60 (7): 773–786.
[16]Merron, G.S. and M.N. Bruton. 1995. Community ecology and conservation of the fishes of the Okavango Delta, Botswana. Environ. Biol. Fish. 43: 109–119.
[17]Simasiku, E.K., C. Hay and J. Abah. 2021. Effects of water level and water quality on small-sized and juvenile fish assemblages in the littoral zones of the Zambezi/Chobe floodplain. Afr. J. Ecol. 2021: 1–13.
[18]Gomes, L.C., C.K. Bulla, A.A. Agostinho, L.P. Vasconcelos and L.E. Miranda. 2012. Fish assemblage dynamics in a neotropical floodplain relative to aquatic macrophytes and the homogenizing effect of a flood pulse. Hydrobiol. 685: 97–107.
[19]Cooke, S.J., E.H. Allison, T.D. Beard, Jr., R. Arlinghaus, A.H. Arthington, D.M. Bartley, I.G. Cowx, C. Fuentevilla, N.J. Leonard, K. Lorenzen, A.J. Lynch, V.M. Nguyen, S. Youn, W.W. Taylor and R.L. Welcome. 2016. On the sustainability of inland fisheries: Finding a future for the forgotten. Roy. Swed. Acad. Sci., Ambio: DOI 10.1007/s13280-016-0787-4.
[20] Linhoss, A.C., R. Munoz-Carpena, M.S. Allen, G. Kiker and K. Mosepele. 2012. A flood pulse driven fish population model for the Okavango Delta, Botswana. Ecol. Model. 228: 27-38.

CHAPTER 2

SYSTEMATICS OF NAMIBIA'S INLAND FISHES

by

Jay R. Stauffer, Jr. and Charles H. Hocutt

I. Introduction

Namibia's freshwater fish fauna was ignored for the most part by European explorers and traders until the discovery of Victoria Falls in 1855 by Dr. David Livingstone. Afterwards, there was a steady increase of hunters, natural history boffins and ichthyologists who travelled the region collecting specimens for museums around the world. Victoria Falls is recognised as a major zoogeographic barrier for fishes with the histories of the Zambezi River above (Upper) and below (Middle-Lower) the falls being unrelated until relatively recently (Chapter 3). Namibia, sharing the borders and rivers above Victoria Falls with Angola, Botswana, Zambia and Zimbabwe, was subsequently criss-crossed many times.

The history of *ichthyology* of southern Africa has been extensively documented.[1,2] The year 1758 represents Year 1 in the modern classification of biota, coinciding with the publication of Carl Linnaeus's classic *Systema Naturae.*[3] His tome included descriptions of the Nile Tilapia (*Oreochromis niloticus*), Goldfish (*Carassius auratus*) and Common Carp (*Cyprinus carpio)*, amongst other species; the latter two are from Europe. The Largemouth Bass (*Micropterus salmoides)* was recognised in North America in 1802 by the French naturalist, Bernard Germain de Lacépède (1756–1825).[4] Ironically, all four species were introduced to Namibia.

Relative to native Namibian inland fishes, William J. Burchell (1822) described the Sharptooth Catfish (*Clarias gariepinus)* and Vaal-Orange Smallmouth Yellowfish (*Labeobarbus aeneus)* from the Orange-Vaal-Senqu River[5], a well studied and much modified ecosystem. The Banded Tilapia (*Tilapia sparrmanii)* was recognised by Andrew Smith in 1840, followed by the Moggel (*Labeo umbratus)* and Orange River Mudfish (*Labeo capensis*) in 1841.[6,7] Wilhelm Peters identified the Manyspined Climbing Perch (*Ctenopoma multispine)* in 1844 and Straightfin Barb (*Enteromius paludinosus)*, Threespot Barb (*Enteromius trimaculatus*), Redeye Labeo (*Labeo cylindricus*), Silver Robber (*Brycinus lateralis)* and Mozambique Tilapia (*Oreochromis mossambicus)* in 1852 from southern African waters[8,9], the latter to be introduced to Namibia. Explorations and the identification of new fish species continued at an accelerated pace through the late 1800s and 1900s with new discoveries continuing until today[10, 11, 12, 13, 14] with 119 described species and six undescribed species known to occur in Namibian waters (Chapter 4).

The Kavango and Kwando-Linyanti Rivers of Namibia are conduits for water from the Angolan-Zambian highlands to the wetlands of Botswana and the Okavango Delta (Chapters 1, 3 and 6). Castelnau (1861) was the first to describe fishes from Lake Ngami near the Delta, including the Southern African Pike (*Hepsetus cuvieri*).[15] Pellegrin (1936) provided the first account of the upper Cubango River fishes in Angola,[16] while three years later the first collection of fishes from the Kavango River in Namibia (then called South West Africa) was obtained by Eedes.[17] The Namibian and Botswana ichthyofauna of the Kavango and Kwando-Linyanti Rivers is well documented[17, 18, 19, 20, 21, 22, 23, 24] as are the fishes of the Upper, Middle and Lower Zambezi River into Zambia, Zimbabwe and Mozambique.[2, 25, 26, 27, 28, 29, 30, 31] The Angolan portions of these rivers are modestly known through several works;[1, 2, 32, 33, 34, 35, 36, 37] however, P. Skelton (*pers. comm.*) and his colleagues with support of *National Geographic* are making exciting inroads into Angola. Approximately 30 species are known from Angolan and Zambian headwaters thus far not collected in Namibia (Chapter 4).[13, 14, 37, 38, 39]

Lakes Guinas, Lake Otjikoto and the Aigamas Cave complex have been researched extensively since the 1936 description of the Otjikoto Tilapia and Cave Catfish.[40] Fisheries studies on the state reservoirs have been conducted since the establishment of the Hardap Fisheries Institute in 1968.

The history of description of inland fishes now known from Namibia is presented in Figure 1; the only non-native species included are the three described in 1758 by Linnaeus. It demonstrates clearly the increased scientific interest in southern African fishes until 1940, followed by nearly a 20-year hiatus that presumably represents the secondary effects of World War II. There was resurgence in the 1960-1980s; not to ignore other ichthyologists, but the master works of R.A. Jubb and M. Poll were both coincidently published in 1967. Another peak occurred after 2000 with 18 new species being described by 2020, dominated by 11 species of morymrids based upon their peculiar *electric organ discharges* (EODs) and four species of new catlets. A revisit of *Serranochromis* spp. in 2020-2021 recognised four new species and elevated a fifth to species status,[13, 14] one of which is known to occur in Namibian waters. As this book goes to press in 2023, 'the Pygmy Topminnow was described in 2021 while the Orange River Chubbyhead Barb was elevated to species status. It is predictable that new species will be described as techniques in physiology, genetics, biochemistry and behaviour are applied to fish systematics in addition to traditional practices.'

Type specimens of fishes from numerous regional expeditions to Namibia, Angola, Botswana, South Africa, Zambia and Zimbabwe have been deposited, housed and analysed in museums globally from the USA to Britain to Austria to Belgium to South Africa and many other countries (Table 1). These collections provide valuable insight to the historical record on fish distributions within the region since the earliest of expeditions.

By far the most important regional repository for historical records of Namibian fishes is the South African Institute for Aquatic Biology (SAIAB), Makhanda (Grahamstown), RSA, from which surveys have been conducted for decades through all the major drainages. It is also the current home of several leading freshwater ichthyologists. The Albany Museum, also in Makhanda and the oldest museum in that country, once was the repository for many fishes in the region, although all of its fish collections have been transferred to SAIAB. The National Museum of Namibia, Windhoek holds important collections with type specimens as well as other species collected with them. The lack of trained personnel and a national depository for wet collections in Angola has hindered the discovery of new species in the upland drainages that might populate Namibia's northern rivers.[37]

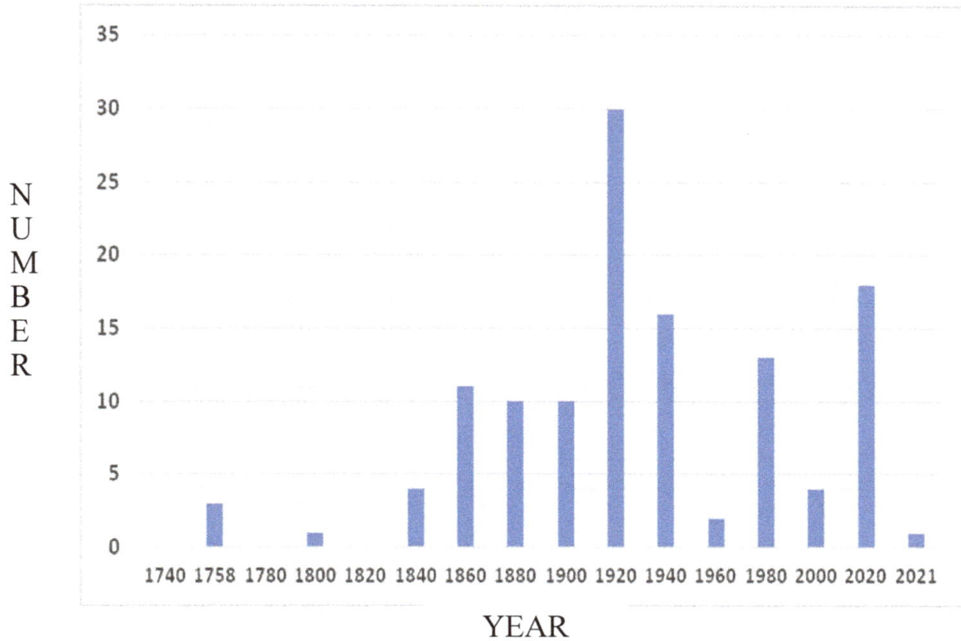

Figure 1. Numbers of native inland species described over time now known in Namibia. Year 1758 represents 'Year 1' in the classification of biota following Carl Linnaeus's *Systema Naturae*[3] that included Nile Tilapia, Goldfish and Common Carp, species introduced to Namibia. The figure does not imply that Namibia was the *type locality* for these species, rather it depicts peak periods of discovery within southern Africa and the presumed secondary influence of World War II after 1940. The far right represents the 2021 description of the Pygmy Topminnow.

Table 1. Examples of inland fish species recorded from Namibia and the respective locations of type specimens.

Cunene River Stonebasher	*Hippopotamyrus longilateralis*	SAIAB, Makhanda, RSA
Southern African Pike	*Hepsetus cuvieri*	British Museum of Natural History, London
Saddled Citharine	*Nannocharax dageti*	Royal Museum for Central Africa, Belgium
Kavango Sand Catlet	*Zaireichthys kavangoensis*	National Museum, Windhoek
Zambezi River Bream	*Pharyngochromis acuticeps*	Naturhistorisches Museum, Vienna
Humpback Largemouth	*Serranochromis altus*	U.S. National Museum, Washington D.C.

II. WHAT IS A FISH?

When one thinks of a fish, an animal that lives in water is envisioned. Fishes are for the most part free-living, aquatic, *poikilotherm* (cold blooded), gill-breathing animals with a hard skull (*craniate*), jaws and fins, rather than having 5-digit limbs (*pentadactyl*). There are, however, many exceptions. A few examples include:

- Walking catfishes (*Clarias* spp.) are not totally aquatic;
- Certain species of marine fishes such as lamnid sharks, tuna and Moonfish are *homeotherms* (warm blooded) that can maintain a stable internal body temperature;
- African Lungfish (*Protopterus* spp.) have both gills and functional lungs to breathe;
- Coelacanths (*Latimeria* spp.) have 5-digit limbs instead of fins (see *Old Four Legs* by J.L.B. Smith 1957);[42] and
- One whole category of very primitive fishes called hagfish that occur in the Southeast Atlantic Ocean off Namibia are vertebrates lacking jaws and a proper vertebral column.

Fishes are the most diverse group of all *vertebrates* (animals with a backbone) and there are more species of fish than all other vertebrate groups (i.e., amphibians, reptiles and mammals. NOTE: technically, birds are reptiles) with an estimated 45,000 species recognised globally. They can be found in all types of aquatic habitats, including :

- water below freezing temperatures (0 °C) in Antarctica to hot springs with temperatures greater than 40 °C;
- elevations 5 km above sea level to 11 km below sea level; and
- freshwater to water with salt content greater than 40 ‰ (parts per thousand) with oceans averaging about 35 ‰.

The most primitive living jawed vertebrates are fishes referred to as coelacanths with one species occurring off the east coast of Africa and the Comoro Islands and another near Indonesia. Modern fishes are separated into two major divisions, commonly referred to as the cartilaginous fish (e.g., sharks, skates and rays) and the bony fish. These groups are as different from each other as mammals are from reptiles. Although no cartilaginous fishes are discussed in this book, they are present off the coast of Namibia and some occasionally enter freshwater.

This book is about freshwater fishes and although available habitat in freshwater lakes and rivers comprise less than 0.0093% of the total water mass on Earth, over 41% of all fish species live in freshwater. Namibia's inland fishes are diverse and occupy a variety of habitats. The mormyrids are primitive fishes capable of generating a weak electrical field. They can detect changes in the field to avoid predators, find prey and identify features of their environment. Barbs, *Labeo* and yellowfish of the family Cyprinidae show a diversity of forms, life history habits and anatomical structures. Catfish found in Namibia also are different from other fishes in that they lack scales and in addition to using gills, can breathe through their smooth skin and/or branchial chambers (*labyrinths*) capable of exchanging atmospheric air for oxygen intake while exhaling carbon dioxide. Also, many catfish have strong sharp *spines* in their dorsal and/or pectoral fins that are used in defence and can inflict severe wounds or give a nasty sting.

To summarise, most fish

- are cold blooded, but some are warm blooded;
- have jaws, but some are jawless;
- have a vertebral column, but some lack one (although with rudimental vertebrae);
- have scales, but some are naked;
- respire by gills, but some have lungs or exchange oxygen through the skin;
- live in water, but some crawl on land while others use their fins for gliding through air;
- are free living, but the Candiru Catfish (*Vandellia cirrhosa*) of the Amazon basin is parasitic; and
- have bony skeletons, while others have cartilaginous skeletons.

As can be seen, fish are difficult to define. They come in all shape and forms, occupy diverse habitats and employ a variety of reproductive strategies.

III. What is a Species?

A. Role of Systematics

Carl Linnaeus, the Father of Taxonomy, published *Systema Naturae* in 1758, a system for the orderly classification of plants and animals based upon their similarities;[3] it is now known as the *Linnaean Hierarchy*. The system (with adaptive changes) is globally used and based upon a series of ranks with *Domain* being highest followed by *Kingdom*, *Class*, *Order*, *Family*, *Genus* and *Species* in descending order.

Systematics is the study of the diversity of organisms and includes taxonomy, classification, nomenclature and identification. *Taxonomy* is the theory and practice of classifying organisms and examines the variation within species, genera, families, etc. *Classification* studies the evolutionary relationship of levels in the Linnaean Hierarchy. For example, why are several species placed in a certain genus, why are selected genera included in one family and why are given families included in a particular order? *Nomenclature* is simply the practice of giving scientific names to particular entities. Finally, *identification* is the ability to assign the proper scientific name to an individual organism. A summary of the features used in systematic studies of fishes, their classification, naming and identification is presented below.

B. Description of a New Species

When describing a new fish species, the *ichthyologist* must define its unique characteristics. These characters may be morphological, genetical, behavioural or physiological differences, e.g., in the case of mormyrids, their electric organ discharge (EODs) patterns are used. Once it is determined that the new form warrants species status, the ichthyologist considers the taxonomic rank in the Linnaean Hierarchy where it belongs, i.e., is it a new species or does it warrant being classified as a new genus or family? Once that is established, the describer must:

1. select one specimen (*holotype:* see *type*) usually from a series of specimens (*paratypes*) upon which to base the species description. The holotype will have been collected from the *type locality*;
2. select a scientific name for the species (see below);

3. provide a description of the new species that may include morphological, genetical, behavioural and life history data;
4. provide information on the range and distribution of the new species;
5. provide a diagnosis of the new species, which supports its placement in a given genus or family and demonstrates why it is unique from all other previously described species; and
6. finally, publish the description in a reputable journal. At that time, the species is formally recognised by the scientific community.

Some new species are easily recognised and acted upon in a timely fashion, while others may await discovery or formal description for decades. Some forms are simply mis-named or mis-identified as new species, this not being corrected until a revision is published. Taxonomy is a long and arduous process that is ever evolving and seldom completed.

C. Nomenclature: Naming of a Species

Species Accounts in Chapters 4 and 5 provide both the common and scientific names of each species. Common names are names of convenience and can change between locations. For instance, small barbs may be called minnows or *ghieliemientjies*. Scientific names are consistent and recognised globally.

The scientific name usually is *binomial,* meaning 'two names' comprised of genus and species, e.g., *Homo sapiens*. On rare occasions, it might contain a third name to recognise subspecies status. Species names are in Latin or Greek. Ideally, the specific name should be Latin if the genus name is Latin. The specific epitaph (name) can be (a) descriptive, (b) named after a person (*matronym* or *patronym*, female or male respectively) or (c) locality (*toponym*). For example, (a) the scientific name of the Redeye Labeo is *Labeo cylindricus*, because its overall body shape is cylindrical; (b) the Southern African Pike, *Hepsetus cuvieri,* was named in honour of French ichthyologist Georges Cuvier; and (c) the Caprivi Killifish, *Nothobranchius capriviensis*, derived its name from the type locality in the Caprivi Region of Namibia (before it was renamed as the Zambezi Region).

Species in the same genus are more alike than species in a different genus. Likewise, genera in the same family are more similar than a genus in a different family, and so forth. This information is lost or confused when using common names. For example, the Slender River Bream, *Chetia gracilis*, is more closely related to the Angolan River Bream, *Chetia welwitschi*, than either is to the Rainbow Bream, *Sargochromis carlottae*, because the first two are in the same genus. If only common names are used, one might think that all three of these species are equally related.

The study of names, and their history is *etymology*, often interesting and informative to laymen and scientists alike. The etymology of Namibian fishes is presented in Chapters 4 and 5, preferably from the original description of the species. Barring that, however, an excellent resource was referred to, i.e., the ETYFish Project that strives to explain the meaning of each fish species name, worldwide.[43]

D. Implications of Systematics to Fisheries Management

Fisheries management in the 21st century will increasingly focus on conserving the biodiversity of fishes and the organisms and environments on which they depend. In theory, proper management will be by those who have sufficient knowledge to identify the species which they are dealing with and whether it is native or non-native in origin. This is not always practical in developing countries such as Namibia, however, which has an environmental assessment protocol under the directive of the Ministry of the Environment, Forestry and Tourism (MEFT) where neither it nor regional consulting firms have qualified *fishery scientists* or ichthyologists.

Given the above limitations, the Linnaean Hierarchy is valuable for fishery scientists and environmentalists to predict that certain strategies used to manage one species may be applied to a different species in the same genus. That is, life histories of phylogenetically related species are often similar (e.g., the various squeaker catfish). There are obvious exceptions (e.g., the Alestiidae contains both the tiny robbers and Tigerfish), but *phylogenetics* provides a starting point for fishery managers.

Namibia has officially recognised several species of 'conservation concern' and their capture and export are expressively prohibited by law. These include the Rock Catfish, Broadhead Catfish, Kunene Broadhead Catfish, Cave Catfish, Cunene Kneria, Shorthead Barb, Namaqua Barb, Largemouth Yellowfish, Caprivi Killifish, Angolan River Bream, Cunene Dwarf Bream, Cunene Bream, Namib River Bream, Thicklipped River Bream and Otjikoto Tilapia.

Biotas are increasingly being placed on the *IUCN* Red List of Threatened Species that tabularises data on species of concern for their management. If a species is sufficiently threatened globally, it may be offered a degree of protection through listing by *CITES*, a multi-lateral treaty of the IUCN. No Namibian fishes are currently CITES-listed although the Cave Catfish, Caprivi Killifish and Otjikoto Tilapia are considered Critically Endangered in Namibia due to their limited distribution and habitat dependence.

IV. Anatomical Features Useful in Fish Systematics

A. General

Most fishes are bilaterally symmetrical with characters on one side mirroring those of the other (Fig. 2). Species accounts often describe fishes from a *lateral* (side) viewpoint, or profile, and the left side is generally used for illustrations and photographs. When viewed from this orientation, *anterior* refers to the forward half toward the head and *posterior* refers to the rear half, or tail end. *Dorsal and dorsum* refer to the top half or back and *ventral* means toward the lower half, belly or underside.

To further delineate regions on the body, these terms may be used together. *Dorsolateral* means the upper sides of a specimen, while *anterioventral* refers to the lower head and breast area. Most fishes found in Namibia are *fusiform* in shape; that is, they are somewhat tapered both anteriorly and posteriorly. When viewed in lateral profile, the body is generally elongate, somewhat slender to moderate in depth or obviously deepened. The spiny eels are extremely elongate, cylindrical and snake-like. When viewed in anterior profile, a fish may be laterally compressed, strongly compressed, dorsoventrally depressed or somewhat rounded (*terete*).

The head of a fish extends from the tip of the snout to the posterior-most margin of the *operculum* that is rigid (i.e., if present the extended soft flap is not measured). The body is divided into the trunk and caudal region. The trunk is that area between the posterior edge of the operculum and the last ribbed vertebra, and the caudal region begins with the first un-ribbed vertebra and extends to the posterior end of the body.

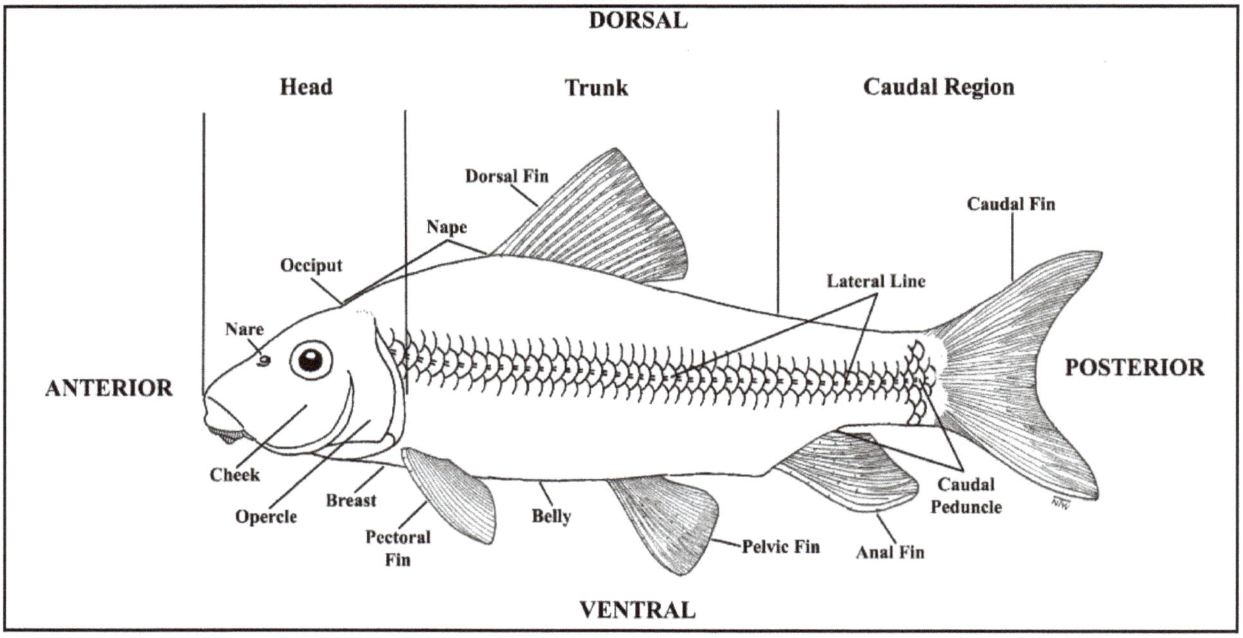

Figure 2. Generalised external fish anatomy with inferior ventral mouth (after Nevin Welte).

The position and orientation of the mouth is nearly as variable as the array of body forms (Fig. 3). From the lateral profile, the anterior end of the mouth may be positioned clearly on the underside of the head (*ventral*), slightly overhung by the snout (*subterminal*), even with the tip of the snout (*terminal*), or clearly opening toward the top of the fish (*superior*). The mouth may be horizontally positioned, or oriented mouthparts from slightly to strongly oblique, i.e., angled.

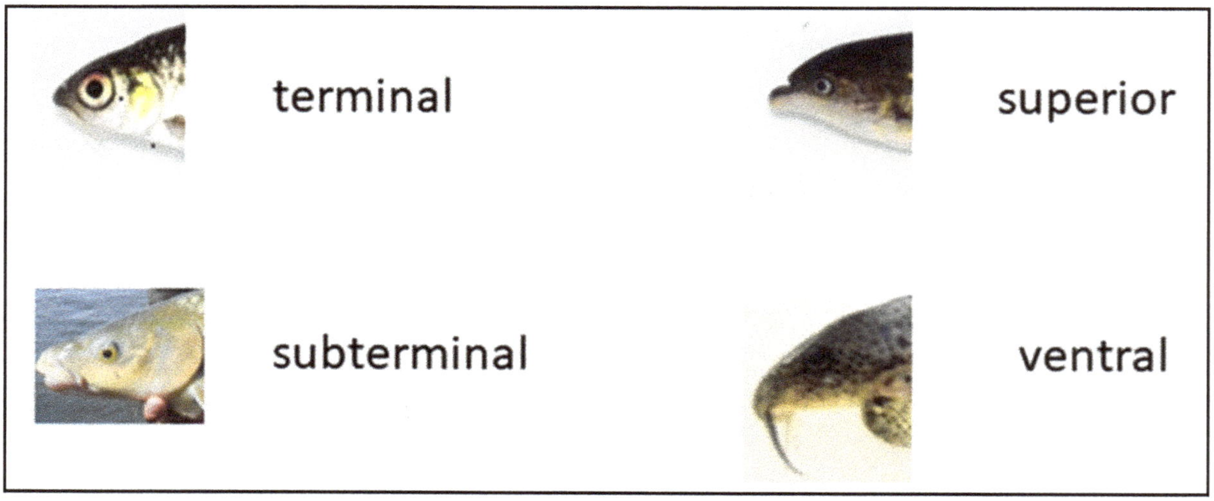

Figure 3. Examples of the position of the mouth in Namibian fishes (terminal: Dashtail Barb; superior: Upper Zambezi Bulldog; subterminal: Vaal-Orange Smallmouth Yellowfish; ventral: Leopard Squeaker) (Photos: B. van der Waal & F. Jacobs).

B. Skeletal Structure

All freshwater fishes in Namibia have a bony skeleton comprised of the *axial skeleton* (includes the bones of the head, vertebral column and median fin structures) and the *appendicular skeleton* (includes the *pectoral* and *pelvic* girdles). The *humeral process*, an outgrowth of the pectoral girdle, is an important character to distinguish the catlets in the family Amphiliidae.

C. Fins

The anterior end of a fin's base is the *origin.* The posterior region is the *distal* area, the middle area is *medial*, and that near the base is *proximal.* Fin margins are variable and may be described as straight, rounded, pointed, *convex* (curved outward or arched), *concave* (curved inward), or *falcate* or *lunate* (sickle-shaped). These terms refer mainly to fins other than the caudal fin. The posterior margin of a caudal fin of Namibian inland fishes may be *truncate* (having a straight vertical profile, e.g., Thicklipped River Bream), rounded (unlobed, with a single rounded margin, e.g., Blackspot Climbing Perch), *emarginate* (shallowly forked, e.g., the sand catlets, *Zaireichthys* spp.), or *forked* (the lobes deeply notched, e.g., Silver Catfish). The relative position of fins on a fishes' body, their length/height and presence/absence are all useful diagnostic features for identifying species.

Fins generally consist of membranes supported by spines or soft rays that are counted to distinguish among species. Spines are usually hard and pointed, but may also be soft, flexible and positioned anterior to rays. Rays are typically jointed, branched and flexible with fused right and left members, but also may be unbranched and inflexible. Spines do not have visible concentric circles throughout their length, while rays do. Some cyprinids (amongst others) have rays that are fused, sharp and commonly called 'spines' as followed here. All Namibian inland species have (a) single *dorsal, anal* and *caudal fins*, termed the *median fins* (except the spiny eels that have a series of individual dorsal spines followed by a soft-rayed dorsal fin), and (b) paired *pectoral* and *pelvic fins* (except the spiny eels lack pelvic fins). Many species have a medial *adipose* fin between the dorsal and caudal fins, but it is fleshy and lacks spines and rays.

D. Skin and Scales

Two layers of skin, consisting of epidermis and dermis, encase the bodies of fishes. The former outer layer has cells that secrete mucus and give most species a slimy look and feel. These secretions protect against infections and reduce friction while swimming.

The skin also contains *chromatophores* and *iridiocytes*, the cells that impart colour or iridescence to a fish, often a distinguishing feature among species and between sexes. Frequent references to the number, pattern and location of chromatophores are made in the Species Accounts (Chapter 4). Colour in nature is important for defence, camouflage and attracting mates. For example, African Pike have colours that blend in with the aquatic vegetation from which they hunt. An important phenomenon is *countershading,* referring to fish being darker dorsally, lighter ventrally, that provides effective camouflage when viewed from above or below, respectively. Fishes that hide in the sand, like the *Zaireichthys* spp., are often a speckled sandy colour. Even stripes and spots can prevent an individual fish from being observed by a predator.

The skin of most fishes is covered by scales. Scale coverage, pattern and arrangement on the head and body describe the *squamation* or *scalation* of a species and are useful diagnostic features. For instance, some Namibian fishes lack scales, such as the catfishes, and are referred to as *naked.* The

Cunene Kneria has so many minute scales it is difficult to count them. Most Namibian fishes are covered with bony ridge scales that are either *cycloid* or *ctenoid* that overlap one another. Ctenoid scales have small *ctenii*, small tooth-like projections, on the exposed margins, while the exposed margins of cycloid scales are smooth. Annual growth rings (*annuli*) are usually evident on ctenoid and cycloid scales, thus facilitating the aging of some species.

E. Jaws and Teeth

The jaws of fishes may be typical or highly modified. The basic bones of the jaw include the (a) *premaxillaries* and the *maxillaries* that form the upper jaw and (b) the *mandible*, or lower jaw. The mandible consists of the *angular* and fused *dentary* bones, as well as the small *articular* bone that is part of the jaw hinge. Bones within the roof of the mouth that are important in the identification of some species include the medial *vomer* and the lateral *palatines*. The ventral region behind the chin and between the angular bones of the mandible is referred to as the *gular region.*

Some jaws are *protractile*; that is, they may be extended forward. Others have a median bridge (*frenum*) that connects the upper lip to the snout, which prevents the upper jaw from being protractile.

Teeth of bony fishes are quite variable in shape and function, and may be located on the jaws, tongue, roof of the mouth and/or on pharyngeal bones, which consist of an upper and lower bone in the pharynx that aids in the mastication or crushing of food. The presence or absence and placement of teeth may be used to distinguish between taxonomic categories such as species or genera.

F. Other Head Elements

Other bones of the head important in distinguishing groups of fishes include the *opercle* (gill cover), *preopercle* (the bone between the cheek and opercle) and *subopercle* (bony plate immediately below and posterior to the opercle). Combined, these bones form the *operculum*, a protective shield for the gills underlying it.

The gills of Namibia's inland fishes consist of five *pharyngeal arches*. The gills exchange carbon dioxide for oxygen when water either (a) is forced over them by movements of the mouth and gill opercle or (b) flows passively over them during swimming. Gill rakers filter and retain food, and depending on the primary prey of a fish species, may be comb-like and densely spaced as in filter feeders or stouter and more widely spaced as in predators.

Barbels, fleshy processes that function as taste and tactile organs, are located on the snout and jaw regions of a number of fishes. Depending on location, they may be considered *nasal* (located near the nares), *maxillary* or *premaxillary* (on the upper jaw), or *mental* (on the chin). They may be elongate and quite obvious, as those of catfishes, or small and barely discernible (some barbs). Their structure, presence/absence, location and colouration may be useful in identification.

The inner ear of fishes functions similarly to that of humans, detecting sounds and gravitational force for maintaining orientation. Three *otoliths* perform these functions in fishes. These calcium carbonate structures, also called "ear stones", are useful in aging many fish species. In some families, including the barbs and catfishes, the inner ear is connected with the swim bladder, which is sensitive to vibrations by a series of bones known as the *Weberian apparatus*. The purpose of this structure is to amplify sound waves.

G. Cephalic-Lateral Line System

This system consists of a network of canals, grooves and pores located on the head (cephalic) and body (lateral). Vibrations in the water from prey, predators and other fishes in a school or obstacles are transferred through the system allowing the fish to assess its environment. Elements of this system are evident externally in many fishes and may aid in the identification of similar species. In some species, they may be interrupted or absent. The lateral line system when present extends onto the body in the form of a *pored lateral line*. It is generally located mid-laterally and if *complete* includes a pore in each scale in the lateral series from behind the opercle to the posterior margin of the *hypural plate*, a bony plate at the base of the caudal fin to which its rays attach. In an *incomplete lateral line*, not all lateral scales will be pored. An *interrupted lateral line* has one or more breaks in the series of pored scales. Some fishes (i.e., cichlids) have incomplete upper and lower lateral lines.

V. Fish Identification

The ability to accurately identify a specimen to the species level is critical to the study of fishes. If specimens are not accurately identified, the results of physiological, behavioural, toxicological, etc. studies will not be reproducible by other investigators, one of the tenants of science. The characteristics of some families and genera make this process rather simple, while for others the identification of an individual to species level can be extremely problematic and frustrating. For instance, there are only minor differences between several cyprinids, while many snoutfishes are morphologically indistinguishable being separated only by their electrical organ discharge (EOD) patterns. Most characters diagnostic for identifying species may be generally classified as *morphometrics* (measurements) and *meristics* (counts).

A. Measurements

Morphometrics deal with a fish's form and the relationships of its various aspects, such as the ratio between length and body depth. Important measurements include *standard length* (SL), *total length* (TL), *fork length* (FL) and *body depth*. Measurements relating to the head include *head length*, *snout length*, *eye diameter* and *postorbital length* (see Glossary).

B. Counts

Meristic counts in this book include those related to scales, fin spines and fin rays. The *lateral line scale count* is commonly used as a diagnostic character. In some species, the pored lateral line is interrupted or absent, in which case the scales in the *lateral series* or *lateral row* are counted. In others, such as cichlids, there are two lateral lines and both rows of pored scales are counted. Another scale count referenced in the book is the total number of *caudal peduncle* (CP) scales, counted around the circumference at the narrowest portion of the tail.

Spines can occur in the dorsal, anal, pectoral and pelvic fins of various fishes, and all are diagnostic in their way. As noted above, some species have hardened fused rays that are commonly treated as spines in the counts. Only spine counts from the dorsal and anal fins are included in the Species Accounts.

Fin-ray counts are important in the identification of many fishes. Potentially, all rays of all fins may be counted, whether branched or unbranched. For most families, however, only the dorsal-fin and anal-fin rays are counted, as followed here. The last two rays of the anal fin are counted as one if they are not branched. If one is branched, they are both counted.

Each Species Account (Chapters 4 & 5) has a spine/ray/scale count that follows the formula: number of dorsal spines (DS), dorsal rays (DR, unbranched, followed by branched); anal spines (AS), anal rays (AR, unbranched, branched); lateral line (LL) and caudal peduncle (CP) scale counts where known. All spine, branched ray and scale counts use Arabic numerals 1, 2, 3 etc. All unbranched ray counts of the dorsal and anal fins use Roman lower case letters i, ii, iii, etc. For instance:

DS 3, DR 10-12; AS 0, AR iii, 14-18; LL 30-34, CP 14

VI. Key to the Families of Namibia's Inland Fishes (*non-native* in green)

1a. Body cylindrical, elongate and serpentine; several separate spiny dorsal finlets; pelvic fins absent – **Mastacembelidae**
1b. Body not serpentine; no separate dorsal finlets; pectoral and pelvic fins present – 2

2a. Opercle bones not observable exteriorly, i.e., beneath the skin; gill slits small; eyes covered with skin – **Mormyridae**
2b. Opercle bones observable; gill slits prominent; eyes not covered with skin – 3

3a. Scales absent – 4
3b. Scales present – 9

4a. Anal fin long, more than 20 rays – 5
4b. Anal fin short, less than 20 rays – 6

5a. Dorsal fin without spines, with more than 20 rays; caudal fin rounded or truncate – **Clariidae**
5b. Dorsal fin with spine, less than 20 rays; caudal fin forked – **Schilbeidae**

6a. Dorsal and pectoral fins without spines; eyes dorsal on head – **Amphiliidae**
6b. Dorsal and pectoral fins with strong spines; eyes lateral on head – 7

7a. Mouth terminal to subterminal – 8
7b. Mouth distinctly ventral – **Mochokidae**

8a. Adipose fin origin forward of anal fin origin; scattered distinct black spots on body and fins, some forming 5-7 vertical bars; northern rivers only – **Claroteidae**
8b. Adipose fin origin over origin of anal fin; rather uniform olive-brown colour, without distinct black spots and bars on body and fins; Orange River only – **Austroglanididae**

9a. Scales very small, >80 in lateral series; difficult to count – **Kneriidae**
9b. Scales distinct, <80 in lateral series; countable – 10

10a. Pelvic fins modified into elongate single rays – **Osphronemidae**
10b. Pelvic fins comprised of numerous rays – 11

11a. Head without scales – **Cyprinidae**
11b. Head covered with scales or bony plates – 12

12a. Anal and/or dorsal fins with true spines – 13
12b. Anal and dorsal fins lack spines – 15

13a. Single lateral line – **Centrarchidae**
13b. Lateral line in 2 series – 14

14a. Anal fin with 3-4 spines – **Cichlidae**
14b. Anal fin with 8+ spines – **Anabantidae**

15a. Scales ctenoid; 2 rows of teeth on both jaws – **Distichodontidae**
15b. Scales cycloid; lacking 2 rows of teeth on both jaws – 16

16a. Adipose fin present – 17
16b. Adipose fin absent – 19

17a. Elongate body; dorsal fin positioned well posterior to pelvic fin origin – **Hepsetidae**
17b. Body not elongate; dorsal fin over or slightly posterior to pelvic fin origin – 18

18a. Presence of a distinct black spot on shoulder (humeral spot) – **Characidae**
18b. No distinct black spot on shoulder – **Alestidae**

19a. Males with elongate gonopodium on anal fin – **Poeciliidae**
19b. Males without elongate gonopodium on anal fin – 20

20a. Origin of dorsal fin anterior to origin of anal fin – **Nothobranchidae**
20b. Origin of dorsal fin posterior to origin of anal fin – **Procatopodidae**

VII. REFERENCES

[1]Jubb, R.A. 1967. The freshwater fishes of southern Africa. AA Balkema, Cape Town: 248 pp.

[2]Skelton, P.H. 2001. A complete guide to the freshwater fishes of southern Africa. Struik Publ., Cape Town: 395 pp.

[3]Linnaeus, C. 1758. Systema Naturae (*translated:* System of nature through the three kingdoms of nature, according to classes, orders, genera and species, with characters, differences, synonyms, places). 10th Ed., Stockholm: Laurentius Salvius.

[4]Lacepede, B.G. 1802. Histoire Naturelle des Poissons 4: 1–728.

[5]Burchell, W. J. 1822. Travels in the interior of southern Africa: Volume 1. Longman, London: 582 pp.

[6]Smith, A. 1840. Illustrations of the zoology of South Africa, Part 9. Smith, Elder and Co., Cornhill, London.

[7]Smith, A. 1841. Illustrations of the zoology of South Africa, Part 14. Smith, Elder and Co., Cornhill, London.

[8]Peters, W.C.H. 1844. Einige neue fische und amphibian aus Angola und Mossambique. Monatsb. Akad. Wiss. Berlin 1844: 32–37.

[9]Peters, W.C.H. 1852. Diagnosen von neuen Flussfischen aus Mossambique. Monatsb. Akad. Wiss. Berlin 1852: 275-276, 681–685.

[10]Braganca, P.H.N., P.H. Skelton, R. Bills, D. Tweddle and A. Chakona. 2021. Revalidation and redescription of "*Lacustricola*" *chobensis* (Fowler, 1935) and description of a new miniature species of "*Lacustricola*" from Southern Africa (Cyprinodontiformes: Procatopodidae). Ichthy. Herpe. 109(1): 123–137.

[11]Braganca, P.H.N., T.G. Smith, E.J.W.M.N. Vreven and A. Chakona. 2020. Integrative taxonomy reveals hidden diversity in the southern African darters genus *Nannocharax* Günther 1867 (Characiformes: Distichodontidae). J. Fish. Biol. 2020: 1–11.

[12]Braganca, P.H.N., R.M. van Zeeventer, R. Bills, D. Tweddle and A. Chakona. 2020. Diversity of the southern Africa *Lacustricola* Myers, 1924 and redescription of *Lacustricola johnstoni* (Günther, 1894) and *Lacustricola myaposae* (Boulenger, 1908) (Cyprinodontiformes, Procatopodidae). ZooKeys 923: 91–113.

[13]Stauffer, J.R., Jr., R. Bills, P.H. Skelton and O.L.F. Weyl. 2020. Re-elevation to species level and redescription of *Serranochromis jallae* and *Serranochromis robustus* (Teleostei: Cichilidae). Zootaxa 4858(1): 126-134.

[14]Stauffer, J.R., Jr., R. Bills and P.H. Skelton. 2021. Four new species of *Serranochromis* (Teleostei: Cichilidae) from the Cuanza and Okavango river systems in Angola, including a preliminary key for the genus. Zootaxa 4908(1): 66–84.

[15]Castelnau, F. 1861. Mémoire sur les poissons de l'Afrique Australe. J-B Baillière et Fils, Paris: 78 pp.

[16]Pellegrin, J. 1936. Contribution a l'ichthyologie de l'Angola. Arq. Mus. Bocage, Lisboa 7: 45–62.

[17]Jubb, R.A. and I.G. Gaigher. 1971. Checklist of the fishes of Botswana. Arnoldia Rhod. 5: 1–22.

[18]Skelton, P.H., M.N. Bruton, G.S. Merron and B.C.W. van der Waal. 1985. The fishes of the Okavango drainage system in Angola, South West Africa and Botswana: Taxonomy and distribution. Bull. J.L.B. Smith Inst. Ichthy. 50: 1–21.

[19]van der Waal, B.C.W. 1991. A survey of the fisheries in Kavango, Namibia. Madoqua 17(2): 113–122.

[20]van Zyl, B.J. 1992. A fish ecological study of the Okavango and Kunene rivers with a special reference to fish production. Ph.D. thesis, Rand Afrikaans University: 380 pp.

[21]Hay, C. J., T. Naesje, J. Breistein, K. Harsaker, J. Kolding, O.T. Sandlund and B.J. van Zyl. 2000. Fish population, gill net selectivity and artisanal fisheries in the Okavango River, Namibia: Recommendations for sustainable fishery. NINA-NIKU Project Report 10: 1–105.

[22]Bruton, M., G. Merron and P. Skelton. 2018. Fishes of the Okavango Delta and Chobe River, Botswana. Penguin Random House, South Africa: 120 pp.

[23]Merron, G.S. 1991. The ecology and management of the fishes of the Okavango Delta, Botswana, with particular reference to the role of the seasonal flood. Ph.D. thesis, Rhodes University, Grahamstown: 171 pp.

[24]Merron, G.S. 2020. Notes on the community structure and biology of fishes in the Okavango Delta. Bots. Notes Rec. 52: 93–99.

[25]Jackson, P.B.N. 1961. The fishes of Northern Rhodesia: A checklist of indigenous species. Dept. Game & Fish., Govern. Print., Lusaka.

[26]Kenmuir, D.H.S. 1983. Fishes of Kariba. Kingstons Ltd., Harare: 131 pp.

[27]Bell-Cross, G. and J.L. Minshull. 1988. The fishes of Zimbabwe. National Museums and Monuments of Zimbabwe, Harare, Zimbabwe: 294 pp.

[28]van der Waal, B.C.W. and P.H. Skelton. 1984. Checklist of fishes of Caprivi. Madoqua 13(4): 303–321.

[29]Hay, C.J., T.F. Naesje, S. Kapirika, J.H. Koekemoer, R. Strand, E.B. Thorstad and K. Harsaker. 2002. Fish populations, gill net catches and gill net selectivity in the Zambezi and Chobe Rivers, Namibia, from 1997 to 2000. NINA Proj. Rept. 17: 1–88.

[30]Tweddle, D. 2010. Overview of the Zambezi River system: Its history, fish fauna, fisheries, and conservation. Aquat. Ecosyst. Health Manag. 13(3): 224–240.

[31]Marshall, B.E. 2011. The fishes of Zimbabwe and their biology. Smithiana Monogr. 3: 291 pp.

[32]Poll, M. 1967. Contribution à la Faune Ichthyologique de l'Angola. Publicações Culturais 75. Companhia dos Diamentes de Angola (DIAMANG), Lisbon: 381 pp.

[33]van der Waal, B.C.W. 1991. Fish life of the Oshana Delta in Owambo, Namibia and the translocation of Cunene species. Madoqua 17(2): 201–209.

[34]Hay, C.J., B.J. van Zyl, F.H. van der Bank, J.T. Ferreira and G.J. Steyn. 1997. A survey of the fishes of the Kunene River, Namibia. Madoqua 19(2): 129–141.

[35]Hay, C.J., T.F. Naesje and E.B. Thorstad. 2008. Fish populations, gill net catches, and gill net selectivity in the Kunene River, Namibia. NINA Project Report 325: 1–98.

[36]Bills, R., P. Skelton and F. Almeida. 2012. A survey of the fishes of Upper Okavango River system in Angola. SAIAB Invest. Rept. No. 73: 62 pp.

[37]Skelton, P.H. 2019. The freshwater fishes of Angola, pp. 207–242. In: B.J. Huntley, V. Russo, F. Lages and N. Ferrand (Eds.). Biodiversity of Angola, Science & Conservation: A Modern Synthesis. Springer Nature, Switzerland.

[38]Mutizwa, T.I., W.T. Kadye and A. Chakona. *in press*. Deep genetic and morphological divergence in the *Hippopotamyrus ansorgii* species complex (Teleostei: Mormyridae) in southern Africa. J. Fish Biol. 2021.

[39]Skelton, P.H., J.R. Stauffer, A. Chakona and J.M. Wisor. 2021. Two new species of African bubble-nesting *Microctenopoma* (Teleostei: Anabantidae) from Angola. Ichthy. Explor. Freshw., IEF 1134: 1-16. doi.org/10.23788/IEF-1134.

[40]Trewavas, E. 1936. Dr. Karl Jordan's expedition to South-West Africa and Angola: The freshwater fishes. Novitates Zool. XL: 63–74.

[41]Kambikambi, M.J., W.T. Kadye and A. Chakona. In press. Allopatric differentiation in the *Enteromius anoplus* complex in South Africa, with the revalidation of *E. cernuus* and *E. oraniensis*, and description of a new species, *E. mandelai* (Teleostei: Cyprinidae). J. Fish Biol.

[42]Smith, J.L.B. 1957. Old Fourlegs: The story of the coelacanth. Readers Union, Longmans, London: 260 pp.

[43]https://etyfish.org

CHAPTER 3
DRAINAGE HISTORY AND THE ORIGINS OF NAMIBIA'S INLAND FISHES

by

Charles H. Hocutt

I. Geological Setting

The southern African landscape has evolved in different ways, in different places and at different times, hence there is no single 'narrative' of landscape evolution.[1] The hydrology is complex and characterised by numerous inland deltas, *palaeo*-lakes, cataracts and *elbows of river capture*.[2] Even for the largest rivers on the continent, questions remain how and when they obtained their present-day form[3] and how fish distributions occurred.[4,5,6]

A. Gondwana

From the perspective of the Southern Hemisphere, it is generally accepted that South America, Africa, the Arabian Peninsula, Madagascar, India, Australia and Antarctica were joined together >500 million years ago (*Ma*) forming a super-continent *Gondwana* (or Gondwanaland), although they were not in their current geographical positions being close to the South Pole. Geological[7] and biological evidence[8] for the existence of the *palaeo*-continent is abundant (Fig.1). The Cape Fold Mountains, an 850-km long belt of parallel ridges on the south and western tip of South Africa, were created as plates collided to form Gondwana.

Gondwana experienced a 70 million year (*My*) period of glaciation ~340–272 Ma known as the Late Palaeozoic Ice Age or LPIA (Fig. 2).[10,11,12,13] The LPIA glaciation originated from multiple glacial centres globally, not necessarily at the same time, and its gradual demise led to a greenhouse effect.[14]

The LPIA is evidenced today in southern Africa by glacial *tillite* deposits known as the Dwyka Group.[15] Southern Africa was not covered by a single ice sheet, but smaller ice sheets, ice caps and alpine glaciers that ebbed and flowed with prevalent climatic conditions.[16] The Dwyka Group in Namibia extends intermittently from the Kunene River southward to the Huab River and beyond to Mariental, Keetmanshoop and Aranos,[7,12,15] continuing into South Africa (Fig. 3).[13]

Phenomena associated with glaciers are ice streams, i.e., channels filled with ice and tillite flowing faster than the surrounding ice sheet and being the main conduits for ice flow from the sheet's centre to the margin. The famed Henno Martin of *The Sheltering Desert* saga and a renowned geologist recognised tillite in the Kunene River basin[10] below the 120-m high Ruacana Falls,[17] noting it was certain to occur in nearby drainages. Such ice streams occupied the *proto*-Huab[12], Hoarusib and Kunene Rivers[102] draining from a 'Kaokoveld ice sheet' likely discharging into a shallow marine environment of Brazil or Uruguay[18] prior to the breakup of Gondwana. The preservation of an intact *fjord* network dating to the LPIA has been described recently for northwest Namibia (Fig. 4), corresponding to an extensive, ~50,000 km^2 glacial landscape. The ice sheet was up to 1.7 km thick.[102] At least two glacial advance-retread events have been identified for the region.[18,19]

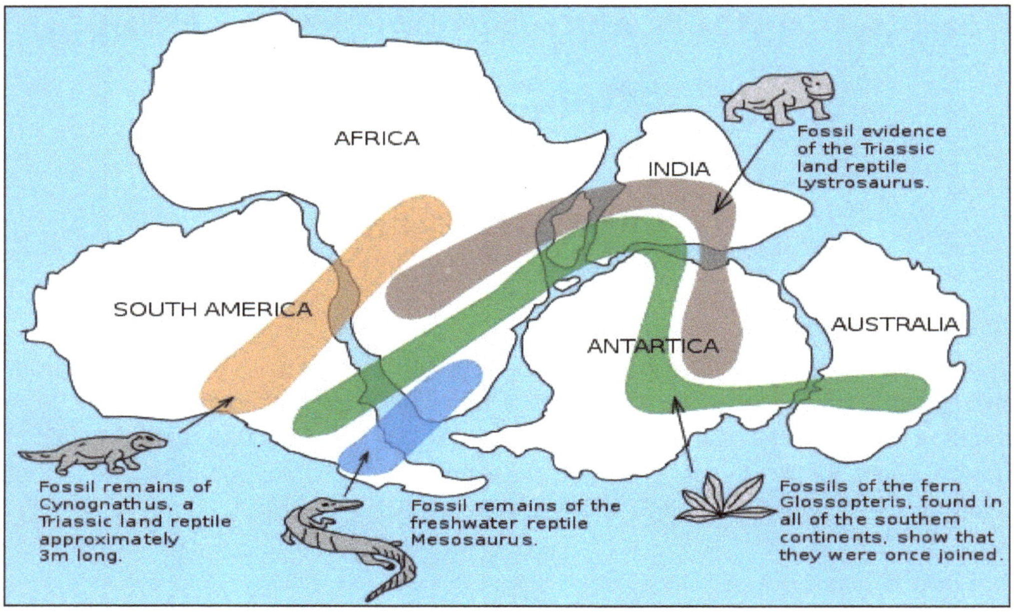

Figure 1. Examples of fauna and flora supporting the geological evidence for the super-continent Gondwana.[9]

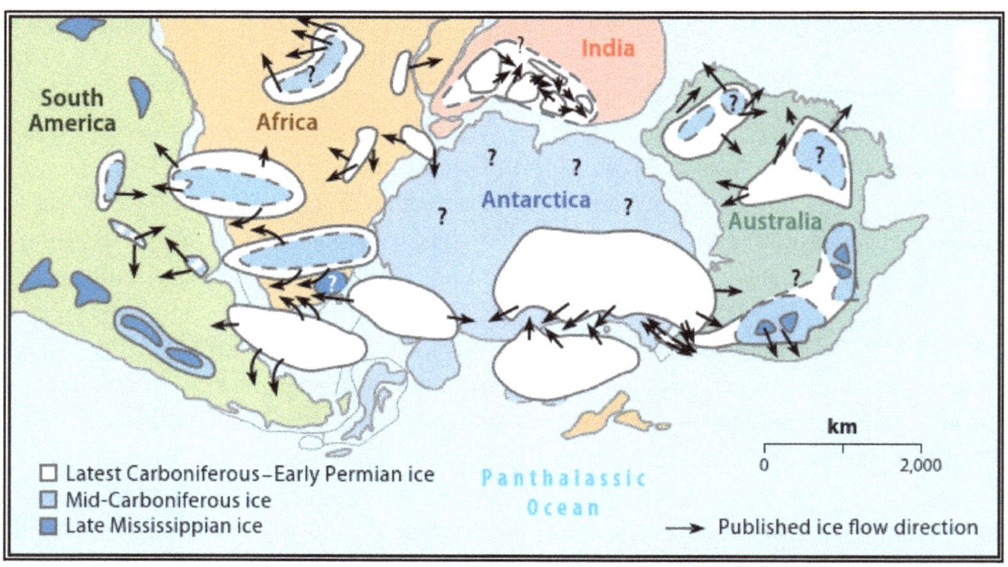

Figure 2. Reconstruction of Late Palaeozoic Ice Age (LPIA) distributed over 21 ice sheets covering 18.6 x 10^6 km^3 of Gondwana, excluding the Arabian Peninsula.[14]

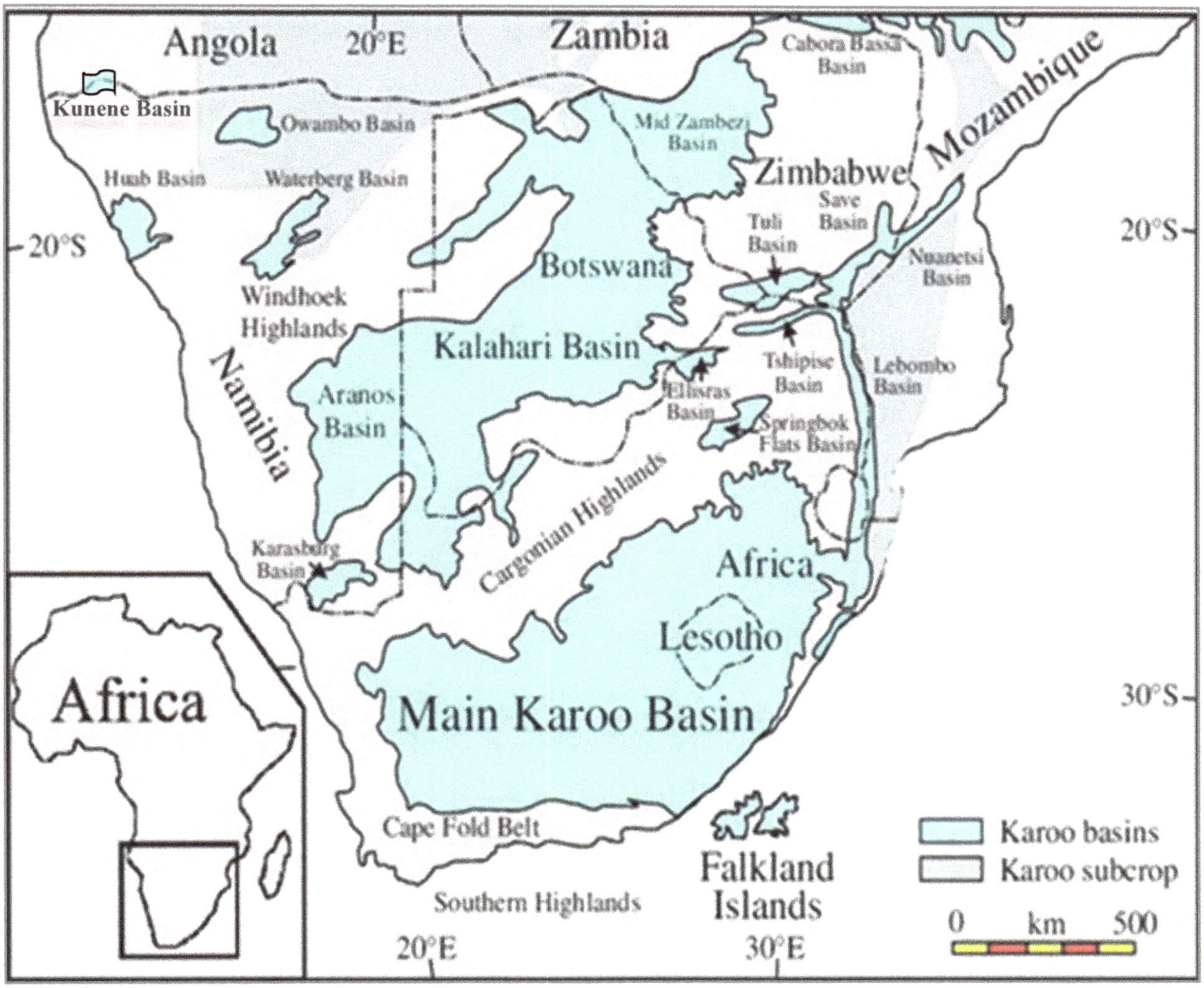

Figure 3. Distribution of Dwyka tillite deposition (blue) in southern Africa.[16] The Falkland Islands are depicted in their Gondwana position.

The development of southern Africa's inland continental waters began about 180 Ma when Gondwana commenced to rift apart. The breakup was in stages with the first major rifting at ~180 Ma essentially dividing Gondwana into western (South America-Africa-Arabian Peninsula) and eastern (Madagascar-India-Australia-Antarctica) portions that drifted apart.[20,21] The next major stage in terms of southern Africa was about 130 Ma when Africa and South America begin to split, forming the *proto*-Southeast Atlantic Ocean followed by an uplift of the continent and regional tilting.[22] Afterwards, Africa experienced a long period of tectonic calmness with events mainly being episodic and concentrated in existing rifts. Another major event was initiated about 30 Ma, continuing today with the Arabian Peninsula spreading away from northern Africa and the formation of the Red Sea, Gulf of Aden and the East African Rift System.[23]

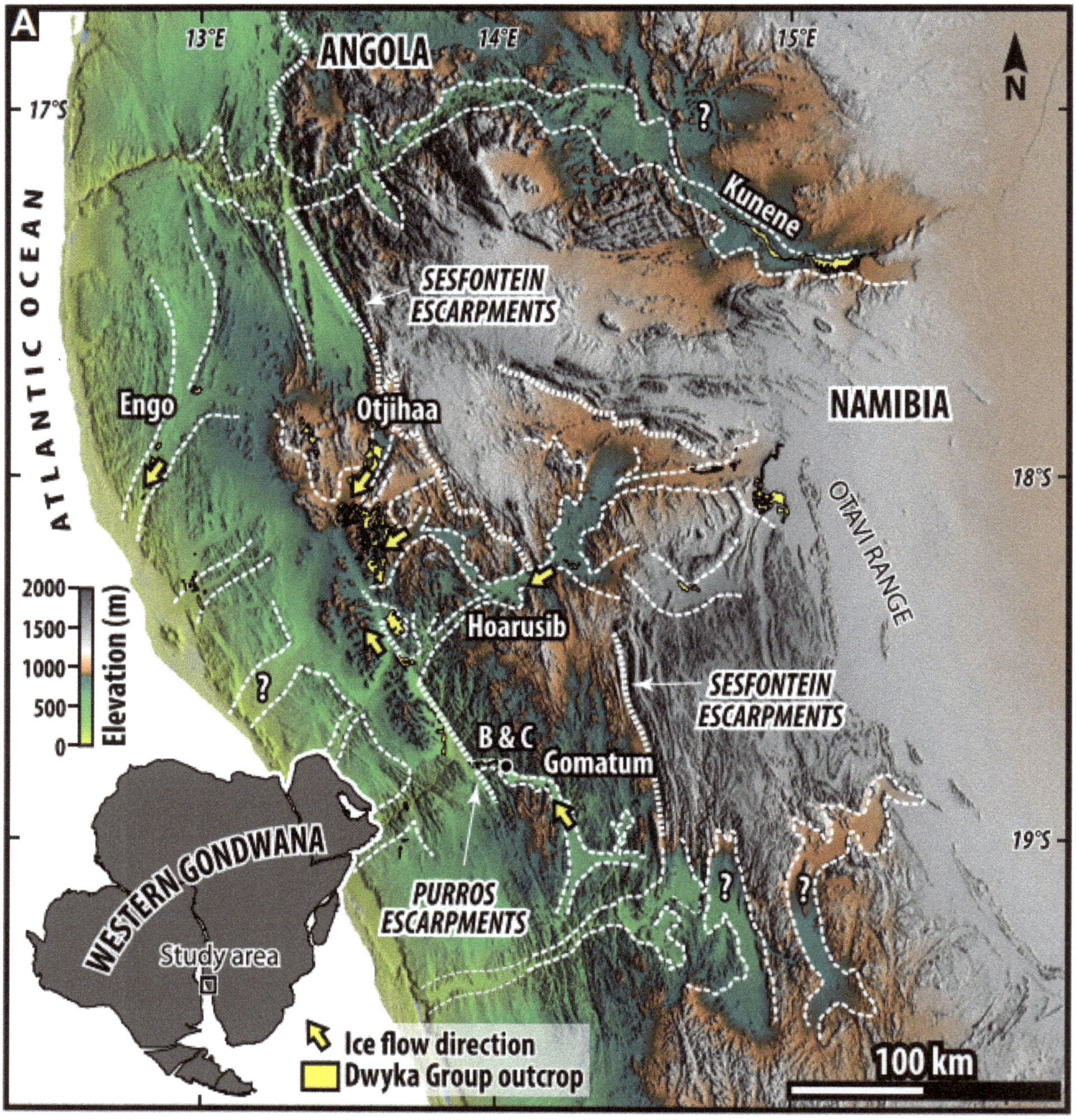

Figure 4. Northwest Namibia with glacial valleys depicted (white dashed lines) dating to the Late Palaeozoic Ice Age (LPIA) ~300 Ma. Northward and westward *palaeo*-ice flows are indicated (yellow arrows). Ice retreat was accompanied by postglacial marine incursions that turned the valleys into *fjords* that were subsequently drowned with rising sea level and covered with non-glaciogenic sediments for ~170 million years (My), leaving an excellent state of preservation.[102]

B. Great Escarpment

The uplift 130 Ma left the almost horseshoe-shaped, 5,000-km long Great Escarpment of Africa (Fig. 5) extending from Angola southward through Namibia to South Africa, eastward around the Cape, then north along the Zimbabwe-Mozambique border.[24] The escarpment was bounded by a narrow band of coastal streams along its flank and a high elevation plateau to the interior. These coastal streams, although short, had steeper gradients and erosive power than interior rivers, thus resulting in (1) incisions cutting deep into the escarpment, (2) gorge and waterfall development and (3) retreat of the escarpment toward the interior.[25] The retreat has been variable depending upon location and the controlling factors of tectonic processes, climate and topography. For instance, retreat near the border region of South Africa and Lesotho has been estimated at 50–95 m/My, but ~16 m/My for Namibia.[1]

The ~2,573-m high Brandberg Mountain, 90 km from the coast and highest point in Namibia; 2,347-m high Gamsberg, 170 km from the coast; and the central highlands of the Khomas Hochland, are all components of the Great Escarpment in Namibia (Fig. 6).[26] The interior plateau in the region of the Gamsberg has a mean elevation of 1,800 m above sea level which is gently inclined southward with a gradient of 1%.

Figure 5. Great Escarpment of southern Africa (gray band) extending from Angola southward through Namibia to South Africa, eastward around the Cape, then north along the Mozambique-Zimbabwe border.[24]

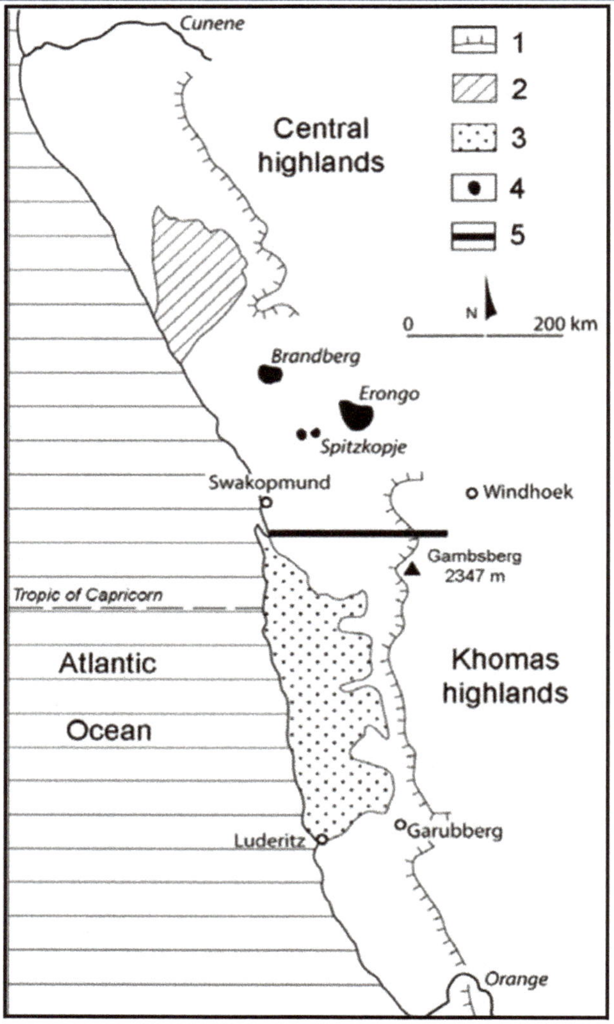

Figure 6. Major geomorphological regions of Namibia: (1) Great Escarpment; (2) Etendeka lava fields; (3) Namib Sand Sea; (4) subvolcanic complexes; and (5) west-east profile across the central Namibian piedmont.[26]

C. Kalahari Basin

The Kalahari Basin (Fig. 7) is key to fish distribution in south-central Africa.[6] The basin formed in stages commencing ~130–100 Ma associated with the rifting of Gondwana and uplift of the continental margins. This was followed by a down-warp of the Earth's crust forming a huge concave depression in south-central Africa,[27] thus creating an unique continental landscape with major catchments draining to the interior rather than to the exterior's adjacent seas.[28,29] The basin is identified by its sediments, the Kalahari Group, deposited over the past 70 million years (My) with their greatest depths today approaching 450 m across the Namibia-Angola border. It reached 2.5 million km^2 in size and extended 2,200 km from the Democratic Republic of the Congo (DRC) with connections to the Congo River,[6] southward through Zambia and Botswana to South Africa, and from Namibia east to Zimbabwe.[30,31] With many regional rivers flowing into it, the basin was a huge wetland *archipelago* that facilitated dispersal of ancestral fish stocks within and around it.[28]

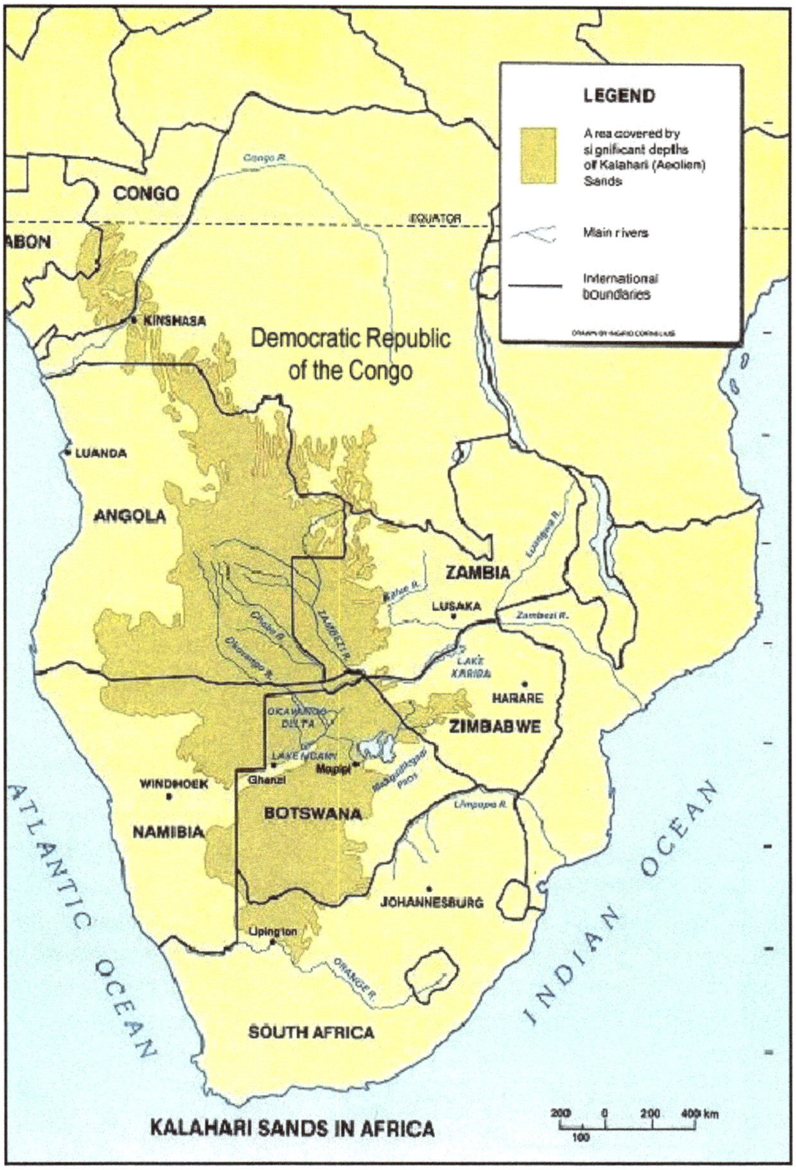

Figure 7. The Kalahari Basin is generally identified by its sediments, the Kalahari Group (tan colour),[32] although the sediments do not fully define the basin.[33]

The modern drainage network of southern Africa developed as the Kalahari Basin evolved, characterised by three river divides, broadly parallel to the coastline (Fig. 8). The drainage divides are associated with three discrete axes of flexure uplift (without mountain building) as a consequence of the breakup of Gondwana and subsequent plate reorganisation surrounding southern Africa.[34,35] The outer arc followed the crest of the Great Escarpment, separating relatively short coastal rivers from the main drainages of the inland plateau. Uplift along the Escarpment Axis was initiated by the opening of the Indian and Atlantic Oceans at ~126 Ma.

The middle divide along the Etosha–Griqualand–Transvaal Axis (EGT) dates to 85–65 Ma. It separates the Orange River basin from the southern Limpopo and cuts across an abandoned former link between the Molopo-Nossob and the Orange Rivers.[34] The EGT Axis is inferred on the basis of fossil wood preserved in the Mahura Muthla *palaeo*-channel astride the axis.

The innermost divide, known as the Ovambo–Kalahari–Zimbabwe (OKZ) Axis, separates (a) the major Limpopo and Zambezi drainage basins in Zimbabwe and (b) the Limpopo from fossil endoreic river lines in Botswana. It dates to 44–33 Ma.

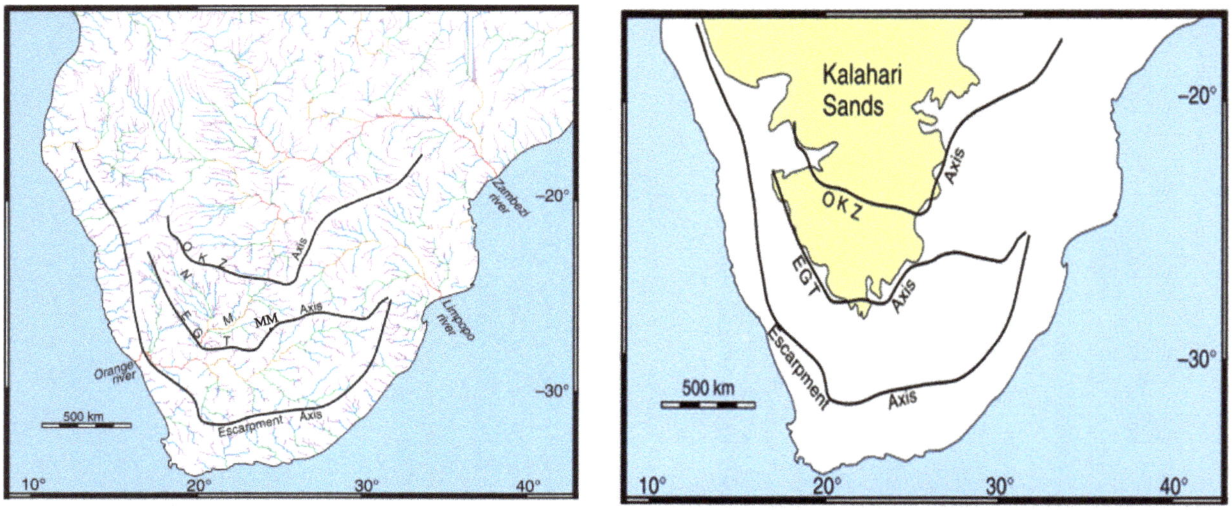

Figure 8. Three drainage divides superimposed upon the current drainage patterns of southern Africa (left). These divides have been interpreted as broad flexure axes of uplift correlated with the breakup of Gondwana and plate reorganisation. The divide associated with the Great Escarpment is the oldest at ~126 Ma, becoming progressively younger to the interior with the EGT Axis (Etosha-Griqualand–Transvaal) dated to 85–65 Ma and the OKZ (Ovambo–Kalahari–Zimbabwe) Axis at 44–33 Ma. The EGT Axis basically is the southern boundary of the Kalahari Basin, while the younger OKZ Axis bisects the basin (right), virtually forming the southern boundary of *palaeo*-Lake Makgadikgadi (M = Molopo River; N = Nossob River; MM = Mahura Muhtla).[34,35]

The Kalahari Basin receded due to the combined forces of continued tectonic activity, climatic change and river captures by the coastal streams cutting through the Great Escarpment on the east, west and south, including the drainage realignments invigorated by the EGT and OKZ Axes uplifts. Smaller sub-basins formed as climates changed and the Kalahari Basin dried; the *palaeo*-Etosha (*aka* Owambo) Basin (Fig. 9) and *palaeo*-Lake Makgadikgadi (Fig. 10) were prominent among them.

The Etosha Basin,[36] *palaeo*-Lake Kunene and *palaeo*-Lake Etosha[37,38] were intermediate stages between the Kalahari Basin and the current Etosha Pan. *Palaeo*-Lake Kunene was fed by the 'southward draining' *palaeo*-Kunene, Cuvelai and Kavango Rivers with the Kavango being the foremost supply before its current channel commenced to form to the north about 7–6 Ma,[36] thus greatly diminishing inflow to the lake. Adding to the demise of Lake Kunene were cycles of inundation during the rainy seasons and extensive weathering and erosion during the increasingly dry climate.[39] Lake Kunene possibly drained west to the *proto*-Southeast Atlantic Ocean by the Hoanib River prior to an uplift[40] of >400 m that closed the route ~3 Ma, giving the lake a surface area[27] of ~82,000 km^2 and being 45 m deep.[40] A more recent uplift rejuvenated westward streams flowing off the Great Escarpment, including the Atlantic coastal *proto*-Kunene River that cut through the Great Escarpment to create Ruacana Falls before eventually capturing the *palaeo*-Kunene River at about 35 ka. Afterwards, the fragmented Cuvelai Basin became the primary source of discharge to Lake Kunene; this inflow fluctuated and diminished through the years and Lake Kunene senesced into a smaller sub-basin, Lake Etosha, predecessor to the current Etosha Pan that is located at the greatest lake depth in Namibia. The latest wet episode at Etosha dates to ~2,500 years ago.[37,38]

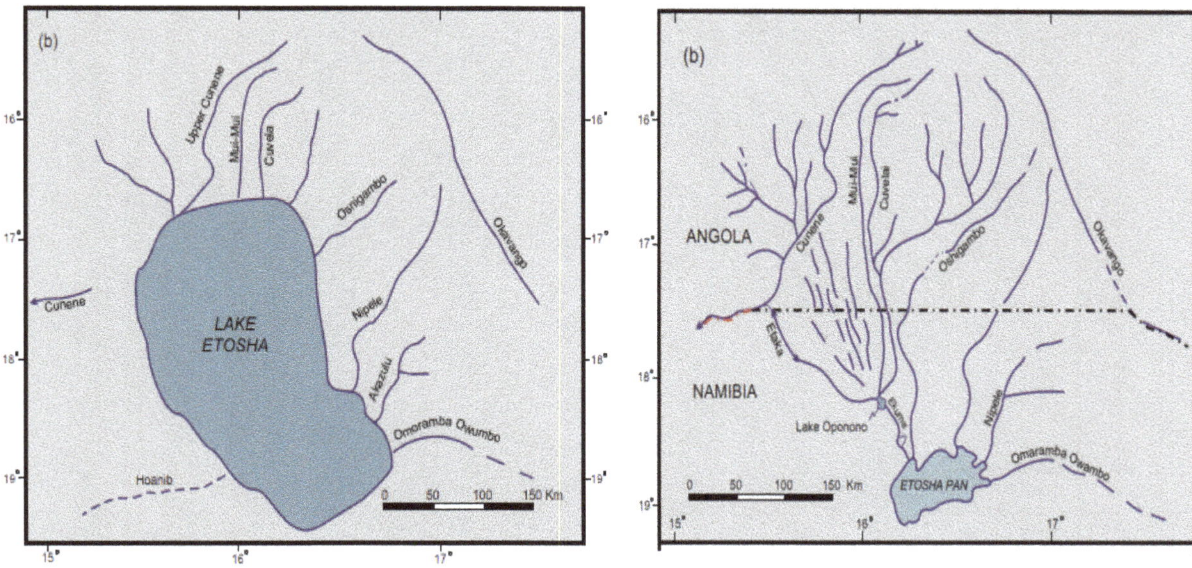

Figure 9. Lake Etosha ~3 Ma (left) and Etosha Pan 35 ka to present (right). *Omuramba* Owambo in the southeast corner was a tributary to the *Omuramba* Omatako of the Kavango River that had meandered toward its current position 7–6 Ma. The Hoanib River (left) was possibly the outlet of *palaeo*-Lake Kunene/Etosha to the *proto*-Southeast Atlantic Ocean before the *proto*-Kunene River captured the *palaeo*-Kunene River ~35 ka. Prior to that capture, the Etaka *oshana* (right) may have been a dispersal route to the *proto*-Kunene River.[27]

Palaeo-Lake Makgadikgadi (Fig. 10) formed ~1.5–0.4 Ma principally in Botswana, but included portions of Angola, Namibia, Zambia and Zimbabwe with its southern boundary limited by the OKZ Axis flexure (Fig. 10). The lake was unstable and filled at least seven times,[41] establishing shorelines at five distinct elevations (~995, 945, 936, 920 and 912 m).[42] It reached up to 175,000 km^2 in size at its highest levels, more than twice the size of Lake Victoria (68,000 km^2), before senescing into Makgadikgadi Pan, Okavango Delta, Lake Ngami, Deception Pan and lesser pans.[42,43] As explained below, the southward-flowing *palaeo*-Upper Zambezi River was the master drainage of the Angolan highlands that supplied the *palaeo*-lake that had outlets through the Limpopo-Shashe River to the Indian Ocean. This changed with regional uplift, diversion of the Upper Zambezi into the Middle Zambezi River and creation of Victoria Falls, or Mosi-oa-Tunya.[44]

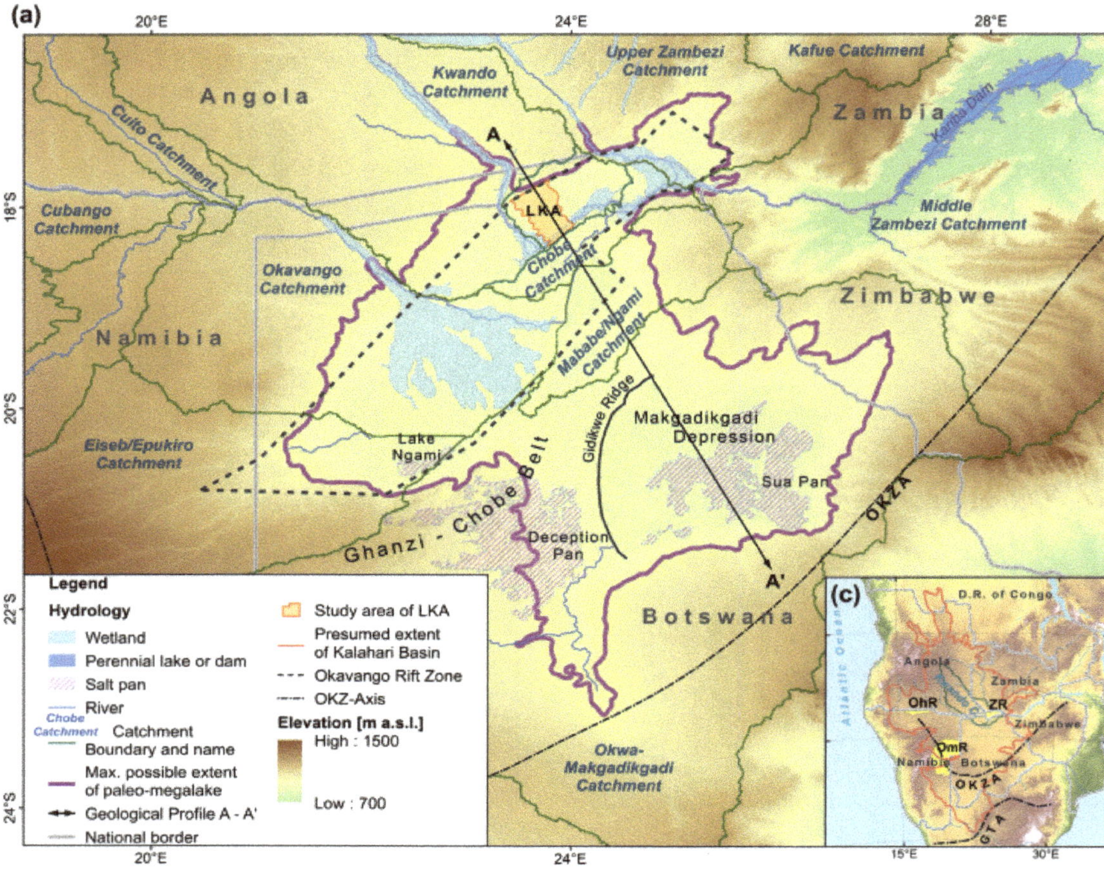

Figure 10. *Palaeo*-Lake Makgadikgadi (outlined in purple) with international boundaries (gray lines) and other local features noted,[41] dating from about 400,000 to 2,000 years ago (YA) when it dried up after having served as a centre of radiation and dispersal of inland fishes. Coincidently, Lake Makgadikgadi was also a base from which modern *hominins* dispersed 130–110 ka.[45,46]

II. Drainage Patterns of Regional Rivers

It is generally agreed that southern Africa received its ancestral fish stocks from the north through headland connections, influenced by the interconnected waterways of the Kalahari Basin. Relative to most of south-central Africa including Namibia, the headwaters of the *palaeo*-Upper Zambezi and *palaeo*-Congo Rivers were intimately linked allowing for the exchange of fishes through contiguous swamps or headwater *stream captures*.[47,48] These connections generally closed in time, but remain visible today through (a) elbows of capture, i.e., sharp changes in stream flow direction, (b) *mekgacha,* or dry 'fossil' valleys found in the Kalahari Desert and (c) shallow drainage divides that overflow during peak flooding. The timing of migrations did not occur simultaneously for all stocks and some representatives of fishes in the same families arrived at different intervals.

The Orange River drainage to the south is ancient. It received ancestral fish stocks very early that (a) spread both within and beyond its current basin, (b) became isolated and (c) evolved into different species in some cases. The routes of dispersal from the Kalahari Basin and *palaeo*-Lake Makgadikgadi, its senesced phase, must have intermittently opened and closed.

A. Upper Zambezi River and Namibia's northern rivers

The *palaeo*-Upper Zambezi River (UZR) evolved into the master stream of the upper Angolan highlands after the breakup of Gondwana and flowed southeast through Namibia beyond today's Okavango Delta in Botswana to the Indian Ocean via the Limpopo-Shashe River valley (Fig. 11). There is evidence that *palaeo*-Lake Etosha had an outlet to the Kavango River via the *Omuramba* Owambo[37] (Fig. 9; Chapter 7, Fig. 1). Thus, all of Namibia's northern water courses from the *palaeo*-(upper) Kunene and Cuvelai Rivers eastward through the current Upper Zambezi River were tributaries to the *palaeo*-UZR. The similarity in the current day fish fauna of these systems, an "Upper Zambezi fauna", supports the geological evidence.[6,49]

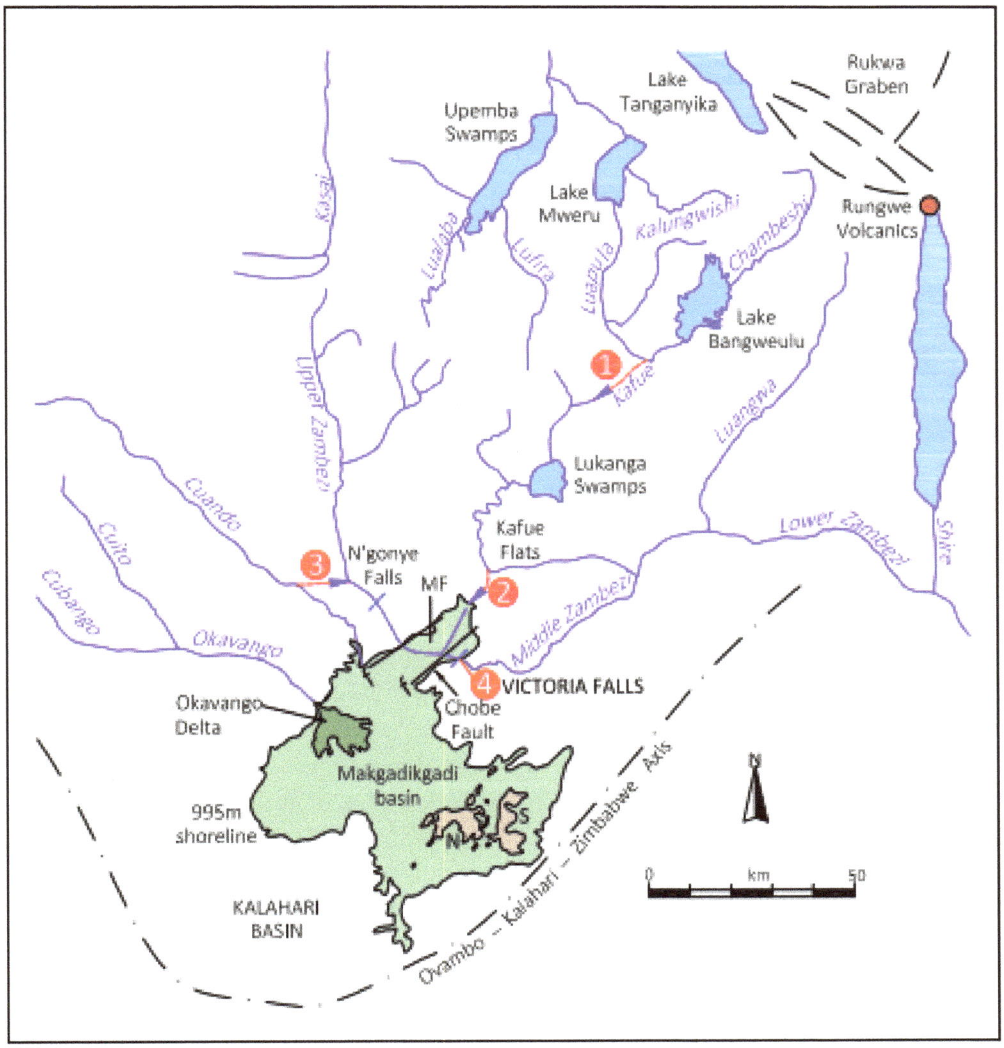

Figure 11. *Palaeo*-Lake Makgadikgadi, *palaeo*-Upper Zambezi River and Victoria Falls. Dashed lines represent former river links that once fed into the Lake Makgadikgadi: (1) abandoned Chambeshi-Kafue link; (2) abandoned Kafue-Zambezi link that flowed through the Machile Flats, MF; (3) abandoned Cuando-Upper Zambezi link; (4) Victoria Falls. Pale green shows the maximum extent of *palaeo*-Lake Makgadikgadi (N = Ntwetwe Pan, S = Sowa Pan).[43]

The Middle and Lower Zambezi Rivers evolved as separate systems,[2] related only in more recent times to the UZR after the creation of Victoria Falls (Fig. 12). With the breakup of Gondwana, the Lower Zambezi was a small coastal river heading up on the Great Escarpment flowing into the Indian Ocean, but it had strong erosive power. To the interior laid the *palaeo*-Middle Zambezi River that may well have flowed northeast, toward an eastward flowing *palaeo*-Congo River,[50] until it was captured by the Lower Zambezi River.[51] In a step-wise progression, the Middle Zambezi, lying about 400–500 m in elevation below the *palaeo*-UZR, captured the latter. The marked difference in elevation would have resulted in rapid headwater erosion as the two sections of the river were united.[43] It has been estimated that the rate of recession of the gorge below Victoria Falls was 42–80 mm/annum.[25]

The *palaeo*-Chambeshi River, originally a component of the Congo basin, seemingly reversed its flow twice. It was captured by the *palaeo*-Upper Zambezi-Kafue system, including the Upemba and Bangweulu districts, Zambia. Its flow was directed via the *palaeo*-Kafue River into the *palaeo*-UZR that fed *palaeo*-Lake Makgadikgadi (Fig. 11). The lake's level dropped from 990 to 945 m above sea level with the recapture of *palaeo*-Chambeshi River by the Congo system.[42,51]

The *proto*-Kafue River was isolated from the *palaeo*-UZR ~500 ka resulting in the formation of *palaeo*-Lake Patrick, which inundated an area of ~17,000 km^2 in the vicinity of the present day Kafue Flats.[52,53] The *proto*-Kafue was subsequently captured by the Middle Zambezi River 300–200 ka, diverting another significant watershed from the *palaeo*-UZR. This resulted in *palaeo*-Lake Makgadikgadi's shoreline receding to 936 m.[42] The similarity of today's fish fauna between the Upper Zambezi and Kafue Rivers affectively demonstrates how drainage evolution has influenced modern-day distributions.[6,28,42,54]

Palaeo-Lake Makgadikgadi continued to diminish to (a) 920 m as headwaters of the UZR were impounded to form *palaeo*-Lake Bulozi and (b) 912 m as the UZR was increasingly diverted through the Batoka Gorge as Victoria Falls evolved (Fig. 12). At this point, Lake Makgadikgadi was supplied only by the Kavango and Kwando Rivers with inflow quantities too little to sustain it.[42]

The separation of the *palaeo*-UZR and Lake Makgadikgadi's outlets to the Indian Ocean through the Limpopo-Shashe, and possibly Save Rivers,[53] is dated approximately to the OKZ regional uplift that created the conditions for the Okavango Delta. Also likely contributing to the lake's termination were (a) reduced regional rainfall,[30] (b) evaporation and transpiration and (c) the final stage of the UZR being diverted over Victoria Falls,[53] perhaps by a cataclysmic flood[55] that overwhelmed the divide at Chimamba Rapids 40.71 km downstream of the modern falls. The erosion of the Batoka Gorge at Chimamba Rapids at ~669 ka was significant in that elevation was a requirement to maintain the level of *palaeo*-Lake Makgadikgadi at 945 m.[44,53,56]

It is estimated that the 100-km long Batoka Gorge below the modern falls eroded at a rate of 0.042–0.080 m/yr over the past 2.54–1.4 My. The distance of the modern falls to the end of the third gorge, 2.96 km, was eroded over the past 71–57 ka at a rate of 0.042–0.052 m/yr.[42]

Thus, all of Namibia's northern rivers historically drained across the continent from northwest to southeast to discharge into the Indian Ocean, excluding the *proto*-Kunene. The *proto*-Kunene River as the exception was at the same time a minor coastal stream heading up on the western edge of an uplifted escarpment and flowing west into the *proto*-Southeast Atlantic Ocean after the rifting apart of Gondwanaland. The *proto*-Kunene eroded through the Great Escarpment and captured the upper *palaeo*-Kunene River, a western tributary to the Cuvelai basin, at a purported date of ~35 ka.[30]

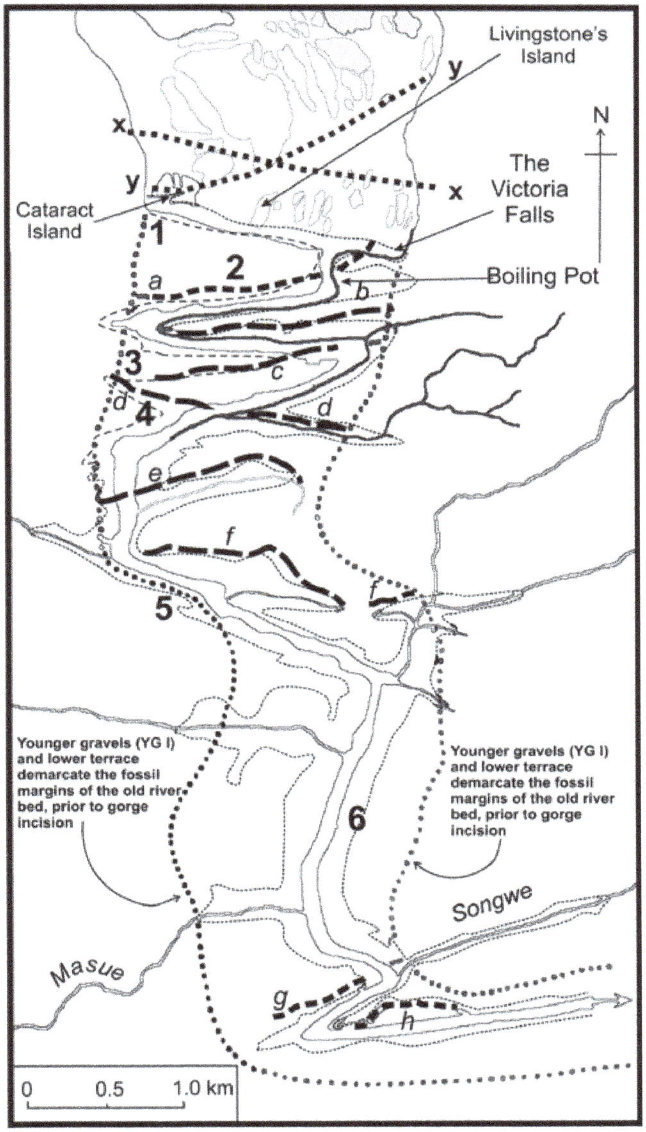

Figure 12. Victoria Falls, its gorges (1–6) and past positions (a–h, dashed lines). The edges of the Batoka Gorge approximated by fine dashed lines. Dotted lines x–x and y–y note where future falls could develop.[44]

Henno Martin[10] recognised that tillite from the Lower Palaeozoic Ice Age ~340–272 Ma below Ruacana Falls had facilitated the *proto*-Kunene in capturing the *palaeo*-Kunene above the Escarpment:

> *In the Kunene valley the hard, basal tillite is overlain by boulder-shale that contains yellow concretions of marl ... the Dwyka beds in the Kunene valley have furthermore been locally folded and contorted by the glaciers. The big Ruacana falls owe their existence to the erosion of the soft boulder-shale, and in those parts of it where it has just been stripped of Dwyka beds ..."*

Although many fishes of today's Kunene River are distinctly related to the UZR fauna, others have presumably followed a north to south route along the coast[6,57] or through headland connections.[47] Many Zambesian species are absent or lack generic or family affinities in the Kunene (Tigerfish; many cyprinids, especially *Labeobarbus* spp.; various catfishes; climbing perch; spiny eels). However, 14 of 65 fishes native (21%) to the Kunene-Cuvelai systems are endemic, a consequence of the isolation of *palaeo*-Etosha (aka Owambo) Basin from the Zambezi influence. With the drying of the *palaeo*-Etosha Basin and surrounding landscape, stocks of Snake Catfish (*Clarias theodorae*) and Banded Tilapia (*Tilapia sparrmanii*) sought refuge and were isolated in the Aigamas Cave System and Lake Guinas, respectively, evolving into the Cave Catfish (*C. cavernicola*) and Otjikoto Tilapia (*T. guinasana*).

The level of endemism of today's Kunene River fishes strongly infers that founder groups of UZR fishes reached the lower *proto*-Kunene River prior to its capture 35 ka of the upper *palaeo*-Kunene River. Such routes could have included lesser tributaries to Lake Kunene/Etosha such as the Etaka *oshana* (Fig. 9), a tributary to the Ekuna *oshana* (Chapter 7, Fig. 9) and known to flow in both directions.[37] Headwater interchanges remain active today (Chapter 7) between the Kunene and Cuvelai basins facilitating dispersal across their divide.

The separation of Kunene mormyrids from UZR stocks has been dated to about 2.7–0.7 Ma,[58,59] whereas Southern Mouthbrooder stock separation was about 0.4 Ma.[60] Sharptooth Catfish of the Kunene are genetically related to the 'southern zone' Kavango, Upper Zambezi and most South African populations including the Orange River,[3] which are thought to have separated from other sister stocks about 0.4 Ma.

B. Orange River

The geological history and drainage pattern of the modern Orange River on the southern border contrast with Namibia's northern rivers. Its headwaters are associated with eastern South Africa and Lesotho within 200 km of the Indian Ocean and drain westward 2,300 km to the Atlantic Ocean with an average gradient of 1.4 m/km. The Orange evolved through time with its mouth shifting northward and its course influenced by continental drift, crustal uplift, tilting of the interior plateau and climatic conditions.[61] There is evidence that the interior plateau was at sea level 290–270 Ma prior to the rifting of east Gondwanaland which started ~130 Ma, but has been terrestrial since then.[62] All *palaeo*-drainage reconstructions for southern-most Africa suggest that the majority of the subcontinent has been westward-draining since that time. Offshore sediment samples record two major periods of erosion in southern Africa, 150–120 Ma and 100–80 Ma.[63,64] This was followed by a minor cycle of uplift of the interior plateau 81–66 Ma and a major event 30 Ma to a combined elevation ≥1,000 m (up to 3,000 m), countered by events of denudation that removed ~4–6 km of continental crust from the coastal regions and <1.5 km from the interior.[62] Most of this stripping occurred by 65 Ma with minimal erosion since then.[33,64]

The prevalent opinion is that the modern Orange River evolved from two rivers lying south of the influence of the *palaeo*-UZR, both flowing east to west (Fig. 13). The northern most was the Kalahari River draining southern Botswana and Namibia to discharge into the Southeast Atlantic in the vicinity of the mouth of the current-day Orange River. The Karoo River, further south, drained a catchment inclusive of the modern Vaal and Orange Rivers and followed a course to discharge near the mouth of the current Olifants River of the Western Cape, South Africa.[65,66,67] At about 60–50 Ma, the Kalahari captured the Karoo River by the intermediate Koa River,[66] but not before the Karoo provided an avenue for fish dispersal to the Atlantic slope.[5,68,69] Certain diamond placer deposits suggest the piracy occurred much earlier at about 90 Ma.[70] All these dates significantly precede the influence of *palaeo*-Lake Makgadikgadi, dated from 400 to 2 ka.

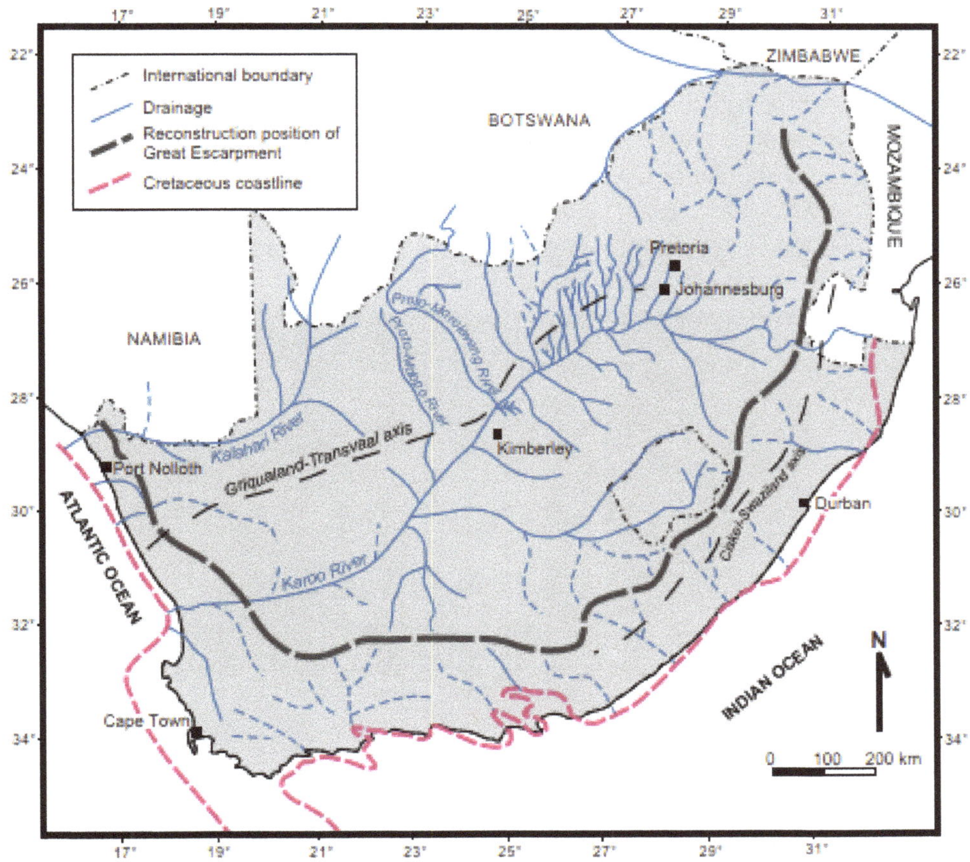

Figure 13. The westward draining Kalahari and Karoo Rivers 130 Ma with their respective mouths approximate to the Orange and Olifants Rivers of today.[27,66] Various interpretations allow for the Molopo River to have flowed into either the Kalahari or Karoo Rivers.

The well known 56-m Augrabies Falls is a consequence of uplift of the area with the Orange channel today following a fault zone that is possibly in a state of reactivation.[71] Data suggest that the interior plateau has undergone minimal lowering (~100+ m) for 60 My with the falls acting as a regional base level for areas that lie upstream.[26] The Orange River is the only major drainage basin in Africa with a history of more than 100 My that flows on an actively rising swell. Thus for south-central Africa, the Orange obtained its modern course earlier than any other major river south of the Congo River.

The Namib Desert began to form by no later than 34 Ma. Dry xeric conditions were maintained by the cold offshore Benguela Current upwelling (Fig. 15), which stabilised the air masses and suppressed year-round formation of precipitating clouds.[72] On the southern tip of the continent, only the Indian Ocean side has experienced truly wet conditions since that time.[23]

Successful *invasive* species of the *proto*-Orange River and Western Cape coast of South Africa would have been *r*-selected, tolerant of unstable environmental conditions and/or ecologically generalists. Habitat specialists would have been handicapped. Once *pioneer* species successfully established themselves in new environs, survival would be dependent upon their ability to evolve with the prevalent, but changing conditions. The earliest invasion wave had to have included ancestral stock of the *tetraploid* barbs, cyprinids virtually confined from the current mouth of the Orange River southward beyond its ancestral mouth, the Olifants River, to east along the Cape Fold Mountains (CFMs) (Fig. 14). Interesting is what did not reach the Orange and/or expired: the speciose mormyrids, citharines, numerous catfish families, various cichlid genera, lampeyes, robbers/tigerfish and spiny eels. This infers a shift away from favourable migratory conditions brought about by altered hydrology, climate and landscapes that severed the routes from the north.

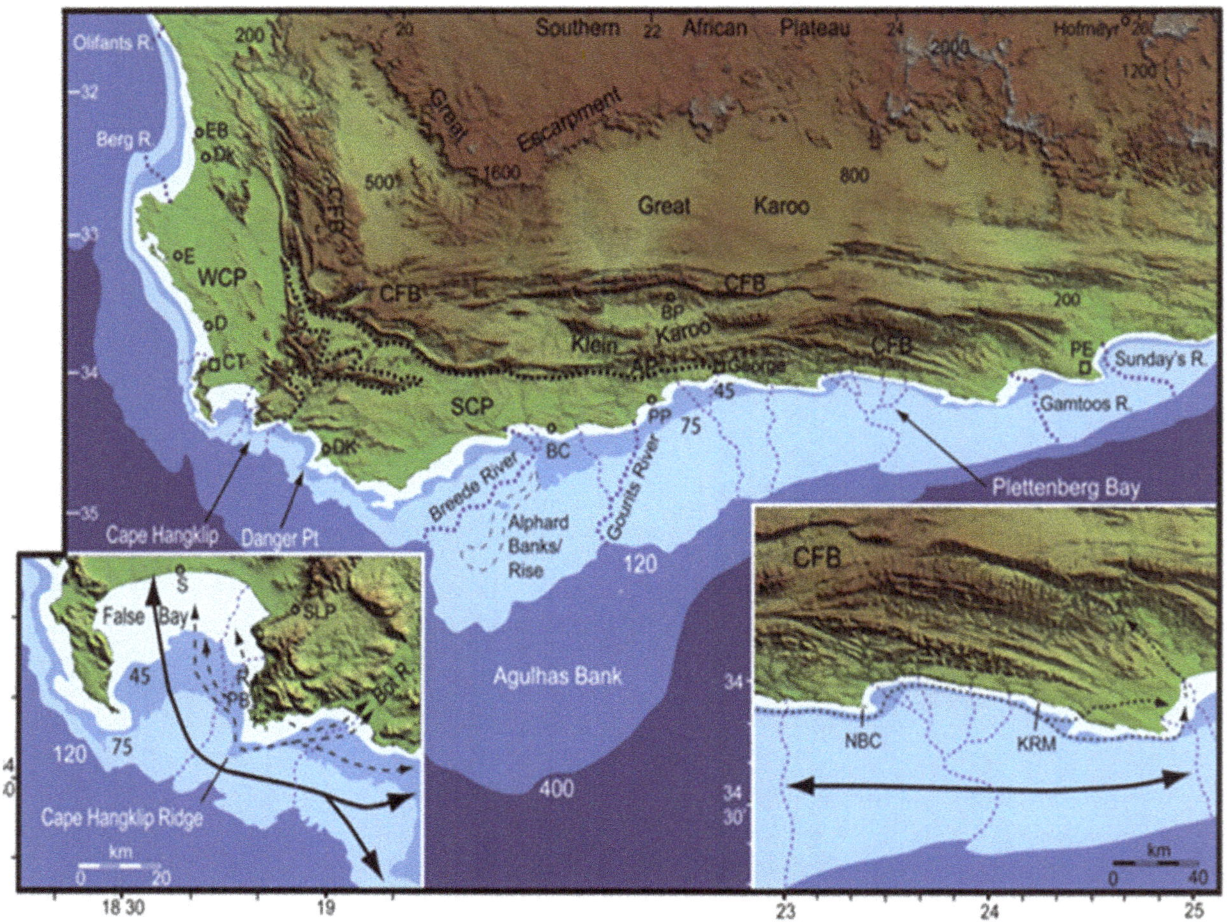

Figure 14. Cape Fold Belt (CFB) of the Western Cape, South Africa, extending along the Western Coastal Plain (WCP) south of the Olifants River and eastward along the Southern Coastal Plain (SCP) to Sundays River. Bathymetry contours at -75 m and -120 m represent Pleistocene interglacial and glacial maxima shorelines, respectively, facilitating human and inland fish migrations alike along the exposed coastal plain.[79]

The tetraploid barbs of South Africa are classified in five different genera[73] with the genus *Namaquacypris* being monotypic with the only representative (Namaqua Barb) being endemic to the lower Orange River below Augrabies Falls. The remaining tetraploid cyprinid genera include *Pseudobarbus, Sedercypris, Cheilobarbus* and *Amatolacypris. Pseudobarbus* has 11 species in it, with one restricted to the Olifants/Clanwilliam system and six endemic to the Western Cape south to the Breede River. *Sedercypris* is represented by two species endemic to the Olifants/Clanwilliam River while two species of *Cheilobarbus* spp. are endemic to the Berg and Breede Rivers.[73] Most of these tetraploid species have distributions from the lower Orange River through the 850-km long Cape Fold Mountain range, but have affinities with *diploid* cyprinids found elsewhere in the Orange-Vaal drainage.

The genus *Amatolacypris* is a sister group to the above tetraploid cyprinids. Its single representative (*A. trevelyani*) is isolated from the remainder, found in the Amatola Mountains of the Eastern Cape in the Kei and Mbashe Rivers that flow into the Indian Ocean. Thus, it too is descended from a very early lineage of cyprinids that were transferred from the *palaeo*-Orange basin by stream capture. Subsequently, they evolved into a distinct genus in isolation.

Although there are several ways tetraploids can occur naturally, hybridisation and environmental change play a large role in the initiation of new species. Tetraploidy has the distinct advantage over closely related diploid organisms by having greater genetic diversity, facilitating adaptation to instable or altered environs.[74] The environmental conditions in and around the Karoo River, *palaeo*-Orange River and the Cape Fold Mountains were far from stable. By analogy, the recent impoundment and alteration of the hydrological regime of Namibia's Fish River very likely promoted hybridisation of labeo and yellowfish as seen in present day stocks.[75]

The CFMs were created ~440–253 Ma with the collision and separation of the plates that formed Gondwana. They are part of a belt also found in Argentina, Falkland Islands, Antarctica and Australia.[76] They pre-date and are separate from the Great Escarpment with unique flora and fauna. Table Mountain, towering over Cape Town at an elevation 1,086 m, was actually a valley of the original CFMs with higher peaks surrounding it, which have now eroded away. This infers that ancestral cyprinid stocks dispersed very early to the Olifants-Clanwilliam River, then used minor stream captures or coastal routes during lowered sea level to spread southward, then eastward, through the CFMs. The Clanwilliam Sandfish (*Labeo seeberi*) is also endemic to the western CFMs, while the Orange River Sardine (*Engraulicypris gariepinus*) is considered endemic to the Orange-Fish River below Augrabies Falls.

The catfish genus *Austroglanis* spp. distribution pattern is similar to the above cyprinids. Only three *Austroglanis* spp. are known, two occurring in the Western Cape, and *A. sclateri* from Orange River drainage. However, their lineage is seemingly very ancient and related to the breakup of Gondwana with closest relatives being either Asian or North American, not African.[104] Southern Africa anabantids[105] and galaxiids[77,106] also have Gondwana origins.

Once stocks arrived in and dispersed through the Cape Fold belt, physical changes in the environment could be expected to promote isolation of populations and speciation through time. Once considered to have a rather depauperate fish fauna of 18 species, it is now thought up to 43 isolated lineages occur in the Cape Fold Ecoregion.[77] The recent events of the Pleistocene from 2.6 Ma to ~11.7 ka demonstrate how dispersal routes were provided with lowered sea levels, then were severed as sea levels rebounded, effectively isolating fish stocks and promoting vicariant speciation events. Sea levels varied from +34.9 m some 4 Ma near the Olifants River[78] prior to the Pleistocene to -75 m during Pleistocene interglacial periods to -120 m during the four glacial maxima, the last being only ~21–11.7 ka. The southern coastal plain of South Africa, including the Cape Fold Mountains, increased five-fold as sea level fell from -75 to -120 m, although less so on the west coastal plain north of Cape Town.[79] *Periglacial* events have occurred in the Cape Fold Mountains since the last glacial maxima advancement ~21–11.7 ka until recent, although decreasingly so, while the historical presence of cirque glaciers and moraines have been identified in the Lesotho Highlands and Drakensburg Mountains of the Great Escarpment, where some peaks reach up to 3,482 m.[80]

To summarise, the merging of the ancestral Kalahari and Karoo Rivers to form the *proto*-Orange River is accepted. More speculative are the route(s) of transfer of fishes from the north into the *palaeo*-Orange system and the timing. Combined, the five tetraploid cyprinid genera, Clanwilliam Sandfish (*Labeo*), Orange River Sardine, Banded Tilapia, Southern Mouthbrooder and the three *Austroglanis* spp. demonstrate that older routes into the *palaeo*-Orange basin were available to a number of species with varying life styles. In this regard, the Trans-Tswana River, Molopo River, Nossob-Auob Rivers and *palaeo*-Lake Makgadikgadi are considered.

C. Trans-Tswana River

It has been hypothesised[23] that the Trans-Tswana River was a very ancient drainage that encompassed much of Namibia and Botswana as a major component of the *palaeo*-Orange River (Fig. 15). It differs from the above proposed Kalahari River (Fig. 13)[26,66] in that its headwaters extended from the Cuvelai to the Upper Zambezi Rivers prior to the down-warp of the Kalahari Basin.[30] Too, it has been postulated that Trans-Tswana River flowed directly into the Kalahari River,[29,30,81] then captured the upper Karoo River ~30 Ma to form the course of the modern Orange River. Alternately, the Trans-Tswana River entered the Karoo River 30 km downstream of the current confluence of the Vaal and Orange Rivers, and this may have been the same for the *proto*-Molopo River.[27] In either case, the *proto*-Orange River entered the Atlantic Ocean in proximity to its mouth today.

Subsequently, crustal warping and uplift of the South African swell ~30 Ma (a) severed the Cuvelai, Kavango, Kwando and Upper Zambezi's southern outlet through the Trans-Tswana River,[23] and (b) infilled the Kalahari Basin and/or diverted these waters southeast through the Limpopo. Thus, the Trans-Tswana scenario included at least the southern half of the Kalahari Basin and, eventually, *palaeo*-Lake Makgadikgadi's regional influence.

The hypothetical formation of the *proto*-Orange by the Trans-Tswana River <30 Ma is far more recent than the proposed capture 60–50 Ma[65,66,67] or ~90 Ma based upon diamond placer deposits.[70] Geological evidence to support the Trans-Tswana model is contentious;[33] however, offshore sediment depositions at the Orange River mouth[62] infer a mature river ~93–86 Ma four times its current size[69] with depositions peaking 90–80 Ma.[23]

Shuttle Radar Topographic Mission (SRTM) digital elevation profiles (Fig. 16) were obtained for 8,354 km along 29 mekgacha (or *omuramba*) and ephemeral rivers of the central Kalahari Desert.[82] These data identified 55 previously unknown *knickpoints* (sharp changes in elevation) and four drainage reversals, some pre-dating 23 My. While the research was not designed to answer the validity of a Trans-Tswana River, the results also did not support its presence. The answer to this may be simple, i.e., the Trans-Tswana is sufficiently old that it has vanished under the sands of time, pre-dating data obtained in the SRTM study and hence having no influence on the distribution of modern fish fauna.[83]

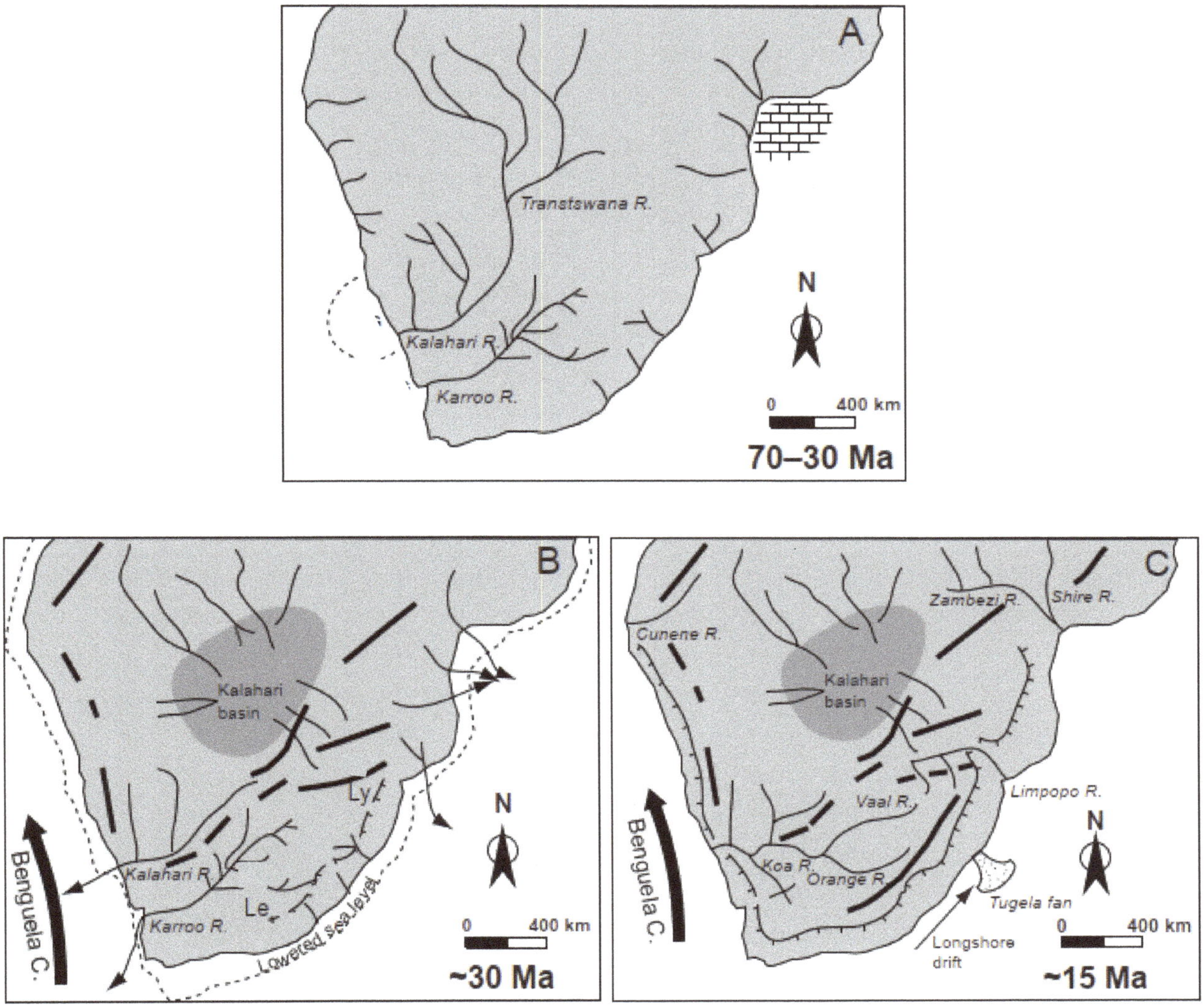

Figure 15. Illustration of *proto*-Orange River 70–30 Ma with Trans-Tswana River comprised of all the drainages from the Cuvelai eastward to the Upper Zambezi Rivers and flowing into the Kalahari River (A). The more southern Karoo River 70–30 Ma flowed into the SE Atlantic Ocean via today's Olifants River; its principal tributaries were the Vaal and Upper Orange Rivers as known today. Break-up of the Trans-Tswana River occurred by 30 Ma (B). All drainages from the Cuvelai to the Upper Zambezi were diverted eastward presumably through the modern Limpopo River valley prior to regional uplift, infilling of the Kalahari Basin and the formation of *palaeo*-Lake Makgadikgadi. (C) Southern Africa rivers 15 Ma with the Kalahari River having captured the Vaal-Upper Orange from the Karoo River.[23]

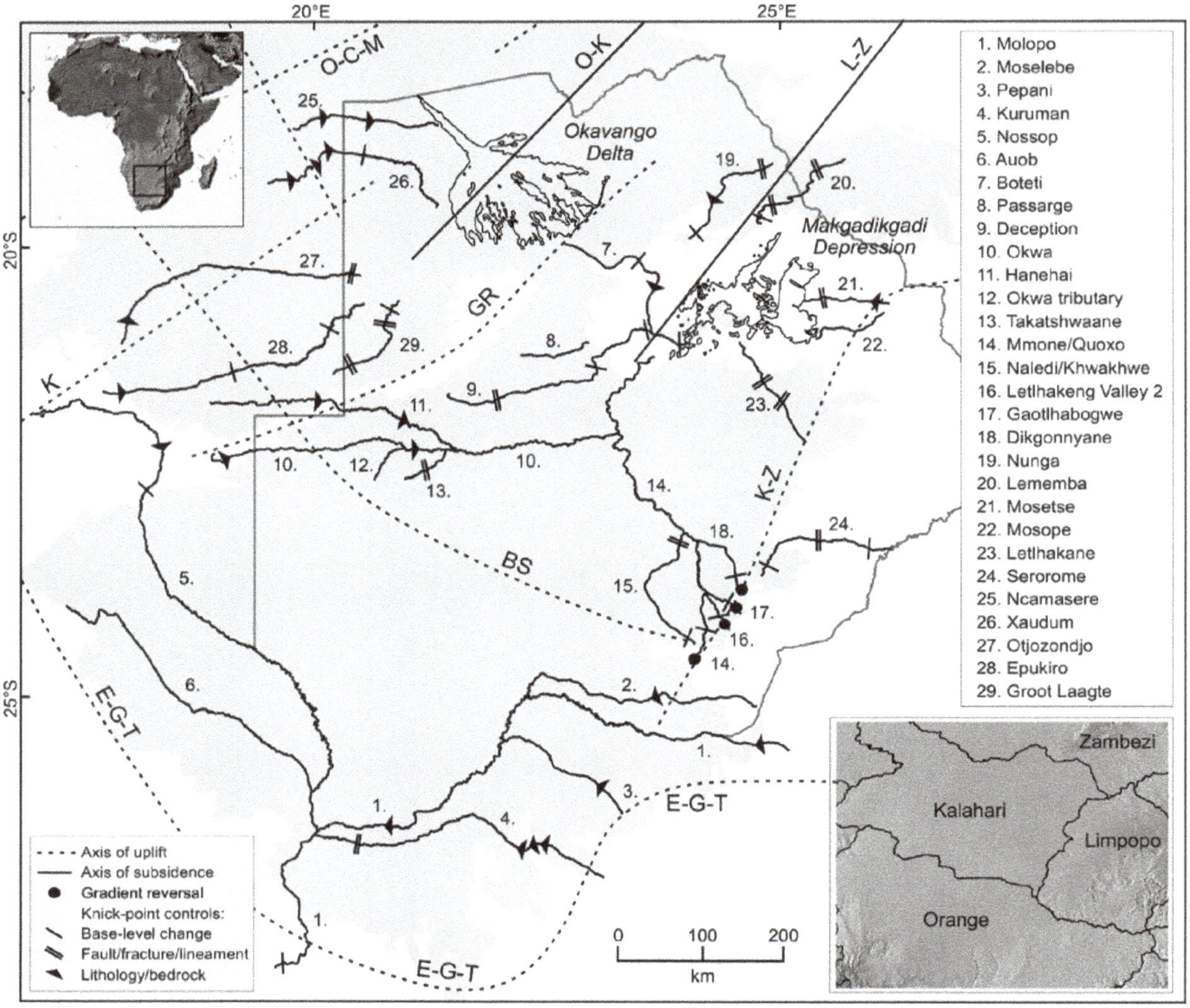

Figure 16. Detailed image derived from Shuttle Radar Topographic Mission (SRTM) digital elevation data for the central Kalahari Desert, covering 8,354 km along 29 *mekgacha* and ephemeral rivers.[82] Depicted are (a) ancient drainage lines; (b) 55 previously unknown *knickpoints* (bars and triangles across the various drainages); (c) four drainage reversals (black circles); and (d) Kalahari sediments (shaded areas).

D. Molopo River

Mekgacha are an important landscape feature in the semi-arid to arid Kalahari Desert. The majority of valleys do not flow, except in their headwater regions, but may contain pools of water on a seasonal basis.[84] The Molopo River is a poorly-studied ancient mekgacha, dating to 300 Ma and being infilled by pre-glacial Dwyka sediments.[55] The Molopo is now the border between Botswana and South Africa, draining west on the southern rim of the Kalahari Basin, its confluence with the Orange River 17 km downstream of Augrabies Falls (Fig. 17). It receives the mainly endorheic Nossob-Auob-Olifants and Kuruman Rivers and other minor tributaries from Namibia, Botswana and South Africa. It has a low gradient of 700 mm/km and has not reached the Orange in recorded history with rains sinking into the Kalahari sands.[85,86]

The Molopo originally flowed northward away from the Karoo River into the Kalahari Basin.[27] Ponding occurred to create *palaeo*-Lake Molopo[55] about 60 Ma and other regional uplifts enhanced the lake that possibly resembled a series of interconnected *playas* differing slightly in salinities and salts as local conditions prevailed. There are also indications of drainage reversals. An increasingly dry climate and stream capture in the Molopo's headlands by the Limpopo River ~18 Ma contributed to its reduced regional importance as a perennial river, even during wetter conditions. The divide has shifted southward some 75–100 km since that time as the Limpopo has encroached upon the Molopo and the Kalahari basins.[81] Eventually, the Molopo was influenced by uplift, a westward tilting of the landscape and a wetter climate about 4 Ma to seek a western outlet. Low rainfall conditions and growth of the Kalahari Desert over the past 2 My has resulted in its demise (Fig. 17).[55]

Although geological evidence is marginal that the above-mentioned Trans-Tswana River existed, and biogeographic support is lacking, there is little question that the Molopo drainage did, and does, exist. The ancient Molopo in its geographical intermediate position must have provided an early entry route for fish into the *palaeo*-Orange and Limpopo systems from the north. Subsequent landscape evolution, altered drainage patterns and dry xeric conditions would have combined to have shifted its northward direction across the Kalahari Basin to a westward-draining route along the basin's southern periphery, its mouth migrating westward along the *palaeo*-Orange River until its current position below Augrabies Falls was achieved.

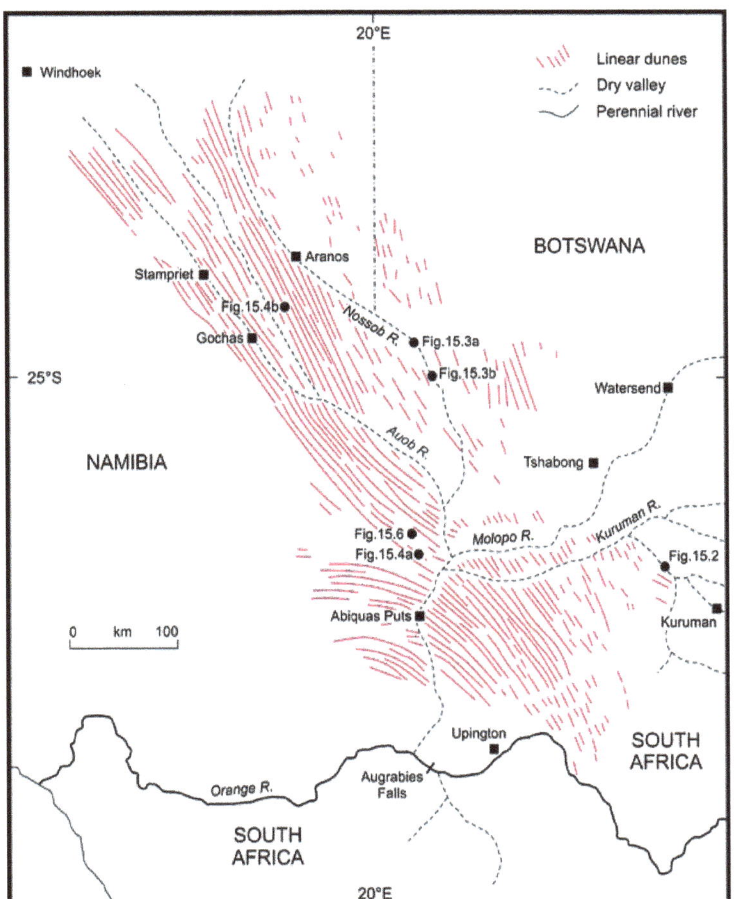

Figure 17. Linear dunes (red) of the Kalahari Desert now blocking and absorbing the waters of the Molopo-Nossob-Auob River basin, historically discharging into the Orange River 17 km downstream below Augrabies Falls.[87]

E. Nossob-Auob River

The upper Nossob-Auob sub-basin appears to have had a complex history independent of the greater Molopo basin. This is based upon geological evidence and the curious occurrence of the Chubbyhead Barb in the Kuiseb River, an Atlantic slope drainage that heads up against the Nossob-Auob River along the Great Escarpment. It currently lacks a mouth or estuary with its lower section covered by mobile sand dunes. Floodwaters breach the dunes and reach Walvis Bay perhaps every 10–15 years.

The Chubbyhead Barb species complex is widely distributed in South Africa with four distinct lineages.[88] One of these (*Enteromius oraniensis)* is considered endemic to the Orange River drainage above Augrabies Falls while the others are endemic to the Olifants River of the Western Cape (*E. cernuus*), Eastern Cape (*E. mandelai*) and the Gouritz River system (*E. anoplus*). The only known specimens of the Chubbyhead Barb (presumably *E.* cf. *oraniensis*) from Namibia are from collections made in 1968–1973 in the ephemeral Kuiseb River and its main tributary, the Gaub River (Fig. 18),[89] that occasionally reaches the Atlantic Ocean. However, its presence was known[90] before 1967 and it was again reported[91] being in 'western drainages' in 1991. It, unfortunately, was not included in later regional references[4,5] and was 'forgotten' until C. Hay 'rediscovered' it in the National Museum, Windhoek during the preparation of this book. Adding to this misfortune, Namibia experienced virtual Covid lockdown for over a year, thus it was not possible to mount an expedition to demonstrate if the population remains viable, although there is no reason to expect otherwise.

C. Hocutt

Figure 18. The Gaub River, tributary to the ephemeral Kuiseb River, that heads up on the western rim of the Great Escarpment at Gamsberg Mountain (Photo: C. Hocutt).

Both the Kuiseb and Gaub Rivers have eroded through the western boundary of the Great Escarpment with the Gaub nearly encircling Gamsberg Mountain, standing at 2,347 m in elevation. Combined, (a) the remoteness of the Kuiseb basin, (b) the harshness of the habitat, (c) the distance to the closest extant populations in the Orange River at least 700 km away (direct air distance to Augrabies Falls) and (d) its known presence in the Kuiseb dating to before 1967, make it unlikely that this species was introduced. Thus, the Kuiseb population is considered a relict population of the

Chubbyhead Barb lineage, similar to the Shorthead Barb restricted to the Kunene River in Namibia and the western Angolan highlands, as well as possible other chubbyhead outliers in the DRC and Mozambique.[83]

The presence of the Chubbyhead Barb in the Kuiseb basin suggests its lineage was present in an old interior drainage and transferred across the Great Escarpment by a stream piracy with past tributaries of the Nossob-Auob River. Such a piracy[27] has been dated at 5–3 Ma, when the upper Fish River north of Keetmanshoop was once part of the Nossob-Auob basin before being captured by the lower Fish River.

Numerous fault lines[92] and a classic elbow of capture of the upper Fish River northwest of Mariental infer that the *palaeo*-drainage direction of the region was to the northeast (Fig. 19). A question based on the above mentioned (Fig. 16) SRTM map is, did the upper Nossob-Auob once continue northeast into the Kalahari Basin/*palaeo*-Lake Makgadikgadi, hence providing a more direct corridor for dispersal of the Chubbyhead Barb (and other species) into both the Kuiseb River and the lower Orange Basin? Afterwards, the northeast-flowing Nossob-Auob Rivers were perhaps re-directed southward along the Kalahari Basin's western margin to join the Molopo, this route known as the *palaeo*-Aranos River[27] that flowed southward along the edge of the Kalahari Basin into the Kalahari River. Climate change, denudation of the landscape and the increased growth of the Kalahari Desert would have masked these ancient drainage affiliations.

Regardless of the route used (Molopo vs Nossob-Auob), the Chubbyhead Barb lineage further dispersed throughout the *palaeo*-Orange River into the Karoo, the Western Cape's Olifants River and by headwater captures into the Limpopo River and South African coastal rivers. This *r*-selected species complex is well known for its adaptability to a variety of environments and colonisation of marginal waters.[93] These routes of dispersal are equally plausible for the early entry of other species stocks into the Orange drainage.

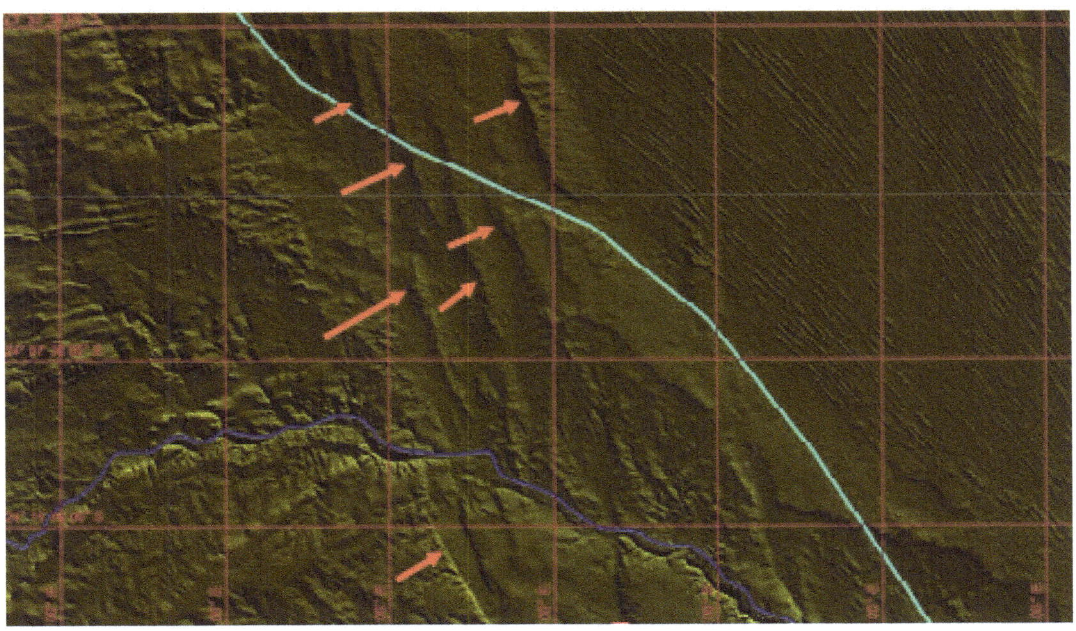

Figure 19. Faults northwest of Mariental (arrows) with the upper Fish River (dark blue) and basin margin (light blue).[92] The inverted 'U-curve' of the west-to-east flowing Fish River strongly infers an elbow of capture.

F. Lake Makgadikgadi

The evolution of *palaeo*-Lake Makgadikgadi and Victoria Falls, discussed above, complements the idea of 'waves' of dispersion of fishes[5] into southern Africa. The lake was known to have filled at least seven times, while Batoka Gorge below Victoria Falls had at least 6–8 different fall lines over the past 1 My. Each refilling of Lake Makgadikgadi would have provided dispersal routes via channels visible on the landscape today as interrupted pans and the low-lying mekgacha valleys and channels.[27,94] The shifting between dry and wet conditions and hydrology during each refill would have presented both *lotic* and *lentic* conditions for colonising fishes to disperse into both the Orange and Limpopo drainages.

Molecular studies[3] demonstrated three distinct genetic markers for the Sharptooth Catfish: (1) a southern zone, inclusive of *palaeo*-Lake Makgadikgadi, all of Namibia's northern rivers and the Orange River; (2) a central zone, the Middle Zambezi, Limpopo, Komati and Maputo Rivers; and (3) an eastern zone, the Lower Zambezi River and Lake Malawi region. The genetic separation of southern stocks was about 1.7–0.4 Ma, with at least two introductions. The most recent introduction (0.4 Ma) corresponds well[95] with the drying of Lake Makgadikgadi about 2 ka, a similar time to Lake Kunene.[37]

The Sharptooth Catfish pattern coincides with the known distributions of other species groups. Essentially the same route of dispersal was postulated for the widely distributed sand catlets (*Zaireichthys* spp.) complex below Victoria Falls to Malawi through to the Save River, Mozambique, now requiring additional examination.[96] Three closely related species are recognised above the falls in the UZR (Pallid Sand Catlet, Blotched Sand Catlet), Kwando (Pallid Sand Catlet), Kavango (Blotched Sand Catlet, Kavango Sand Catlet) and Kunene River (Kunene Sand Catlet) basins;[97] however, Pallid Sand Catlet populations differ between the Kwando and UZR drainages.[96]

The similarity in chromosome structure of the cichlids *Serranochromis macrocephalus* from the UZR and *Chetia brevicauda* of the Mussapa River, Mozambique, also is consistent with an ancient drainage of the UZR through the Limpopo River basin.[95] Lake Makgadikgadi is considered as the centre of radiation of the (a) cichlid *Serranochromis* and *Thoracochromis* spp. stocks that dispersed east to the Indian Ocean slope, northward into the Congo basin and west into Atlantic flowing rivers, and (b) *Synodontis* spp. stocks, themselves derived from two different invasions from the north[98] that reached the UZR ~5 Ma via the Congo basin, then radiated and diversified over the past 2.3–0.8 Ma.[99]

The dispersal, evolution and speciation of southern African mormyrids supports the impression of Lake Makgadikgadi as a centre of fish radiation with distinct faunas above and below Victoria Falls and along the Indian Ocean coast.[59,103] Once in the Limpopo River, stocks of *Hippopotamyrus*, *Marcusenius*, *Mormyrus*, *Petrocephalus* and *Pollimyrus* likely spread by waifing and 'river hopping' along the coast during periods of cyclonic storms. A western clade of churchills (*Petrocephalus longicapitis* and *P. magnoculis*) from the Kunene, Kavango and UZR is distinct from an eastern clade (*P. catostoma* and *P. wesselsi*) near the Indian Ocean. *Marcusenius altisambesi* (Kavango) and *M. multisquamatus* (Kunene) separated from *M. macrolepidotus* (Lower Zambezi) 5–9 Ma and from one another 0.7–2.7 Ma.[59] It is thought that *Pollimyrus castelnaui* (Kunene, Cuvelai and Kavango Rivers) and *P. marianne* (Upper Zambezi) separated 3.8-7.6 Ma, while differentiation of the geographically intermediate *P. cuandoensis* (Kwando-Linyanti) was about 2.1 Ma. *Hippopotamyrus longilateralis* (Kunene) became isolated from the more eastern

(Kavango, Kwando, UZR) *H. ansorgii* 1.2–2 Ma.[58] However, there are at least seven undescribed lineages of *H. ansorgii* with relationships unresolved.[47]

Within the barb genus *Enteromius*, three Namibian species stand out: Straightfin Barb, Threespot Barb and Chubbyhead Barb, the latter discussed above. The Straightfin Barb is unique in that it is the only cyprinid found in all of Namibia's perennial rivers, as well as various ephemeral systems. The distribution of the Threespot Barb is peculiar. It is known from the Kunene and Cuvelai Rivers in northern Namibia/Angola and the Orange River system to tributaries discharging into the Indian Ocean as far north as Tanzania. It is unknown from the geographical-intermediate Kavango, Kwando-Linyanti and Upper Zambezi-Chobe Rivers, where it is replaced[96] by the Dashtail Barb (*E. poechi*), some with three spots. Lake Makgadikgadi was certainly a dispersal route/radiation centre until 2,000 years ago when it dried, severing courses to the west, i.e., Etosha Basin and the Kunene River.

III. SYNOPSIS

The inland fishes of Namibia are unquestionably linked to the evolution of the Kalahari Basin, its sub-basins and the *palaeo*-Upper Zambezi River. Indeed, all of Namibia's northern rivers were physically joined within the Kalahari Basin until tectonic uplift, drainage diversions and climate change altered the region resulting in the (relative) isolation of the Kunene and Cuvelai basins. The Kavango, Kwando-Linyanti and Upper Zambezi-Chobe Rivers retain connectivity today, especially during major flooding, with high similarity in their ichthyofaunas.

The complex geological history and the fishes of the Kunene River infer ichthyofaunal transfer from multiple sources: coastal, headwater and Upper Zambezi links. The UZR affiliation was open, then closed with the shifting of the Kavango drainage line and demise of Lake Etosha. Four mormyrid species are either endemic to the Kunene or shared with the closely-affiliated ephemeral Cuvelai River. Mormyrid stocks are, however, derived from the UZR as are the ancestral stocks of the Kunene Sand Catlet (*Zaireichthys kunenensis*) and undescribed Cunene Broadhead Catfish *Clariallabes* sp. The Cunene Kneria (*Kneria maydelli*) is endemic to the Kunene with no other kneriids found elsewhere in Namibia, but relatives occur to the north in the Atlantic coastal streams of Angola; the Papermouth Barb (*Enteromius mattozi*) exhibits a similar pattern. The Slender River Bream (*Chetia gracilis*), Angolan River Bream (*C. welwitschi*), various largemouth bream (*Serranochromis* spp.), Thicklipped River Bream (*Thoracochromis albolabris*) and Namib River Bream (*T. buysi*) are likely derived from the Lake Makgadikgadi centre of radiation,[95] reinforcing the importance of the *palaeo*-UZR basin. The Shorthead Barb (*E. breviceps*) in the Kunene River is of a Chubbyhead Barb lineage that seemingly is very old, based upon various widely disjunct populations of related species throughout southern Africa. The level of endemism in Kunene River fishes supports the multiple entry of stocks into the *proto*-river prior to its capturing the upper *palaeo*-Kunene River 35 ka, which retains seasonal connections with the Cuvelai basin. The *palaeo*-Upper Zambezi/Lake Makgadikgadi also had undeniable links with the Limpopo River, followed by faunal dispersal south along the Indian Ocean coast.

The meagre fish fauna of Orange-Vaal-Senqu River on Namibia's southern border is perplexing, despite it being the oldest drainage in all of Africa south of the Congo River. The system's long established east to west drainage evolved largely independent of the *palaeo*-UZR. Nonetheless, the fishes of the Orange basin are derived from more northern stocks that took advantage of changes in drainage divides and had the characteristics of successful invasive species. Only 12 indigenous species occur in Namibia's portion of the drainage, with six of these being endemic. Of the 12 native species, four are shared with Namibia's northern perennial rivers: Straightfin Barb, Sharptooth Catfish, Southern Mouthbrooder and Banded Tilapia. These species are highly adaptable to changing environmental conditions, exhibit *r*-selectivity or are nest guarders of offspring, all of which are desirable characteristics of successful colonising pioneer species. Amongst the dispersal outlets to the Orange basin were the Molopo River, *palaeo*-lake Makgadikgadi and very possibly the *palaeo*-Aranos River, the latter represented today by the Nossob-Auob basin. Perhaps the controlling factor of low species richness in the Orange-Vaal portions of the drainage was ecological, i.e., a predominately shifting sand substratum between well defined banks without rock outcroppings, productive littoral embayments or floodplains.

There is ample geological and biological evidence that the Karoo River discharged through the Western Cape's Olifants-Clanwilliam River mouth allowing for ancient fish lineages to disperse southward into the Cape Fold Mountains, then eastward along the exposed coastal plain during the 2.58 My Pleistocene. Variable environmental conditions promoted isolation and speciation of certain fishes with tetraploidy being exhibited amongst the cyprinids. Subsequently, the Karoo River was captured by the Kalahari River, diverting the entire drainage over Augrabies Falls to the current mouth of the Orange River.

There is evidence for a very old drainage realignment of the upper Nossob-Auob basin that facilitated dispersal of the Chubbyhead Barb into the ephemeral, Atlantic slope Kuiseb-Gaub Rivers that head up on the Great Escarpment. This is supported by the museum holdings of the Gobabeb Training & Research Centre that contain a single specimen each of River Sardine and unidentified cyprinid. While provenance is lacking, they most likely were collected within the Kuiseb basin.[107] The Kuiseb drainage is considered a linear oasis for wildlife in the harsh Namib Desert environment,[108] thus it may harbour other aquatic relicts. (See Addendum, P. 365)

Drainage rearrangements through time have been the primary mechanism that shaped the distribution patterns of southern Africa's inland fishes. However, recent research has demonstrated that many species that were previously considered to have broad geographic ranges across multiple isolated river systems are actually complexes of historically isolated lineages, many representing new species.[47] Examples include studies on the tetraploid cyprinids,[73] galaxids,[77] stonebashers[47] and sand catlet[96] complexes. Moreover, electric organ discharges (EODs) have been demonstrated as not only integral to mormyrid systematics, but also are key to their speciation with data indicating that EODs of mormyrid populations may differentiate before there are morphological distinctions.[47,103] Thus, it is not an understatement to say that there is currently a new age of discovery of southern Africa's inland fishes with many more species to be described in the future as new technologies are employed to complement traditional morphometric and meristic methodologies. Understanding the routes and timing of dispersal(s) will be facilitated as *palaeo*-drainage lines are more accurately defined employing remote sensing tools.[85,100,101]

IV. REFERENCES

[1]Knight, J. and S.W. Grab. 2018. Chapter 2: The geomorphic evolution of southern Africa during the Cenozoic, pp. 6–28. In: P.J. Holmes and J. Boardman (Eds.). Southern African landscapes and environmental change. Routledge, London.

[2]Goudie, A.S. 2005. The drainage of Africa since the Cretaceous. Geomorp. 67: 437–456.

[3]Van Steenberge, M.W., M.P.M. Vanhove, A.C. Manda, M.H.D. Larmuseau, B.L. Swart, F. Khang'Mate, A. Arndt, B. Hellemans, J.V. van Houdt, J-C. Micha, S. Koblmuller, R. Roodt-Wilding and F.A.M. Volckaert. 2020. Unravelling the evolution of Africa's drainage basins through a widespread freshwater fish, the African sharptooth catfish *Clarias gariepinus.* J. Biogeog. 47: 1739–1754.

[4]Skelton, P.H. 1993. A complete guide to the freshwater fishes of southern Africa. Southern Book Publ., Harare: 388 pp

[5]Skelton, P.H. 2001. A complete guide to the freshwater fishes of southern Africa. Struik Publ., Cape Town: 395 pp.

[6]Skelton, P.H. 2019. The freshwater fishes of Angola, pp. 207–242. In: B.J. Huntley, V. Russo, F. Lages and N. Ferrand (Eds.). Biodiversity of Angola, Science & Conservation: A Modern Synthesis. Springer Nature, Switzerland.

[7]Ziegera, J., J. Rotheb, M. Hofmannc, A. Gartnerc and U. Linnemann. 2019. The Permo-Carboniferous Dwyka Group of the Aranos Basin (Namibia) - How detrital zircons help understanding sedimentary recycling during a major glaciation. J. Afr. Earth Sci. 158: 1–19.

[8]McLoughlin, S. 2001. The breakup history of Gondwana and its impact on pre-Cenozoic floristic provincialism. Aust. J. Bot. 49(3): 271–300.

[9]USGS. https://pubs.usgs.gov/gip/dynamic/continents.html

[10]Martin, H. 1953. Notes on the Dwyka succession and on some pre-Dwyka valleys in South West Africa. Trans. Geol. Soc. S. Af. 56: 37–41.

[11]Godderis, Y., Y. Donnadieu, S. Carretier, M. Aretz, G. Dera, M. Macouin and V. Regard. 2017. Onset and ending of the late Palaeozoic ice age triggered by tectonically paced rock weathering. Nat. Geosci. 10:1-5.

[12]Andrews, G.D., A.T. McGrady, S.R. Brown and S.M. Maynard. 2019. First description of subglacial megalineations from the late Paleozoic ice age in southern Africa. Plos One 14(1): 1–10.

[13]Dietrich, P. and A. Hofmann. 2019. Ice margin fluctuation sequences and grounding zone wedges: The record of the Late Palaeozoic Ice Age in the eastern Karoo Basin (Dwyka Group, South Africa). J. Interntl. Assoc. Sedim. 5(2): 247–271.

[14]Montanez, I.P. and C.J. Poulsen. 2013. The Late Paleozoic Ice Age: An evolving paradigm. Ann. Rev. Earth Planet. Sci. 41 (24): 1–28.

[15]Martin, H. 1981. The Late Paleozoic Dwyka Group of the South Kalahari Basin in Namibia and Botswana and the subglacial valleys of the Kaokoveld in Namibia, pp. 61–66. In: M.J. Hambrey and W.B. Harland (Eds.). Earth's pre-Pleistocene glacial record. Cambridge University Press, Cambridge.

[16]Isbell, J.L., D.I. Cole and C. Octavian. 2008. Carboniferous-Permian glaciation in the main Karoo Basin, South Africa: Stratigraphy, depositional controls, and glacial dynamics. Spec. Pap. Geol. Soc. Am. 441: 71–82.

[17]de Moor, F.C., H.M. Barber-James, A.D. Harrison and C.R. Lugo-Ortiz. 2000. The macroinvertebrates of the Cunene River from the Ruacana Falls to the river mouth and assessment of the conservation status of the river. Afr. J. Aq. Sci. 25: 105–122.

[18]Fedorchuk, N.D., J.L. Isbell, N.P. Griffis, F.F. Vesely, E.L.M. Rosa, I.P. Montanez, Q-Z Yin, M.H. Huyskens, M.E. Sanborn, R. Mundil, F.F. Vesely and R. Iannuzzi. 2021. Provenance of late Paleozoic glacial/post-glacial in the eastern Chaco-Parana basin, Uruguay, and southernmost Parana basin, Brazil. J. S. Am. Earth Sci. 106: doi.org/10.1016/j.jsames.2020.102989.

[19]Rosa, E.L. and J.L. Isbell. 2019. Glaciogenic sedimentary infill of late Paleozoic glacial paleovalleys of the Kaokoveld, northwest Namibia. Geol. Soc. Am. Abst. 51(5): doi: 10.1130/abs/2019AM-340908.

[20]Hocutt, C.H. 1987. Evolution of the Indian Ocean and the drift of India: A vicariant event. Hydrobiologia 150: 203–223.

[21]Heads, M. 2005. Towards a panbiogeography of the seas. Biol. J. Linn. Soc. 84: 675–723.

[22]Earthworks. 2021. A plate tectonics primer. http://www.reeves.nl/gondwana.

[23]Burke, K. and Y. Gunnell. 2008. The African erosion surface: A continental-scale synthesis of geomorphology, tectonics, and environmental change over the past 180 million years. Geol. Soc. Am. Mem. 201: 66 pp.

[24]Clark, V.R., N.P. Barker and L. Mucina. 2011. The Great Escarpment of southern Africa: A new frontier for biodiversity exploration. Biodiv. Conserv. 20(20): 2543–2561.

[25]Goudie, A.S. 2020. Waterfalls: Forms, distribution, processes and rates of recession. Quaest. Geog.39(1): 59–77.

[26]Peulvast, J-P., F. Betard and Y. Lageat. 2009. Long-term landscape evolution and denudation rates in shield and platform areas: A morphostratigraphic approach. Géomorphologie 2: 95–108.

[27]Haddon, I.G. 2005. The sub-Kalahari geology and tectonic evolution of the Kalahari Basin, southern Africa. Ph.D. Diss., University of the Witwatersrand, Johannesburg: 343 pp.

[28]Cotterill, W. and S. Goodier. 2008. DNA clocks for dating landforms. Quest 4(2): 19–21.

[29]Jones, M.J. 2010. The groundwater hydrology of the Okavango basin. Okacom/FAO Tech. Rept.: 82 pp.

[30]Haddon, I.G. and T.S. McCarthy. 2005. The Mesozoic–Cenozoic interior sag basins of Central Africa: The Late-Cretaceous–Cenozoic Kalahari and Okavango basins. J. Afr. Earth Sci. 43(1–3): 316–333.

[31]Thomas, D.S.G. and P.A. Shaw. 1990. The deposition and development of the Kalahari Group sediments, central southern Africa. J. Afr. Ear. Sci. 10 (1–2): 187–197.

[32]Pinard, M.I. 2010. The use of sands in road construction. 1st AFCAP Practitioners Conf., Addis Ababa, Ethiopia, 23–25 Nov.

[33]de Wit, M.C.J. 2021. *pers. comm.*

[34]Moore, A.E. 1999. A reappraisal of epeirogenic flexure axes in southern Africa. S. Afr. J. Geol. 102: 363–376.

[35]Moore, A., T. Blenkinsop and F. Cotterill. 2009. Southern African topography and erosion history: Plumes or plate tectonics? Terra Nova 21: 310–315.

[36]Miller, R. McG., M. Pickford and B. Senut. 2010. The geology, palaeontology and evolution of the Etosha Pan, Namibia: Implications for terminal Kalahari deposition. S. Afr. J. Geol. 113 (3): 307–334.

[37]Hipondoka, M.H.T. 2005. The development and evolution of Etosha Pan, Namibia. Ph.D. Diss., Universität Wuerzburg: 152 pp.

[38]Hipondoka, M.H.T., H.J. Jousseb, J. Kempf and D. Busche. 2006. Fossil evidence for perennial lake conditions during the Holocene at Etosha Pan, Namibia. S. Afr. J. Sci. 102: 93–95.

[39]Buch, M.W. 1997. Etosha Pan – the third largest lake in the world? Madoqua 20(1): 49–64.

[40]Stuart-Williams, V. 1992. Etosha: Third largest lake in the world. In: A. Marsh and M. Seely (Eds.). Oshanas – sustaining people, environment and development in central Owambo, Namibia. Windhoek: Typoprint: 13.

[41]Baumle, R., T. Himmelsbach and U. Noell. 2018. Hydrogeology and geochemistry of a tectonically controlled, deep-seated and semi-fossil aquifer in the Zambezi Region (Namibia). Hydrogeol. J. 27: 885–914.

[42]Moore, A.E., T. Blenkinsop and F. Cotterill. 2012. Dynamic evolution of the Zambezi-Limpopo watershed, Zimbabwe. S. Afr. J. Geol. 115: 551–560.

[43]Moore, A.E. 2016. Victoria Falls and its gorges, pp. 16–20. In: R. Viljoen, M. Viljoen, C. Annhauser (Eds.). Africa's top geological sites. Penguin Random House, South Africa.

[44]Moore, A.E. and F. Cotterill. 2010. Victoria Falls: Mosi-oa-Tunya – The smoke that thunders, pp. 143–153. In: P. Migon (Ed.). Geomorphological landscapes of the world. Springer Science+Business Media.

[45]Reeves, C.V. 2019. Gravity survey and the origin of mankind. www.reeves.nl/news/item/ id/77.

[46]Chan, E.K.F., A. Timmermann, B.F. Baldi, A.E. Moore, R.J. Lyons, S-S Lee, A.M.F. Kalsbeek, D.C. Petersen, H. Rautenbach, H.E.A. Fortsch, M.S.R. Bornman and V.M. Hayes. 2019. Human origins in a southern African palaeo-wetland and first migrations. Nature 575: 185–189.

[47]Mutizwa, T.I., W.T. Kadye and A. Chakona. 2021. Deep genetic and morphological divergence in the *Hippopotamyrus ansorgii* species complex (Teleostei: Mormyridae) in southern Africa. J. Fish Biol. 2021: 1–14.

[48]Skelton, P.H., J.R. Stauffer, A. Chakona and J.M. Wisor. 2021. Two new species of African bubble-nesting *Microctenopoma* (Teleostei: Anabantidae) from Angola. Ichthy. Explor. Freshw., IEF 1134: 1–16. doi.org/10.23788/IEF-1134.

[49]Tweddle, D., R. Bills, E. Swartz, W. Coetzer, L. Da Costa, L. Engelbrecht, J. Cambray, B. Marshall, D. Impson, P.H. Skelton, W.R.T. Darwall and K.G. Smith. 2009. The status and distribution of

freshwater fishes, pp. 21–37. In: W.R.T. Darwall, K.G. Smith, D. Tweddle and P.H. Skelton (Eds.). The status and distribution of freshwater biodiversity in southern Africa, IUCN, Gland, Switzerland and SAIAB, Grahamstown, South Africa.

[50]Stankiewicz, J. and M.J. de Wit. 2006. A proposed drainage evolution model for Central Africa-Did the Congo flow east? J. Afr. Earth Sci. 44: 75–84.

[51]Tweddle, D. and P.H. Skelton. 2008. New species of '*Barbus*' and *Labeobarbus* (Teleostei: Cyprinidae) from the South Rukuru River, Malawi, Africa. Smithiana Bull. 8: 25–39.

[52]Simms, M.J. 2000. Preliminary report on the sediments of Lake Patrick, pp. 279–282. In: L.S. Barham (Ed.). The Middle Stone Age in Zambia. West. Acad. Special. Press.

[53]Moore, A.E. and P.A. Larkin. 2001. Drainage evolution in south-central Africa since the breakup of Gondwana. S. Afr. J. Geol. 104: 47–68.

[54]Cotterill, F.P.D. 2004. Drainage evolution in south-central Africa and vicariate speciation in swamp-dwelling weaver birds and swamp flycatchers. The Honeyguide 50 (1): 7–25.

[55]Bootsman, C.S. 1998. The evolution of the Molopo drainage. Ph.D. Diss., Witwatersrand Univ., Johannesburg: 262 pp.

[56]Grove, A.T. 1969. Landforms and climactic change in the Kalahari and Ngamiland. Geograph. J. 135: 191–212.

[57]Poll, M. 1967. Contribution à la Faune Ichthyologique de l'Angola. Publicações Culturais 75. Companhia dos Diamentes de Angola (DIAMANG), Lisbon. 381 pp.

[58]Kramer, B. and E.R. Swartz. 2010. A new species of Slender Stonebasher within the *Hippopotamyrus ansorgii* complex from the Cunene River in southern Africa (Teleostei: Mormyriformes). J. Nat. Hist. 44: 2213–2242.

[59]Kramer, B. and M. Wink. 2013. East-west differentiation in the *Marcusenius macrolepidotus* species complex in southern Africa: The description of a new species for the lower Cunene River, Namibia (Teleostei: Mormyridae). J. Nat. Hist. 47(35–36): 2327–2362.

[60]Koblmuller, S., U.K. Schliewen, N. Duftner, K.M. Sefc, C. Katongo and C. Sturmbauer. 2008. Age and spread of the haplochromine cichlid fishes in Africa. Mol. Phylogen. Evol. 49: 153–169.

[61]Kounov, A., G. Viola, M.J. de Wit and M. Andreoli. 2009. Denudation along the Atlantic passive margin: New insights from apatite fission-track analysis on the western coast of South Africa, pp. 287–306. In: F. Lisker, B. Ventura and U.A. Glasmacher (Eds). Thermochronological methods: From palaeotemperature constraints to landscape evolution models. Geol. Soc. Lond. Spec. Publ. 324.

[62]Baby, G., F. Guillocheau, J. Morin, J. Ressouche, C. Robin, O. Broucke and M. Dall'Asta. 2018. Post-rift stratigraphic evolution of the Atlantic margin of Namibia and South Africa: Implications for the vertical movements of the margin and the uplift history of the South African Plateau. Mar. Petrol. Geol. 97: 169-191.

[63]Stanley, J.R. and R.M. Flowers. 2020. Mesozoic denudation history of the lower Orange River and eastward migration of erosion across the southern African Plateau. Geol. Soc. Am., Lithosphere. 12(1): 74–87.

[64]de Wit, M. 2007. The Kalahari epeirogeny and climate change: Differentiating cause and effect from core to space. S. Af. J. Geol.110(2-3): 367–392.

[65]Dingle, R.V. and Q.B. Hendry. 1984. Late Mesozoic and Tertiary sediment supply to the eastern Cape Basin (SE Atlantic) and palaeo-drainage systems in southwestern Africa. Mar. Geol. 56(1-4): 13–26.

[66]de Wit, M.C.J. 1999. Post-Gondwana drainage and the development of diamond placers in western South Africa. Econ. Geol. 94: 721–740.

[67]Kounov, A., G. Viola, M.J. de Wit and M. Andreoli. 2008. A Mid Cretaceous paleo-Karoo River valley across the Knersvlakte plain (northwestern coast of South Africa): Evidence from apatite fission-track analysis. G. Soc. S. Afr. 111: 409–420.

[68]Jubb, R.A. 1964. Freshwater fishes and drainage basins. S. Afr. J. Sci. 60: 17–21.

[69]Barber-James, H.M. 2003. The biogeography of the Prosopistomatidae, with a particular emphasis on southern African species. Biogeog. Biodiv. Ecol. 263–270.

[70]Phillips, D., J.W. Harris, M.C.J. de Wit and E.L. Matchan. 2018. Provenance history of detrital deposits, west coast of Namaqualand, South Africa. Mineral. Petrol. 112: 259–273.

[71]Madi, K. 2016. Seismicity and neotectonic uplift in the Augrabies Falls National Park, Namaqualand, Northern Cape, South Africa. Earthquake Sci. 29: 281–289.

[72]Grodek, T., E. Morin, D. Helman, I. Lensky, O. Dahan, M. Seely, G. Benito and Y. Enzel. 2019. Eco-hydrology and geomorphology of the largest floods along the hyperarid Kuiseb River, Namibia. J. Hydrol.: https://doi.org/10.1016/j.jhydrol.2019.124450.

[73]Skelton, P.H., E.R. Swartz and J. Vreven. 2018. The identity of *Barbus capensis* Smith, 1841 and the generic status of southern African tetraploid cyprinids (Teleostei, Cyprinidae). Eur. J. Tax. 410: 1–29.

[74]Leggatt, R.A. and G.K. Iwama. 2003. Occurrence of polyploidy in the fishes. Rev. Fish Biol. Fisher. 13: 237–246.

[75]Van Vuuren, N.G., P.F.S. Mulder, J.T. Ferreira and F.H. van der Bank. 1989. The identification of hybrids of *Barbus aeneus* X *B. kimberleyensis* and *Labeo capensis* X *L. umbratus* in Hardap Dam, SWA/Namibia. Madoqua 16: 27–34.

[76]Blewett, S.C.J. and D. Phillips 2016. An overview of Cape Fold Belt chronology: Implications for sediment provenance and the timing of orogenesis, pp. 45–55. In: B. Linol and M.J. de Wit (Eds.). Origin and evolution of the Cape Mountains and Karoo basin. Springer Internatl. Publ., Switzerland.

[77]Chakona, G., E.R. Swartz and A. Chakona. 2017. The status and distribution of a newly identified endemic galaxiid in the eastern Cape Fold Ecoregion, of South Africa. Aq. Conserv. Mar. Freshw. Ecosyst: 1–13.

[78]Hearty, P., M. Raymo, M. Sandstrom, A. Rovere and M. O'Leary. 2016. Timing and estimates of Plio-Pleistocene sea-level highstands from the Republic of South Africa (RSA). Geophys. Res. Abst. 18.

[79]Compton, J. 2010. Pleistocene sea-level fluctuations and human evolution on the southern coastal plain of South Africa. Quartern. Sc. Rev. 30(5–6): 506–527.

[80]Mark, B. G. and H.A. Osmaston. 2008. Quaternary glaciation in Africa: key chronologies and climatic implications. J. Quaternary Sci. 23: 589–608.

[81]Partridge, T.C. and R.R. Maud. 1987. The geomorphic evolution of southern Africa since the Mesozoic. S. Afr. J. Geol. 90: 179–208.

[82]Nash, D. and F. Eckardt. 2015. Drainage development, neotectonics and base-level change in the Kalahari Desert, southern Africa. S. Afr. Geograph. J. 98(2): 308–320.

[83]Skelton, P.H. 2021. *pers. comm.*

[84]Nash, D.J. 1996. On the dry valleys of the Kalahari: Documentary evidence of environmental change in central southern Africa. Geograph. J. 162(2): 154–168.

[85]Nash, D.J. and G.H. Endfield. 2002. Historical flows in the dry valleys of the Kalahari identified from missionary correspondence. S. Afr. J. Sci. 98: 244–248.

[86]Van Veelen, M. and T. Baker. 2009. Feasibility study of the potential for sustainable water resources development in the Molopo-Nossob watercourse. ORASECOM Rept. 007/2009: 106 pp.

[87]Nash, D.J. 2015. Of dunes, depressions and dry valleys: The arid landscapes of the Kalahari Desert, pp. 129–138. In: S. Grab and J. Knight (Eds.). Landscapes and landforms of South Africa. Springer Internatl. Publ., Switzerland.

[88]Kambikambi, M.J., W.T. Kadye and A. Chakona. 2021. Allopatric differentiation in the *Enteromius anoplus* complex in South Africa, with the revalidation of *E. cernuus* and *E. oraniensis*, and description of a new species, *E. mandelai* (Teleostei: Cyprinidae). J. Fish Biol. 99(3): 931-954

[89]Dixon, J.E.W. and M.J. Blom. 1973. Some aquatic vertebrates from the Namib Desert, South West Africa. Madoqua, Series II, 3(69–73): 31–32.

[90]Jubb, R.A. 1967. The freshwater fishes of southern Africa. AA Balkema, Cape Town: 248 pp.

[91]Bethune, S. and K.S. Roberts. 1991. Checklist of the fishes of Namibia for each wetland region. Madoqua 17 (2): 193–199.

[92]Swart, R. 2008. An earth science review of the Orange-Fish River Basin. Report produced for the ephemeral river basins in southern Africa (ERB) project. Desert Research Foundation of Namibia (DRFN), Windhoek: 46 pp.

[93]Cambray, J.A. and M.N. Bruton. 1985. Age and growth of a colonizing minnow, *Barbus anoplus*, in a man-made lake in South Africa. Environ. Biol. Fish 12(2): 131–141.

[94]Shaw, P.A., D.S.G. Thomas and D.J. Nash. 1992. Late Quaternary fluvial activity in the dry valleys (mekgacha) of the middle and southern Kalahari, southern Africa. J. Quart. Sci. 7: 273–281.

[95]Joyce, D.A., D.H. Lunt, R. Bills, G.F. Turner, C. Katongo, N. Duftner, C. Sturmbauer and O. Seehausen. 2005. An extant cichlid fish radiation emerged in an extinct Pleistocene lake. Nature 435: 90–95.

[96]Tweddle, D. 2021. *pers. comm.*

[97]Eccles, D.H., D. Tweddle and P.H. Skelton. 2011. Eight new species in the dwarf catfish genus *Zaireichthys* (Siluriformes: Amphiliidae). Smithiana Bull. 13: 3–28.

[98]Day, J.J., R. Bills and J.P. Friel. 2009. Lacustrine radiations in African *Synodontis* catfish. J. Evol. Biol. 22(4): 805–817.

[99]Day, J.J., C.R. Peart, K.J. Brown, J.P. Friel, R. Bills and T. Moritz. 2013. Continental diversification of an African catfish radiation (Mochokidae: *Synodontis*). Syst. Biol 62(3): 351–365.

[100]Persendt, F.C. and C. Gomez. 2015. Assessment of drainage network extractions in a low-relief area of the Cuvelai Basin (Namibia) from multiple sources: LiDAR, topographic maps, and digital aerial orthophotographs. Geomorphology (2015): http://dx.doi.org/10.1016/j.geomorph.2015.06.047.

[101]Paillou, P., S. Lopez, E. Marais and K. Scipal. 2020. Mapping paleohydrology of the ephemeral Kuiseb River, Namibia, from radar remote sensing. Water 12: 1–13.

[102]Deitrich, P., N.P. Griffs, D.P. Le Heron, I.P. Montanez, C. Kettler, C. Robin and F. Guillocheau. 2021. Fjord network in Namibia: A snapshot into the dynamics of the late Paleozoic glaciation. Geology 49: https// doi.org/10.1130/G49067.7.

[103]Kramer, B. 2020. Communication and wavelength analysis in weakly electric fishes with special emphasis on mormyrids, pp. 255-280. In: B. Fritzsch (Ed.) and H. Bleckman (Vol. Ed.). The senses: A compre–prehensive reference. Vol. 7, Elsevier, Academic Press.

[104]Schedel, F.D.B., A. Chakona, B.L. Sidlauskas, M.O. Popoola, N.U. Wingi, D. Neumann, E.J.W.M.N. Vreven and U.K. Schliewen. 2022. New phylogenetic insights into the African catfish families Mochokidae and Austroglanididae. J. Fish Biol. 100: 1171-1186.

[105]Wu, F., D. He, G, Fang and T. Deng. 2019. Into Africa via docked India: A fossil climbing perch from the Oligocene of Tibet helps solve the anabantid biogeographical puzzle. Sci.Bull. 64: 455-463.

[106]Wallis, G.P. 2021. Evolutionary genetics and biogeography of galaxiid fishes (Teleostei:Galaxiiformes: Galaxiidae). Diversity 13: 1-6.

[107]Marais, E. 2021. *pers. comm.*

[108]Kok, O.B. and J.A.J. Nel. 1996. The Kuiseb River as a linear oasis in the Namib Desert. Afr. J. Ecol. 34(1):39-47.

CHAPTER 4
NAMIBIA'S NATIVE INLAND FISHES

by

Charles H. Hocutt, Clinton J. Hay, Francois J. Jacobs,
Glenn S. Merron, Ben C.W. van der Waal, Ben J. van Zyl
and Sarah K. Tribelhorn

I. Diversity of Namibia's Inland Fishes

A. General

A total of 137 species of inland fishes is known to Namibia (Table 1) based on a compendium of

all known historical records,[e.g., 1,2,3,4,5]
taxonomic revisions, [e.g., 6,7,8,9,10,11]
species descriptions [e.g., 12,13,14,15]
and personal records since the 1970s (Hay, Hocutt, Jacobs, Merron, van der Waal, van Zyl).

Of this total, 125 are native species, six of which are undescribed, plus 12 non-native species addressed in Chapter 5. Accounts are provided for each species except where (a) sister species (e.g., many mormyrids) are considered indistinguishable based on their morphological overlap, or (b) insufficient information exists for recognised, but undescribed species, i.e., "Kavango Suckermouth" (*Chiloglanis* sp.), "Climbing Perch" (*Microctenopoma* sp.), "Sand Catlet" (*Zaireichthys* sp.), "Topminnow" (*Lacustricola* sp.) and "Bream" (*Sargochromis* sp.). The lone exception to the latter is the Cunene Broadhead Catfish (*Clariallabes* sp.) for which more information is available and was featured on a postage stamp in 2003; this example demonstrates the often slow pace of taxonomic studies.

B. Diversity by Drainage

Fish diversity is highest in rivers along the northern border with Angola; however, river names as well as common and scientific names of fishes can be confusing depending upon whether the published literature uses Namibian or Angolan lexicon. By example, the *formally described* Cunene Dwarf Bream, Cuando Dwarf Stonebasher and Cubango Stonebasher go by the conventional spelling in Angola, as opposed to typical Namibian spelling of Kunene, Kwando and Kavango, e.g., Kunene Sand Catlet. River and species names used here follow those published in the literature. Otherwise, Namibian spelling is given preference. Some species lacking common names were assigned names based on their morphology after conferring with authorities, e.g., the Saddled and Broad-striped Citharines.[6]

The native fishes of Namibia are distributed among 17 families, with three families accounting for over 63% of the total known species. The Cyprinidae (labeos, yellowfishes, barbs and minnows) is the most speciose family, represented by seven genera containing 36 species. The cyprinids are followed by the Cichlidae (river bream, largemouth breams and tilapia) with 26 species (including one undescribed form) and the cryptic Mormyridae (snoutfishes) with 15 species. Six families are represented by single species (Kneriidae, Hepsetidae, Austroglanididae, Claroteidae, Schilbeidae and Nothobranchiidae). Combined, there are 27 species of catfish distributed amongst six families in nine genera.

The Orange-Fish River along the Namibia/South African border has the most *depauperate* fauna, with only 12 *indigenous* species. Five cyprinids and a catfish are considered *endemic* to this portion of the drainage representing a very high level of endemism (50%), although some have been introduced elsewhere. Disturbingly, there are at least nine non-native species in the lower basin linked to aquaculture (e.g., Common Carp, Grass Carp, Mozambique Tilapia, Nile Tilapia) or found in private (e.g., Goldfish, Koi), state (Largemouth Bass) and municipal waters (e.g., Mosquitofish, Guppy). The Orange system is highly modified in both countries for water supply and irrigation, which has led to alterations in the natural environment and hybridisation amongst closely related yellowfish and labeo species.

As reported in Chapter 6, there is a 1,500+ km hiatus of perennial rivers between the Orange River outlet into the Southeast Atlantic Ocean and the Kunene River to the north along the Angola border. In between, there are several westward-directed ephemeral drainages that seldom enter the Atlantic. Of distinct interest is the Kuiseb River, that harbours the only known Chubbyhead Barbs in Namibia, considered a *relict* population.

The Kunene has 65 known species in Namibia; however, an additional 18 inland species are known from Angola (Table 2) that presumably could be expected to occur further downstream in Namibia.[4] Eastward from the Kunene basin in northern Namibia are (in order) the Cuvelai, upper Kavango (Cubango in Angola), Kwando (Cuando in Angola) and Upper Zambezi Rivers. Chapters 3 and 7 discuss in detail the intimate link between the Kunene and Cuvelai Rivers. To date, 28 species are known to occur in the Cuvelai drainage, but 18 more have been collected in the water supply canal and storage dam (Calueque-Oshakati Canal and Olushandja Dam) between the two rivers, all of Kunene River origin. In effect, they are *introduced* species. Thus, 47 species are recognised in the accounts as being present in the Cuvelai basin. The Line-spotted Barb has been reported only in the *iishana* at this date.

The Kavango River and Upper Zambezi-Chobe Rivers in Namibia have 84 and 85 native species as compared to 104 and 93, respectively, reported for Angola[4] (Table 2). Additionally, two new species of largemouth bream occurring in the upper Kavango (*Serranochromis cacuchi*) and Cuito (*S. alvum*) Rivers of Angola are not yet known from Namibia.[14] The contrast of numbers of species between Namibia and Angola for the Kavango and Upper Zambezi Rivers can be attributed to their significantly larger drainages in Angola (and Zambia) presenting a diversity of upland habitats absent from Namibia. While the Kavango and Upper Zambezi-Chobe Rivers nearly have the same number (84 vs. 85) of species in Namibia, the compositions differ. The Large-bodied Churchill, Okavango Churchill and Kavango Sand Catlet are endemic to the Kavango, while an undescribed climbing perch is shared only with the neighbouring Kwando-Linyanti River system. Species known only in Namibia from the Upper Zambezi-Chobe drainage are Szabo's Stonebasher, Long-head Churchill, Zambezi Dwarf Stonebasher, Saddled Citharine, Caprivi Killifish and an undescribed form of sand catlet.

For ***INLAND FISHES AND FISHERIES OF NAMIBIA***, the division between the Upper and Middle Zambezi River is Victoria Falls, a major barrier to fish dispersal. The head of Cahora Bassa Dam in Mozambique separates the Middle and Lower Zambezi River. FAO's FishBase lists 199 species for the entire basin.[17] A few species found below the falls, but not above, are the Zambezi Parrotfish (*Cyphomyrus discorhynchus),* Vundu (*Heterobranchus longifilis*), Eastern Sand Catlet (*Zaireichthys monomotapa*) and Yellow-belly Largemouth (*Serranochromis robustus*).

The Kwando (Cuando) River heads up in Angola between the Kavango (Cubango) and Upper Zambezi Rivers, flows south to join the eastward flowing Linyanti River to supply Lake Liambezi (Chapter 6). The Linyanti River is a rather continuous low-gradient extension of the Kwando, thus they are considered together in this work. The intermediate-positioned Kwando-Linyanti system is a much smaller drainage than either the Kavango or Upper Zambezi with less habitat diversity in Namibia. Still, 74 species are known from the drainage. The Cuando Dwarf Stonebasher and Pallid Sand Catlet are considered endemic.

The entire area of 40,000 km^2 between the Okavango Delta panhandle, Kwando River and the Upper Zambezi-Chobe system is a network of interconnected waterways and ancient channels (Chapters 3). The Chobe River is a virtual backwater of the Upper Zambezi River and may flow in either direction depending upon the severity of flooding of the latter; thus they are discussed together in the Species Accounts. To further demonstrate the low relief and interconnectedness of the entire region, the Kavango River feeds Botswana's Okavango Delta and overflow during very high floods may fill Lake Ngami or pass water into the Kwando-Linyanti swamps via the Selinda Spillway (aka Magweggana River), a depression between two geological faults that is expected to deflect the Kavango River into the Upper Zambezi drainage in time. Some 87 species are known from the Okavango Delta and Chobe River,[5] while 93 are recognised in Namibia from the Kavango, Kwando-Linyanti and Upper Zambezi-Chobe Rivers, combined.

Table 1. Checklist and distribution of all native and *non-native* inland fishes by major river systems and national waterbodies in Namibia (phylogeny follows Betancur-R et al. 2017).[18] Abbreviations are: Ku = Kunene; CV = Cuvelai; Ka = Kavango; KL = Kwando-Linyanti; UZC = Upper Zambezi-Chobe; OR = Orange-Fish; Kb = Kuiseb; NW = National Waterbodies; cf. = similar to; and sp. = recognised, but not formally described species.

COMMON NAME BY FAMILY	SPECIES	DISTRIBUTION
Mormyridae: Snoutfishes		
Cubango Stonebasher	*Cyphomyrus cubangoensis*	Ka; KL; UZC
Slender Stonebasher	*Hippopotamyrus ansorgii*	Ka; KL; UZC
Cunene River Stonebasher	*Hippopotamyrus longilateralis*	Ku; CV
Szabo's Stonebasher	*Hippopotamyrus szaboi*	UZC
Upper Zambezi Bulldog	*Marcusenius altisambesi*	Ka; KL; UZC
Skeleton Coast Bulldog	*Marcusenius desertus*	Ku
Cunene Bulldog	*Marcusenius multisquamatus*	Ku; CV
Western Bottlenose	*Mormyrus lacerda*	Ku; CV; Ka; KL; UZC
Long-head Churchill	*Petrocephalus longicapitis*	KL; UZC
Large-bodied Churchill	*Petrocephalus magnitrunci*	Ka
Cunene Churchill	*Petrocephalus magnoculis*	Ku; CV
Okavango Churchill	*Petrocephalus okavangoensis*	Ka
Dwarf Stonebasher	*Pollimyrus castelnaui*	Ku; CV; Ka
Cuando Dwarf Stonebasher	*Pollimyrus cuandoensis*	KL
Upper Zambezi Dwarf Stonebasher	*Pollimyrus marianne*	UZC
Kneriidae: Knerias		
Cunene Kneria	*Kneria maydelli*	Ku
Cyprinidae: Labeos, Yellowfishes, Barbs and Minnows		
Goldfish	*Carassius auratus*	NW
Upjaw Barb	*Coptostomabarbus wittei*	CV; Ka; KL; UZC
Grass Carp	*Ctenopharyngodon idella*	OR
Common Carp	*Cyprinus carpio*	OR; NW
Amur Carp	*Cyprinus rubrofuscus*	NW
Zambezi River Sardine	*Engraulicypris brevianalis*	Ka; KL; UZC
Orange River Sardine	*Engraulicypris gariepinus*	OR
Kunene River Sardine	*Engraulicypris howesi*	Ku; CV
Spottail Barb	*Enteromius afrovernayi*	Ku; Ka; KL; UZC
Blackback Barb	*Enteromius barnardi*	Ku; CV; Ka; KL; UZC

Barotse Barb	*Enteromius barotseensis*	Ku; Ka; KL; UZC
Hyphen Barb	*Enteromius bifrenatus*	Ku; CV; Ka; KL; UZC
Shorthead Barb	*Enteromius breviceps*	Ku
Dwarf Barb	*Enteromius brevidorsalis*	Ka; KL; UZC
Orange-fin Barb	*Enteromius eutaenia*	Ku; Ka; KL; UZC
African Banded Barb	*Enteromius fasciolatus*	Ku; Ka; KL; UZC
Sickle-fin Barb	*Enteromius haasianus*	Ka; KL; UZC
Redspot Barb	*Enteromius kerstenii*	Ku; CV; Ka; KL; UZC
Line-spotted Barb	*Enteromius lineomaculatus*	Ku; CV; Ka; UZC
Papermouth	*Enteromius mattozi*	Ku; CV
Zigzag Barb	*Enteromius miolepis*	Ka; KL; UZC
Copperstripe Barb	*Enteromius multilineatus*	Ku; CV; Ka; KL; UZC
Orange River Chubbyhead Barb	*Enteromius* cf. *oraniensis*	Kb
Straightfin Barb	*Enteromius paludinosus*	Ku; CV; Ka; KL; UZC; OR; NW
Dashtail Barb	*Enteromius poechii*	Ka; KL; UZC
Beira Barb	*Enteromius radiatus*	Ku; CV; Ka; KL; UZC
Thamalakane Barb	*Enteromius thamalakanensis*	Ku; CV; Ka; KL; UZC
Threespot Barb	*Enteromius trimaculatus*	Ku; CV; OR
Longbeard Barb	*Enteromius unitaeniatus*	Ku; CV; Ka; KL; UZC
Cunene Labeo	*Labeo ansorgii*	Ku; CV
Orange River Mudfish	*Labeo capensis*	OR; NW
Redeye Labeo	*Labeo cylindricus*	Ka; KL; UZC
Upper Zambezi Labeo	*Labeo lunatus*	Ka; KL; UZC
Silver Labeo	*Labeo ruddi*	Ku; CV
Moggel	*Labeo umbratus*	OR; NW
Vaal-Orange Smallmouth Yellowfish	*Labeobarbus aeneus*	OR; NW
Upper Zambezi Yellowfish	*Labeobarbus codringtonii*	Ka; UZC
Vaal-Orange Largemouth Yellowfish	*Labeobarbus kimberleyensis*	OR; NW
Namaqua Barb	*Namaquacypris hospes*	OR
Northern Barred Minnow	*Opsaridium zambezense*	Ka; KL; UZC

Distichodontidae: Citharines

Saddled Citharine	*Nannocharax dageti*	UZC
Broad-striped Citharine	*Nannocharax* cf. *lineostriatus*	Ka; KL; UZC
Dwarf Citharine	*Nannocharax machadoi*	Ku; CV; Ka; KL; UZC
Multibar Citharine	*Nannocharax multifasciatus*	Ku; Ka; KL; UZC

Alestidae: Robbers and Tigerfish

Striped Robber	*Brycinus lateralis*	Ku; CV; Ka; KL; UZC
Tigerfish	*Hydrocynus vittatus*	Ka; KL; UZC
Silver Robber	*Micralestes acutidens*	Ku; CV; Ka; KL; UZC
Slender Robber	*Rhabdalestes maunensis*	Ku; CV; Ka; KL; UZC

Characidae: Characins

Largespot Tetra	*Astyanax orthodus*	NW

Hepsetidae: African Pike

Southern African Pike	*Hepsetus cuvieri*	Ku; CV; Ka; KL; UZC

Amphiliidae: Mountain Catfish and Sand Catlets

Stargazer Mountain Catfish	*Amphilius uranoscopus*	Ka; UZC
Blotched Sand Catlet	*Zaireichthys conspicuus*	Ka; UZC
Kavango Sand Catlet	*Zaireichthys kavangoensis*	Ka
Kunene Sand Catlet	*Zaireichthys kunenensis*	Ku
Pallid Sand Catlet	*Zaireichthys pallidus*	KL
"Sand Catlet"	*Zaireichthys* sp.	UZC

Austroglanididae: Rock Catfish

Rock Catfish	*Austroglanis sclateri*	OR

Clariidae: Air-breathing Catfish

Broadhead Catfish	*Clariallabes platyprosopos*	Ka; UZC
Cunene Broadhead Catfish	*Clariallabes* sp.	Ku
Cave Catfish	*Clarias cavernicola*	Aigamas Cave complex
Sharptooth Catfish	*Clarias gariepinus*	Ku; CV; Ka; KL; UZC; OR; NW
Smoothhead Catfish	*Clarias liocephalus*	Ku; Ka; UZC
Blunttooth Catfish	*Clarias ngamensis*	Ku; CV; Ka; KL; UZC
Blotched Catfish	*Clarias stappersii*	Ku; Ka; KL; UZC
Snake Catfish	*Clarias theodorae*	Ku; CV; Ka; KL; UZC

Claroteidae: Claroteid Catfishes

Zambezi Grunter	*Parauchenoglanis ngamensis*	Ka; KL; UZC

Mochokidae: Squeakers and Suckermouths

Okavango Suckermouth	*Chiloglanis fasciatus*	Ka; KL; UZC
Neumann's Suckermouth	*Chiloglanis neumanni*	Ku
"Kavango Suckermouth"	*Chiloglanis* sp.	Ka
Leopard Squeaker	*Synodontis leopardinus*	Ku; CV; Ka; KL; UZC
Largespot Squeaker	*Synodontis macrostigma*	Ku; CV; Ka; KL; UZC
Largemouth Squeaker	*Synodontis macrostoma*	Ku; Ka; KL; UZC
Spotted Squeaker	*Synodontis nigromaculatus*	Ka; KL; UZC
Bubblebarb Squeaker	*Synodontis thamalakanensis*	Ka; KL; UZC
Finetooth Squeaker	*Synodontis vanderwaali*	Ku; Ka; UZC
Upper Zambezi Squeaker	*Synodontis woosnami*	Ku; CV; Ka; KL; UZC

Schilbeidae: Butter Catfish

Silver Catfish	*Schilbe intermedius*	Ku; CV; Ka; KL; UZC

Anabantidae: Labyrinth Fishes

Manyspined Climbing Perch	*Ctenopoma multispine*	CV; Ka; KL; UZC
Blackspot Climbing Perch	*Microctenopoma intermedium*	Ka; KL; UZC
"Climbing Perch"	*Microctenopoma* sp.	Ka; KL

Osphronemidae: Gouramies

Threespot Gourami	*Trichogaster trichopterus*	NW

Mastacembelidae: Spiny Eels

Longtail Spiny Eel	*Mastacembelus frenatus*	Ka; KL; UZC
Ocellated Spiny Eel	*Mastacembelus vanderwaali*	Ka; UZC

Cichlidae: River Bream, Largemouth Breams and Tilapia

Slender River Bream	*Chetia gracilis*	Ku
Angolan River Bream	*Chetia welwitschi*	Ku
Redbreast Tilapia	*Coptodon rendalli*	Ku; CV; Ka; KL; UZC; NW
Banded Jewelfish	*Hemichromis elongatus*	Ka; KL; UZC
Threespot Tilapia	*Oreochromis andersonii*	Ku; CV; Ka; KL; UZC
Greenhead Tilapia	*Oreochromis macrochir*	Ku; CV; Ka; KL; UZC; NW
Mozambique Tilapia	*Oreochromis mossambicus*	OR; NW
Nile Tilapia	*Oreochromis niloticus*	UZC; OR; NW
Cunene Dwarf Bream	*Orthochromis machadoi*	Ku; CV
Zambezi River Bream	*Pharyngochromis acuticeps*	Ka; KL; UZC

Southern Mouthbrooder	*Pseudocrenilabrus philander*	Ku; CV; Ka; KL; UZC; OR
Rainbow Bream	*Sargochromis carlottae*	Ka; KL; UZC
Dusky Bream	*Sargochromis codringtonii*	Ka; KL; UZC
Cunene Bream	*Sargochromis coulteri*	Ku; CV
Pink Bream	*Sargochromis giardi*	Ka; KL; UZC
Green Bream	*Sargochromis greenwoodi*	Ka; KL; UZC
"Bream"	*Sargochromis* sp.	Ku; CV
Humpback Largemouth	*Serranochromis altus*	Ku; Ka; KL; UZC
Thinface Largemouth	*Serranochromis angusticeps*	Ku; CV; Ka; KL; UZC
Nembwe	*Serranochromis jallae*	Ka; KL; UZC
Longfin Largemouth	*Serranochromis longimanus*	KL; UZC
Purpleface Largemouth	*Serranochromis macrocephalus*	Ku; CV; Ka; KL; UZC
Brownspot Largemouth	*Serranochromis thumbergi*	Ku; CV; Ka; KL; UZC
Thicklipped River Bream	*Thoracochromis albolabris*	Ku; CV
Namib River Bream	*Thoracochromis buysi*	Ku; CV
Otjikoto Tilapia	*Tilapia guinasana*	NW
Okavango Tilapia	*Tilapia ruweti*	Ka; KL; UZC
Banded Tilapia	*Tilapia sparrmanii*	Ku; CV; Ka; KL; UZC; OR; NW

Nothobranchidae: Killifishes

Caprivi Killifish	*Nothobranchius capriviensis*	UZ – Restricted

Poeciliidae: Livebearers and Topminnows

Mosquitofish	*Gambusia affinis*	NW
Guppy	*Poecilia reticulata*	NW
Swordtail	*Xiphophorus helleri*	NW

Procatopodidae: African Lampeyes

Meshscaled Topminnow	*Lacustricola chobensis*	Ka; KL; UZC
Johnston's Topminnow	*Lacustricola johnstoni*	Ku; CV; Ka; KL; UZC
Striped Topminnow	*Lacustricola katangae*	Ku; Ka; KL; UZC
Pygmy Topminnow	*Lacustricola pygmaeus*	Ka; KL; UZC
"Topminnow"	*Lacustricola* sp.	Ku

Centrarchidae: Sunfishes

Largemouth Bass	*Micropterus salmoides*	OR; NW

Table 2. Inland fishes known from the Cunene (Ku), Cubango (Ka) and Upper Zambezi (UZ) Rivers, Angola, but not thus far recorded from the same Namibian drainages.[4,14,19] Abbreviations follow Table 1.

Mormyridae	
Marcusenius angolensis	Ku; Ka; UZ
Paramormyrops jacksoni	UZ
Petrocephalus longicapitis	Ka
Petrocephalus okavangoensis	UZ
Kneriidae	
Parakneria fortuita	Ka; UZ
Cyprinidae	
Clypeobarbus bellcrossi	Ka; UZ
Enteromius argenteus	Ku
Enteromius breviceps	Ka
Enteromius brevidorsalis	Ku
Enteromius chicapaensis	Ka
Enteromius chiumbeensis	Ka
Enteromius dorsolineatus	Ku
Enteromius evansi	Ka
Enteromius greenwoodi	Ka
Enteromius kessleri	Ka; UZ
Enteromius lujae	Ku; Ka; UZ
Enteromius mocoensis	Ka
Enteromius trimaculatus	Ka
Enteromius cf. *viviparus*	Ku; Ka; UZ
Distichodontidae	
Nannocharax macropterus	Ka; UZ
Alestidae	
Micralestes argyrotaenia	Ku
Amphiliidae	
Amphilius cubangoensis	Ka; UZ
Zaireichthys pallidus	Ka

Clariidae

Clarias dumerilii	Ku

Mochokidae

Chiloglanis angolensis	Ku
Chiloglanis spp.	Ka; UZ
Synodontis nigromaculatus	Ku
Synodontis thamalakanensis	Ku

Schilbeidae

Schilbe yangambianus	UZ

Anabantidae

Ctenopoma multispine	Ku
Microctenopoma steveboyesi	Ka

Cichlidae

Chetia gracilis	Ka
Pharyngochromis acuticeps	Ku
Sargochromis codringtonii	Ku
Sargochromis giardi	Ku
Sargochromis greenwoodi	Ku
Serranochromis alvum	Ka
Serranochromis cacuchi	Ka
Serranochromis jallae	Ku
Serranochromis longimanus	Ka

Procatopodidae

Lacustricola luluae	Ka
Lacustricola macrurus	Ku
Lacustricola mediolateralis	Ka
Lacustricola nigrolateralis	Ka

II. Species Accounts

A. Family Mormyridae (Snoutfishes and Elephantfishes)

The family Mormyridae is endemic to tropical Africa. *Phylogenetically*, snoutfishes are Namibia's most primitive inland fishes, currently represented by six genera and 15 species; however, their taxonomy is unsettled. They are soft-bodied fish with small scales, being silvery, brown, black or grey in colour. The snout of some species is extended into a bulging upper lip or proboscis, whereas the lower jaw is longer than the snout/upper jaw in others. The dorsal and anal fins are often extended with many soft rays. The caudal peduncles are narrow and the caudal fins are forked, some with rounded lobes. Mormyrids usually feed at night on similar food items. Some species migrate in shoals and spawn at the same time of the year.

Many of the species within the genera *Petrocephalus, Pollimyrus* and *Hippopotamyrus* are morphologically very difficult to distinguish, but vary in distinctive *electric organ discharges* (EODs) produced by a specialised organ consisting of modified nerve cells and skeletal muscles located in the caudal peduncle area. Mormyrids can send and receive electric currents to detect small bottom-living invertebrates on which to feed and to communicate with other individuals. The organs are also the source of 'courtship songs' during mating.

Petrocephalus catostoma was formerly considered widespread throughout northern Namibian waters, but is now regarded as restricted to Malawi and Mozambique with its three sister species (*P. okavangoensis, P. magnitrunci* and *P. longicapitis*) indigenous to Namibia. The Zambezi Parrotfish (*Cyphomyrus discorhynchus*) is a valid species below Victoria Falls, but is replaced by the Cubango Stonebasher (*C. cubangoensis*) above the falls in the Upper Zambezi River, Kavango (Cubango) and Kunene Rivers. In other words, the famed "parrotfish run" in the Upper Zambezi River and Impalila Island after heavy rains in June and July is now the "stonebasher bash"!

Within northern Namibian rivers, the diminutive forms of *Pollimyrus* spp., *Petrocephalus* spp. and *Marcusenius* spp. are the most abundant fishes in this family. These species complexes are fairly common throughout the Kunene, Cuvelai, Kavango, Kwando-Linyanti and Upper Zambezi-Chobe systems including Lake Liambezi. Mormyrids do not occur in the Orange-Fish River. Determination of complete distributional ranges and abundance of the mormyrids by field biologists is constrained by both (a) inaccessibility of habitats and (b) species within a complex are often very difficult to distinguish morphologically.

Although some of Namibia's six mormyrid genera look superficially the same, there are some characteristics to assist in identifying them. For instance, *Mormyrus* has a much longer dorsal than anal fin, a very long snout and a small terminal mouth. All other mormyrids have short dorsal fins with only *Cyphomyrus* having a dorsal fin origin anterior to the anal fin. The mouth is superior in *Marcusenius*; inferior in *Petrocephalus*; and subterminal in both *Pollimyrus* and *Hippopotamyrus*, with *Pollimyrus* having a much shorter rounded snout (Fig. 1).

F. Jacobs

Figure 1. Representatives of the *Pollimyrus* (top: subterminal mouth), *Petrocephalus* (middle: inferior mouth) and *Marcusenius* (bottom: superior mouth) species complexes.

Genus *Cyphomyrus*

There is one species recognised in Namibia.

Cubango Stonebasher (*Cyphomyrus cubangoensis*)

F. Jacobs

Etymology: ***Cyphomyrus* = *cypho***, hunchback, referring to convex predorsal profile of back + ***-myrus***, conventional termination for generic names of elephantfishes, derived from ***Mormyrus*** (see below); ***-ensis***, suffix denoting Cubango River, Angola, origin of type specimens

DS 0, DR 28-33; AS 0, AR 21-24; LL 59-71, CP 10-12

Summary: This deep-bodied laterally compressed species has a long dorsal fin, head with a rounded snout and fleshy bulging chin. Its colour is typically dark brown with a dark band between the dorsal and anal fins. The origin of the dorsal fin is anterior to the base of the anal fin. The caudal peduncle is slender and caudal fin forked with broadly rounded lobes. It reaches 100 mm SL.

This nocturnal shoaling snoutfish inhabits river margins and lagoons with a soft bottom and fringing vegetation. Adults are also found in deeper river channels and in rocky environments. Spawning occurs during the onset of the rainy season and/or arrival of annual floodwaters. This species undergoes lateral migrations from river channels to newly inundated floodplain areas and backwater lagoons to spawn.

It is found in the Kavango, Kwando-Linyanti and Upper Zambezi-Chobe River systems of Namibia and also the Okavango Delta. The scientific and common names are both derived from the Cubango River, Angola.

Genus *Hippopotamyrus*

Three species of *Hippopotamyrus* are represented in Namibia, each being morphologically very similar. Each has an elongated laterally compressed body, dark brown to black in colour and with a vertical band between the dorsal and anal fins. They have a distinct rounded caudal peduncle and a forked caudal fin with rounded lobes. Their heads have a rounded snout with a subterminal positioned mouth, the lower jaw protruding less forward than the upper. These three species have overlapping meristic counts.

These fishes feed at night (*nocturnal*) on small invertebrates, especially insect larvae taken from the bottom and off aquatic plant stems. They spawn with the onset of the summer rainy season and/or arrival of annual flood waters in vegetated margins of slow-flowing river channels and floodplains. They are not readily caught due to the inaccessibility of their preferred habitat.

Slender Stonebasher (*Hippopotamyrus ansorgii*)

B. Kramer

Etymology: ***Hippopotamyrus*** = Greek, ***ippos***, horse + Greek, ***potamos***, river; ***ansorgii***, in honour of explorer William John Ansorge (1850–1913) who collected type

DS 0, DR 15-19; AS 0, AR 24-26; LL 67-72, CP 15-19

Summary: This rare current-loving (*rheophilic*) mormyrid inhabits fast-flowing rocky habitats with overhanging well-vegetated banks and root structure where it takes refuge during the day. It is found in the Kavango, Kwando-Linyanti and Upper Zambezi-Chobe Rivers, Namibia, into Angola and Zambia. The systematics of this complex are under review with the Kavango form perhaps being distinct. It is not thus far known from the Okavango Delta. They obtain 150 mm SL.

Cunene River Stonebasher (*Hippopotamyrus longilateralis*)

B. Kramer

Etymology: ***Hippopotamyrus*** = Greek, ***ippos***, horse + Greek, ***potamos***, river; ***longus***, long, referring to long lateral line

DS 0, DR 18-21; AS 0, AR 22-25; LL 74-80, CP 19-22

Summary: This species was originally collected along well-vegetated and steep banks of the Kunene River and is now known from the canal leading to the Cuvelai basin. It appears to be morphologically and physiologically (EODs) adapted to strong currents in its escarpment habitat. It reaches 172 mm SL. No other information is available.

Szabo's Stonebasher (*Hippopotamyrus szaboi)*

B. Kramer

Etymology: ***Hippopotamyrus*** = Greek, ***ippos***, horse + Greek, ***potamos***, river; ***szaboi*** = in honour of Dr. Thomas Szabo, one of the founding fathers of the field of electroreception

DS 0, DR 18-21; AS 0, AR 21-25; LL 61-73, CP 17-22

Summary: The range of Szaboi's Stonebasher is limited in Namibia to the rocky habitat of the main channel of the Upper Zambezi River, which is also the southern limit of its distribution. It is not recorded to date from other Namibian systems. They reach 146 mm SL.

Genus *Marcusenius*

A review of the *Marcusenius* spp. complex revealed three species in Namibia based on species-specific electric organ discharges (EODs). These three species are morphologically similar with overlapping meristic counts. They inhabit similar habitats, have similar feeding preference for aquatic invertebrates and reproduce seasonally among aquatic vegetation in newly inundated floodplains and slow-flowing river channels in relation to the annual flooding regime. Each of these species contributes to the subsistence fishery.

Upper Zambezi Bulldog (*Marcusenius altisambesi*)

F. Jacobs

Etymology: ***Marcusenius*** = in honour of J. Marcusen, ichthyologist, who wrote first mormyrid accounts; ***alti-***,high + ***-sambesi***, Zambezi, referring to distribution in Upper Zambezi River

DS 0, DR 20-26; AS 0, AR 26-30; LL 49-60, CP 12-14

Summary: This mormyrid has a long laterally-compressed body, light golden brown to dark olive and grey with bronze flecks and dark brown blotches. Its chin is a prominently elongated lobe, the mouth superior. The dorsal and anal fins are set far back with the base of the former being posterior to the base of the anal. Lengths reach 195 mm SL.

This widespread and common mormyrid is tolerant of low oxygen and high turbidity levels and is often one of the last remaining species in drying waterbodies. This shoaling fish is common in well-vegetated, muddy-bottomed habitats in slow-flowing rivers as well as in lagoons and floodplains. It is active at night feeding on small aquatic invertebrates, especially insect larvae taken from the bottom and off aquatic plant stems. They spawn during the summer rainy season and/or arrival of annual flood waters in vegetated margins of slow-flowing river channels and newly inundated floodplains.

This species is distributed throughout the Kavango, Kwando-Linyanti and Upper Zambezi-Chobe River systems in Namibia and southward to the Okavango Delta. It is caught by subsistence fishermen using hook-and-line and traditional fish traps; it is represented in the 'stonebasher bash' at Impalila Island. Little is known about the biology and ecology of this mormyrid and additional research is needed.

Skeleton Coast Bulldog (*Marcusenius desertus*)

B. Kramer

Etymology: ***Marcusenius*** = in honour of J. Marcusen, ichthyologist, who wrote first mormyrid accounts; ***desertus***, refers to the Skeleton Coast Desert, surrounding environment of the type locality

DS 0, DR 21-25; AS 0, AR 26-32; LL 55-62 , CP 11-15

Summary: This bulldog is known only from the Kunene River mouth of the Skeleton Coast/Namib Desert from which the common name is derived. This is a harsh wind-swept environment with the river having a sand/gravel/rocky substrate and minor submerged vegetation and reed banks. It is distinguished morphologically from the Cunene Bulldog by having a shorter snout length, smaller eye diameter and shorter distance between the nares. It reaches up to 200 mm SL.

Cunene Bulldog (*Marcusenius multisquamatus*)

B. Kramer

Etymology: ***Marcusenius*** = in honour of J. Marcusen, ichthyologist, who wrote first mormyrid accounts; ***multisquamatus*** = ***multi-***, many + ***-squamatus***, in reference to high number of scales in lateral line

DS 0, DR 23-25; AS 0, AR 28-31; LL 56-64, CP 12-16

Summary: The Cunene Bulldog is reported from the Kunene and Cuvelai drainages of Namibia and Angola, the former lending its common name. It occurs sympatrically with the Skeleton Coast Bulldog at the Kunene mouth. Distinguishing features are noted in the account of the latter; however, the Cunene Bulldog has a high number of lateral line scales as compared to other southern African bulldog species. Its type locality above Epupa Falls has erratic annual discharge from severe flooding to no flow. It can grow to 200 mm SL.

Genus *Mormyrus*

This genus is distinguished by its exceptionally long dorsal fin and extended snout. There are about 22 species distributed through the continent, only two of which occur in southern Africa with one in Namibia.

Western Bottlenose (*Mormyrus lacerda*)

F. Jacobs

Etymology: ***Mormyrus*** = according to Cuvier, Greek name of a "littoral fish variously coloured," probably Striped Sea Bream, ***Lithognathus mormyrus*** (Linnaeus 1758), "applied by Linnaeus, not very happily, to freshwater fishes of a uniform hue"; ***lacerda*** = in honour of the Portuguese explorer and astronomer Francisco José Maria de Lacerda (1753–1798)

DS 0, DR 62-68; AS 0, AR 18-21; LL 80-90, CP 20-22

Summary: The Western Bottlenose has an elongate compressed body coloured grey-brown above, lighter and yellowish below. Its long snout tapers downwards to a small terminal mouth. The dorsal fin is noticeably long, extending from behind the pectoral fins well onto the caudal peduncle. The caudal fin is forked with rounded lobes.

This is the largest mormyrid in Namibian waters though not especially common based on the infrequency of occurrence. It obtains 500 mm TL and 2 kg. It is widely distributed and found in the Kunene, Cuvelai, Kavango, Kwando-Linyanti and Upper Zambezi-Chobe River systems, south to the Okavango Delta and in Angola, Zambia and Zimbabwe. It is also known from the Kafue River, Zambia. It is replaced in the Lower Zambezi by the larger Eastern Bottlenose (*M. longirostris*) that obtains 750 mm SL, 10 kg.

The Western Bottlenose inhabits slower-flowing river channels, backwater lagoons and floodplains with aquatic vegetation and soft bottom substrate. It is nocturnal, feeding on insect larvae, aquatic invertebrates, snails, crabs and small fish. It spawns during the rainy season in inundated shallows amongst aquatic vegetation. It is caught by subsistence fishermen using a variety of methods including hook and line.

Genus *Petrocephalus*

A review of the *Petrocephalus catostoma* complex in 2012 revealed five new species based on species-specific electric organ discharges (EODs), four of which are considered to be in Namibian waters with *P. catostoma* now considered absent. These species are morphologically similar with overlapping meristic counts.

These four species inhabit similar habitats, have similar feeding preference for aquatic invertebrates and reproduce seasonally among aquatic vegetation in newly inundated floodplains and slow-flowing river channels in relation to the annual flooding regime. Each of these species is considered to contribute to the subsistence fishery.

Their respective bodies are compressed, silvery-grey dorsally and creamy white ventrally. Their heads are rounded with an inferior mouth. Their dorsal and anal fins sit posterior on the body and are crescent shape with the anal fin being longer. The caudal peduncle is slender and the caudal fin is forked with rounded lobes.

Long-head Churchill (*Petrocephalus longicapitis*)

B. Kramer

Etymology: ***Petrocephalus*** = ***petro-***, stone + ***-cephalus***, head; ***longicapitis*** = ***longus***, long + ***capit***, head

DS 0, DR, 21-25; AS 0, AR 26-30; LL 37-39, CP 12

Summary: Little is known about this mormyrid and additional research is needed on its biology and ecology. It appears to prefer quiet shallow floodplains, backwaters and oxbows with aquatic vegetation. It is associated with vegetation mats, root balls and undercut banks during the day. Females are larger than males. Length reaches 110 mm SL.

The Long-head Churchill is found in the Kwando-Linyanti system and the Upper Zambezi-Chobe River of Namibia northward to Lumwana River, Zambia. This species is a component of the 'stonebasher bash' in the Upper Zambezi and is taken in the subsistence fishery with traditional and modified gear.

Large-bodied Churchill (*Petrocephalus magnitrunci*)

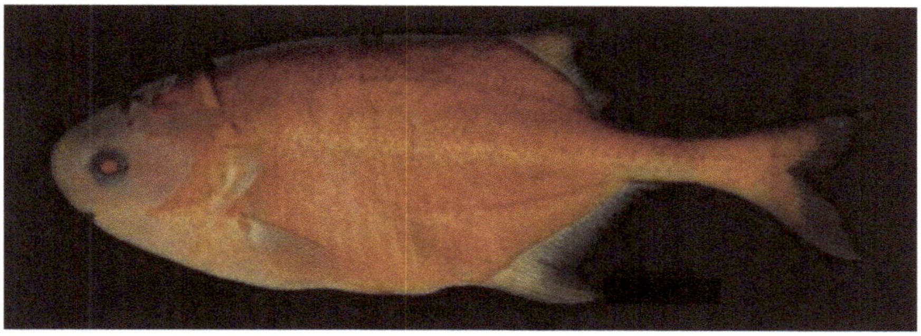

B. Kramer

Etymology: *Petrocephalus* = ***petro-***, stone + ***-cephalus***, head; ***magnitrunci*** = ***magnus***, great + ***truncus***, trunk, i.e., referring to rounded oval body

DS 0, DR 19-22; AS 0, AR 27-31; LL 39-41, CP 11-12

Summary: Their distribution is apparently restricted to the lower Kavango River and the Okavango Delta. The type was collected over a shallow floodplain with mud substrate, dense emergent grasses and water lilies. They reach 89 mm SL. Little else is known about this mormyrid.

Cunene Churchill (*Petrocephalus magnoculis*)

B. Kramer

Etymology: ***Petrocephalus*** = ***petro-***, stone + ***-cephalus***, head; ***magnoculis*** = ***magnus***, great + ***-oculis***, eye

DS 0, DR 20-24; AS 0, AR 26-31; LL 38-42, CP 11-13

Summary: This species has an oval-shaped body with a small inferior mouth. Its eye is proportionally larger than other species within this genus from which its specific name is derived. It was originally described from the Kunene River, hence the common name, but has subsequently been collected in the Cuvelai drainage. It grows to 105 mm SL.

Okavango Churchill (*Petrocephalus okavangoensis*)

F. Jacobs

Etymology: ***Petrocephalus*** = ***petro***, stone + ***cephalus***, head; ***okavangoensis*** = ***-ensis***, suffix denoting place, Okavango River, type locality

DS 0, DR 20-24; AS 0, AR 27-32; LL 37-38, CP 12

Summary: The Okavango Churchill is grey-silver to brown-silver in colour with a distinctly pointed snout and small inferior mouth. The back is dark, lighter ventrally, and the paired fins transparent. Some specimens have a dark spot at the base of the dorsal fin that is absent in the Large-bodied Churchill. The head and body are dorsolaterally compressed. It is known to reach 95 mm SL.

This species seems to prefer swampy, stagnant or slow-flowing waters with papyrus and other emergent vegetation. No other life history information is available. It is found in the Kavango River, Namibia; upstream in the Cubango River, Angola; and downstream in the Okavango Delta. It may reach Lake Ngami and the Makgadikgadi Pans.

Genus *Pollimyrus*

A study of electric signals generated by dwarf stonebashers (*Pollimyrus* spp.) in 2003 and 2013 recognised a complex of three species distributed in Namibia among different watersheds: Kavango River system inhabited by *P. castelnaui*; Kwando River by *P. cuandoensis*; and the Upper Zambezi River by *P. marianne.* All are virtually indistinguishable in terms of their general appearance, but differ in EODs.

Pollimyrus spp. are the smallest mormyrids and share several life history attributes: they inhabit similar habitats, forage on aquatic invertebrates and reproduce annually amongst aquatic vegetation in newly inundated floodplains and slow-flowing river channels in relation to the annual flooding cycle.

Their bodies are compressed and mottled brown with a small subterminal mouth and chin without a swelling. The dorsal fin base is shorter and originates posterior to the anal fin base, both situated far posterior on the body. The caudal fin is forked with rounded lobes.

Dwarf Stonebasher (*Pollimyrus castelnaui*)

F. Jacobs

Etymology: ***Pollimyrus*** = in honour of Belgian ichthyologist Max Poll (1908–1991); ***castelnaui*** = in honour of French naturalist Francis de Castelnau (1810–1880)

DS 0, DR 14-18; AS 0, AR 20-24; LL 46-55, CP 12-14

Summary: The Dwarf Stonebasher is known from the Kunene, Cuvelai and Kavango River systems, Namibia, south to the Okavango Delta.

It is a *eurytopic* species capable of adapting to a wide range of habitats found in densely vegetated margins of moderately-flowing river channels, lagoons, floodplains and backwaters. They feed at night on aquatic insect larvae including a large component of Chironomidae (flies and midges). The common name refers to its small size. This species is caught by subsistence fishermen using a variety of traditional methods. It obtains 70 mm SL.

Cuando Stonebasher (*Pollimyrus cuandoensis*)

B. Kramer

Etymology: ***Pollimyrus*** = in honour of Belgian ichthyologist Max Poll (1908–1991); ***cuandoensis*** = ***-ensis***, suffix denoting place, Cuando River drainage, type locality

DS 0, DR 15-18; AS 0, AR 21-24; LL 45-55, CP 12-16

Summary: This mormyrid is known only from the lower section of the Kwando River. The Cuando Stonebasher differs from the Upper Zambezi Dwarf Stonebasher and the Dwarf Stonebasher in the nature of its EODs and aspects of its anatomy, e.g., a shorter thicker caudal peduncle evolved for manoeuvring in the sluggish waters of the Kwando River as compared to longer thinner caudal peduncles for swimming power in the latter two. It reaches 65 mm SL.

Little is known about this mormyrid's biology and ecology. Its preferred habitat seems to be within floating and emergent vegetation, thus it is possibly caught by subsistence fishermen. Both the common and scientific names refer to the Kwando (Cuando) River that passes through the Zambezi Region, the only known area of occurrence in Namibia.

Upper Zambezi Dwarf Stonebasher (*Pollimyrus marianne*)

B. Kramer

Etymology: ***Pollimyrus*** = in honour of Belgian ichthyologist Max Poll (1908–1991); ***marianne*** = in honour of Marianne Elfriede Kramer, mother of B. Kramer, authority on mormyrid electrolocation

DS 0, DR 14-19; AS 0, AR 21-26; LL 50-59 , CP 13-18

Summary: The Upper Zambezi Dwarf Stonebasher is virtually indistinguishable from other dwarf stonebashers by general morphology, but it has an electric organ discharge (EOD) of longer duration than either the Dwarf or Cuando Stonebasher. Currently, all dwarf stonebashers from the Upper Zambezi are recognised as Upper Zambezi Dwarf Stonebasher, but its overall distribution is unknown. Little is known of the biology and ecology of this species. Length reaches 73 mm SL.

B. Family Kneriidae (Knerias)

This family of fishes is endemic to Africa, containing four genera and about 30 recognised species, only one of which is known to Namibia. They are *rheophilic*, i.e., current loving fishes inhabiting upland headwater streams and rivers. These are small streamlined fish with minute scales. Males develop disc-like contact organs on each gill cover used to maintain contact with the female during spawning. Hence, the genus is often referred to as "earfishes". The family is under review, thus it is likely that more species will be described in the future. (See Addendum, P. 365)

Cunene Kneria (*Kneria maydelli*)

SAIAB

Etymology: ***Kneria***, in honour of Austrian ichthyologist Rudolph Kner (1810–1869) who studied fishes in the Imperial Museum of Vienna; ***maydelli***, in honour of German explorer Gustav Adolf von Maydell (1919–1959) who co-led an expedition to Angola 1952–54

DS 0, DR ii-iii, 6-7; AS 0, AR ii-iii, 5-6; LL 80-100

Summary: This species possesses an elongated body that becomes progressively compressed posterior to the caudal peduncle. It has a pointed head with a smooth snout and ventral mouth. Its body is coloured yellowish brown above and white to silvery below, usually with a horizontal mid-line bar or series of dark brown spots. The back and belly have rows of irregular shaped spots; 3–4 overlapping dark spots on the forked caudal fin base; black spots at the bases of the pectoral and pelvic fins; and the 1st and 2nd rays of the dorsal and anal fins are blackish. The distinctive opercular organ overlies the anterior part of the lateral line and is much larger in diameter than the eye. Males often have tubercles on the cheeks or around the eyes. It obtains 57 mm SL.

These small fish typically shoal in rapid stretches of the Kunene River and at the mouths of inflowing streams. They also inhabit aquatic vegetation lining the river channel. They feed by grazing on *aufwuchs* living on the stems and roots of aquatic plants, and capture insects floating with the river current. Spawning occurs during the summer rainy season.

This species is endemic to Kunene River upstream into Angola, from which the common name is derived.

C. Family Cyprinidae (Barbs, Minnows, Labeos and Yellowfishes)

The Cyprinidae is a large family of freshwater fishes with 36 native species recorded countrywide. Their common name 'barbs' refers to the fact that most species within the family have a pair of barbels on their mouths, which they use to 'taste' the substrate for food at the bottom of rivers, lagoons, floodplains and reservoirs.

Barbs have a wide range of sizes, shapes, life styles and habitats, and many are strong swimmers living in fast-flowing water. They generally have a *fusiform*-shaped body with a single dorsal fin, forked caudal fin and paired pectoral and pelvic fins, the latter positioned mid-body. They are typically the numerically dominant group of fishes in a community with a high diversity in the northern perennial rivers, including the Kunene, Kavango, Kwando-Linyanti and Upper Zambezi-Chobe systems. They are also numerically dominant in the *iishana* of the Cuvelai system and all the National reservoirs such as Hardap, Von Bach, Swakoppoort and Omatako.

Many barbs are opportunistic feeders that live and forage near the bottom as well as in midwater and near the surface, termed *benthopelagic.* Barbs feed on algae, diatoms, detritus, aquatic plants and *periphyton* (or *aufwuchs*), the latter which refers to the small invertebrates and algae that adhere to stems of aquatic plants, rocky surfaces and submerged trees in the water. Spawning typically takes place amongst submerged vegetation during the late spring and summer rainy season and/or arrival of annual floodwaters. Parental care of offspring is lacking in this family.

Upjaw Barb (*Coptostomabarbus wittei*)

B. van der Waal

Etymology: ***Coptostomabarbus***, Greek, ***koptos***, ***-e***, ***-on***, bruise + Greek, ***stoma***, mouth + Latin, ***barbus***, barbel; ***wittei***, in honour of herpetologist Gaston François De Witte (1897–1980), who collected the type specimen

DS 0, DR ii, 7; AS 0, AR ii-iii, 5; LL 26-29, CP 12

Summary: The Upjaw Barb is a small light-pinkish barb with a silvery belly and grey to light-brown stripe along its body. Generally, its fins are bland and translucent. The mouth has no barbels and is sharply upturned, the basis of the common name. It reaches 40 mm SL.

This small common barb is found in well-vegetated margins of river channels, lagoons and floodplains. It feeds on *aufwuchs*. The Upjaw Barb migrates to floodplains to spawn during late spring and summer coinciding with the rainy season and/or arrival of floodwaters. Females lay eggs among aquatic plants.

This species is distributed throughout the Kavango, Kwando-Linyanti and Upper Zambezi-Chobe River systems in Namibia, although it is often rare. There are no definitive MFMR records for this species in the Kunene River in Namibia; however, it is recorded from the canal linking the Kunene to the Cuvelai system and Olushandja Dam, thus it is expected to occur. It is also known from the Okavango Delta as well as the Kafue, Upper Zambezi and Congo River systems, Zambia. It is caught by subsistence fishers in traps, fishing baskets and mosquito seine nets.

Zambezi River Sardine *(Engraulicypris brevianalis)*

M. Riddin

Etymology: ***Engraulicypris*** = Greek, ***engraulis***, anchovy + Greek, ***kyprinos***, carp; ***brevis***, short + ***analis***, anus or anal fin, referring to smaller number of anal-fin rays

DS 0, DR ii, 8; AS 0, AR iii, 13-15; LL 53-57, CP 18

Summary: This slender species has a silvery sheen with a light silver lateral band running along the mid-line. The rounded snout is dark dorsally with the operculum entirely shiny. The dorsal fin sits over the anal fin, far posterior relative to the pelvic fin origin. The large terminal mouth extends behind the anterior border of the large orbit. Spawning males have tubercles on their head and snouts and are smaller than females. The caudal fin is vivid yellow at its fork with the lower lobe longer. They can reach 75 mm SL.

The Zambezi River Sardine is a pelagic shoaling species found in slack-water areas such as eddies and pools below riffles, into open water of the Kavango, Kwando-Linyanti and Upper Zambezi-Chobe Rivers in Namibia to the Okavango Delta. They are otherwise known from the Middle-Lower Zambezi River from Zambia and Zimbabwe through Mozambique southward along the Indian Ocean coast to the Mkuze River, Pongolo River and St. Lucia, South Africa. Populations from different river systems show subtle differences in morphology or colouration, but are not considered different species at this time.

Orange River Sardine *(Engraulicypris gariepinus)*

M. Riddin

Etymology: ***Engraulicypris*** = Greek, ***engraulis***, anchovy + Greek, ***kyprinos***, carp; ***gariepinus***, ***gariep***, a San name meaning "Great Water", now known as the Orange River + ***-inus***, suffix belonging to

DS 0, DR ii, 7-8; AS 0, AR iii, 14-16; LL 49-51, CP 14-16

Summary: This slender silvery-white species has a silver mid-line lateral band running to the forked caudal fin that is clear to pale orange. The rounded snout has dense dark spotting on its tip and the operculum is entirely shiny silver. The large terminal mouth extends to the anterior margin of a large orbit. The dorsal fin sits far posterior on the body originating over the anal fin. Spawning males have tubercles on their lower jaw and lower head. They reach 46 mm SL.

The Orange River Sardine is a pelagic shoaling species found in slack waters such as pools below riffles, eddies and open water of the lower Orange-Fish Rivers, where it is considered endemic. Upstream movement of populations found in the lower Orange and Fish Rivers is limited by Augrabies and Fish River Falls. This species feeds on planktonic crustaceans and insects, such as midges and ants. They are thought to prefer turbid waters, which provide protection from visual predators.

Kunene River Sardine *(Engraulicypris howesi)*

F. Jacobs

Etymology: ***Engraulicypris*** = Greek, ***engraulis***, anchovy + Greek, ***kyprinos***, carp; ***howesi*** = in honour of Gordon John Howes (1938-2013)

DS 0, DR ii, 6-8; AS 0, AR iii, 13-15; LL 51-52, CP 14

Summary: This very slender species is silvery white with a light silver lateral band running along the mid-line. The snout is dark dorsally with the operculum entirely metallic silver. All fin membranes are clear, with the origin of the anal fin having a few dark melanophores. The terminal mouth is large, reaching behind the anterior border of the orbit. The dorsal fin sits far posterior relative to the pelvic fin origin. Spawning males have tubercles on their lower jaw, head and snout and are smaller than females. They can reach 43 mm SL.

Very little is known of the biology of the Kunene River Sardine. It is a pelagic shoaling species, appearing to favour turbid, rocky, river regions where they can gather in pockets of recirculating currents. It is endemic to the Kunene River, Namibia and Angola, but presumably has been introduced into the Cuvelai basin by the Calueque-Oshakati Canal and Olushandja Dam. This species feeds on planktonic crustaceans and insects, such as midges and ants.

Spottail Barb (*Enteromius afrovernayi*)

F. Jacobs

Etymology: ***Enteromius***, Greek, ***enteron***, intestine + Greek, ***myo***, ***mys***, muscle; ***afrovernayi***, ***afro-***, African + ***vernayi*** in honour of Arthur S. Vernay (1877–1960), naturalist-explorer, who funded and led an American Museum of Natural History expedition to Angola

DS 3, DR 7; AS 0, AR iii, 5; LL 27-33, CP 12

Summary: The Spottail Barb has a slender body. It is translucent grey-brown in colour with an iridescent lilac-yellow mid-lateral stripe and silvery to transparent belly. There is a large black spot at the base of the forked caudal fin from which the common name is derived. It has a small upturned terminal mouth with no barbels. The first dorsal spine is serrated. It reaches 57 mm SL.

This small barb is common in vegetated river channels, floodplains and backwater lagoons in the Kunene, Kavango, Kwando-Linyanti and Upper Zambezi-Chobe River systems in Namibia. It also occurs from the Democratic Republic of Congo (DRC) to the Kafue-Middle Zambezi River, Zambia.

It feeds on small aquatic invertebrates and grazes on plant surfaces consuming *aufwuchs*. Spawning begins in late spring and lasts throughout the summer months coinciding with the seasonal rains and/or arrival of floodwaters. Females deposit eggs on aquatic plants.

Spottail barb are caught by subsistence fishers using fishing baskets, weirs, foldable corral traps and seine nets in rivers and lagoons. It is an attractive aquarium species.

Blackback Barb (*Enteromius barnardi*)

B. van der Waal

Etymology: ***Enteromius***, Greek, ***enteron***, intestine + Greek, ***myo***, ***mys***, muscle; ***barnardi***, in honour of Keppel Harcourt Barnard (1887–1964), South African Museum, for his "remarkable contributions" to the taxonomy of South African fishes

DS 0, DR iii, 8; AS 0, AR iii, 5; LL 29-33, CP 12

Summary: This slender species is silver-olive in colour fading to a white belly. There is a black mid-lateral stripe running around the tip of the snout (moustache like) along the body to the caudal fin and irregular black spots along the dorsal line from which its common name is presumably derived. A black spot occurs at the base of the anal fin. The small mouth is flanked by a pair of short barbels. It reaches 70 mm SL.

This small barb occurs in shallow well-vegetated margins of river channels, floodplains and lagoons of the Kunene, Cuvelai, Kavango, Kwando-Linyanti and Upper Zambezi-Chobe River systems of Namibia. It is also known from the Okavango Delta, Kafue-Middle Zambezi and Zambian Congo River systems.

The Blackback Barb feeds on *aufwuchs*. It undertakes potamodromic migrations onto floodplains to spawn in late spring and summer during the rainy season and/or arrival of annual floodwaters. Eggs are laid among aquatic vegetation. Spawning males have tubercles on their snouts. This species is commonly caught by subsistence fishers.

Barotse Barb (*Enteromius barotseensis*)

B. van der Waal

Etymology: ***Enteromius***, Greek, ***enteron***, intestine + Greek, ***myo***, ***mys***, muscle; ***barotseensis***, ***–ensis***, suffix denoting place = type locality: Barotsés, Zambia

DS 0, DR iii, 7-8; AS 0, AR iii, 5-6; LL 25-29, CP 12

Summary: The fusiform body is overall light brown dorsally, silvery below. There are 3-4 black spots along the side with an additional spot at the base of the anal fin. Two pairs of very small barbels flank a terminal mouth. It reaches 50 mm SL.

This small barb is found in well-vegetated margins of river channels and backwaters. It feeds on *aufwuchs*. Spawning adults undertake upstream potamodromic migrations in spring and summer coinciding with the rainy season and/or arrival of annual floodwaters. It is commonly caught by subsistence fishers.

The Barotse Barb inhabits flowing channels of the Kunene, Kavango, Kwando-Linyanti and Upper Zambezi-Chobe River systems in Namibia. It is widely distributed in the DRC, Zambia, Zimbabwe and Angola. The common name refers to the Barotse Floodplain, Zambia.

Hyphen Barb (*Enteromius bifrenatus*)

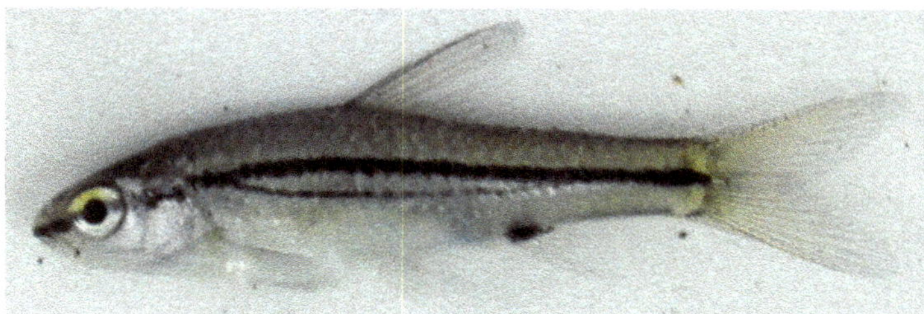

B. van der Waal

Etymology: ***Enteromius***, Greek, ***enteron***, intestine + Greek, ***myo***, ***mys***, muscle; ***bifrenatus***, ***bi-***, two + ***frenatus***, bridled, referring to two distinct black lateral stripes running from snout through eye and gill cover down the body

DS 0, DR iii, 8; AS 0, AR iii, 5; LL 28-32, CP 12

Summary: The Hyphen Barb has a robust body coloured olive green above and silver below. Two distinct black stripes occur along the body: one starts at the snout and runs mid-laterally across the body and the other curves downward convex from behind the head to about the caudal fin following the lateral line and outlined with black spots resembling hyphens. There is a black spot at the anal fin base. A small terminal mouth is flanked by two pairs of barbels. It grows to 70 mm SL.

This small barb is found in well-vegetated, slow- to moderately-flowing river margins, floodplains and lagoons of the Kunene, Cuvelai, Kavango, Kwando-Linyanti and Upper Zambezi-Chobe River systems in Namibia. It is otherwise known from the DRC south to the Limpopo River system in South Africa.

This barb feeds on *aufwuchs* and undergoes spawning migrations during the rainy season into flooded grasses and aquatic plants. It is commonly caught by subsistence fishers.

Shorthead Barb (*Enteromius breviceps*)

J.R. Stauffer

Etymology: ***Enteromius***, Greek, ***enteron***, intestine + Greek, ***myo***, ***mys***, muscle; ***breviceps***, ***brevis***, short + ***-ceps***, head

DS 0, DR iii, 7; AS 0, AR iii, 5; LL 33, CP 13

Summary: The Shorthead Barb has a slender body, light brown in colour above the lateral line, silvery below. Two pairs of barbels occur at the corners of the mouth. An indistinct stripe extends laterally along the body ending in a small black spot at the base of the caudal fin. The dorsal, caudal, anal and pelvic fins are tinted orange in colour. The dorsal fin is short. This species grows to 77 mm SL.

Little is known about this species, but it is likely similar to other barb species: benthopelagic feeding on *aufwuchs* and undergoing potamodromic spawning migrations.

It has been recorded from fontein-fed mountain streams of the Kunene River in Namibia, often occurring with the Cunene Kneria. It is also known from the Cubango and Longa Rivers, Angola.

Dwarf Barb (*Enteromius brevidorsalis*)

B. van der Waal

Etymology: ***Enteromius***, Greek, ***enteron***, intestine + Greek, ***myo***, ***mys***, muscle; ***brevidorsalis***, ***brevis***, short + ***dorsalis***, dorsal, referring to shorter second dorsal ray

DS 0, DR iii, 7; AS 0, AR iii, 5; LL 24-27, CP 11-14

Summary: The body of the Dwarf Barb is short and stocky, olive-yellow above the lateral line and silvery below. It has a small subterminal mouth with either a pair of small barbels or none present. A black stripe runs along the body and ends in a black spot at the base of the forked caudal fin. It has a short dorsal fin and reaches 46 mm TL.

This small barb inhabits river channels, floodplains and backwater lagoons of the Kavango, Kwando-Linyanti and Upper Zambezi-Chobe River systems in Namibia. It is known elsewhere from the Cunene (Angola), Kafue, Quanza and Congo River systems, but is unknown from the Okavango Delta.

The Dwarf Barb feeds on detritus, algae and small aquatic invertebrates. It spawns with the onset of the rainy season and/or arrival of floodwaters amongst aquatic vegetation. It is commonly caught by subsistence fishers.

Orangefin Barb (*Enteromius eutaenia*)

B. van der Waal

Etymology: ***Enteromius***, Greek, ***enteron***, intestine + Greek, ***myo***, ***mys***, muscle; ***eutaenia***, ***eu-***, very + ***taenia***, band, referring to a "sharply defined" black lateral band passing through the eyes and extending to the caudal fin

DS 3, DR 8; AS 0, AR iii, 5; LL 24-27, CP 12

Summary: The body of the Orangefin Barb is robust and coloured dark olive to brown above and silvery white below with yellow to orange fins, the latter feature yielding its common name. A broad black band runs from the snout to the caudal fin with shadow stripes on either side. Large black scales occur at the base of the dorsal fin which has a serrated spine. Two pairs of barbels flank a relatively large mouth for its size. It obtains 140 mm SL.

This is not a common barb although it is distributed in the Kunene, Kavango, Kwando-Linyanti and Upper Zambezi-Chobe River systems of Namibia. It is also known from the northern Okavango Delta; Zambian Congo system; and Lake Tanganyika.

The Orangefin Barb favours well-vegetated margins of moderately-flowing waters and lagoons, especially within papyrus and reed root stocks; it feeds on *aufwuchs*. It is known to move onto recently inundated floodplains to spawn during the rainy season. It is uncommon in the subsistence fishery.

African Banded Barb (*Enteromius fasciolatus*)

F. Jacobs

Etymology: ***Enteromius***, Greek, ***enteron***, intestine + Greek, ***myo***, ***mys***, muscle; ***fasciolatus***, ***fascia***, meaning belt, an allusion to the ±12 blackish-blue vertical bands on its sides

DS 0, DR iii, 8; AS 0, AR iii, 5; LL 25-30, CP 12

Summary: This slender barb is prominently rose coloured with 10-15 pronounced black vertical bars along the body and a silvery-coloured belly. The bars lend its name; some also call it the Red Barb due to its rosy colour. Its fins are translucent rose. Two pairs of barbels flank a terminal mouth. The species reaches 60 mm SL.

This colourful barb is common in well-vegetated waters, such as fringed edges of river channels and permanent lagoons. They are *omnivorous* and feed on *aufwuchs*. This species undergoes lateral migrations from river channels to floodplains to spawn with the arrival of annual floodwaters. Eggs are laid among submerged aquatic vegetation.

The African Banded Barb is widely distributed throughout the Kunene, Kavango, Kwando-Linyanti and Upper Zambezi-Chobe River systems of Namibia. It is also known throughout the Okavango Delta, Middle Zambezi below Victoria Falls, Kafue and Congo River systems. It is caught in the subsistence fishery using fishing baskets, foldable corral traps and mosquito seine nets.

Sickle-fin Barb (*Enteromius haasianus*)

B. van der Waal

Etymology: ***Enteromius***, Greek, ***enteron***, intestine + Greek, ***myo***, ***mys***, muscle; ***haasianus***, ***-anus***, belonging to; in honour of malacologist Fritz Haas (1886–1969), who collected the type specimen

DS 0, DR ii, 8; AS 0, AR ii, 5; LL 35-38 in lateral series

Summary: This small robust barb is silvery brown in colour with a prominent black spot at the base of the caudal fin and a faint spot at the base of the anal fin. Breeding males display a brilliant orange-red hue and the sickle shaped anal fin, which lends the common name, is particularly prominent. A thin black stripe runs along each side of the body (more obvious in live fish). The mouth is small. This is one of the few species of barbs with no barbels and no lateral line. It obtains a length of 32 mm SL.

The Sickle-fin Barb inhabits well-vegetated margins of rivers, backwater lagoons and floodplains. It feeds on *aufwuchs*. They migrate onto floodplains to spawn during the rainy season with the arrival of floodwaters. Eggs are laid among aquatic plants. Males take on a rosy-red hue when spawning. This species is commonly caught by subsistence fishers at the end of the flood cycle.

This is a widely distributed species found throughout the Kavango, Kwando-Linyanti and Upper Zambezi-Chobe River systems in Namibia. It is also known from the Okavango Delta; Kafue and Congo River systems, Zambia; Bangweulu swamps; and Lower Zambezi and Pungwe River drainages in Zimbabwe and Mozambique.

Redspot Barb (*Enteromius kerstenii*)

R. Bills

Etymology: ***Enteromius***, Greek, ***enteron***, intestine + Greek, ***myo***, ***mys***, muscle; ***kerstenii***, in honour of Otto Kersten (1839–1900), early explorer who collected the type specimen

DS 3, DR 7; AS 0, AR iii, 5; LL 23-27, CP 12

Summary: The Redspot Barb is deep bodied, coloured light olive-silver dorsally and silver ventrally. Midbody scales portray a black zigzag pattern and the lateral line is with distinct chevrons. A distinctive red or orange-yellow spot occurs on the gill cover, hence the common name. The mouth is terminal and flanked by two pairs of barbels. The species can each 75 mm SL.

This is a shoaling species found in sheltered well-vegetated habitats and rocky substrates bordering large rivers and channels along the Kunene, Kavango, Kwando-Linyanti and Upper Zambezi-Chobe Rivers in Namibia. It has been collected only once from the canal connecting the Kunene to the Cuvelai basin, thus it has the potential to occur in the latter. It occurs widely throughout Central Africa and is known from Lakes Victoria, Tanganyika and Kivu; the Congo, Kafue-Middle Zambezi and Lower Zambezi Rivers, Zambia; and East Coast rivers in Kenya, Tanzania and Mozambique.

The Redspot Barb feeds on insects, snails, plants, seeds and algae. It migrates upstream during the rainy season and moves onto recently inundated floodplains to lay its eggs on aquatic vegetation. It is caught by subsistence fishers using traditional gear.

Line-spotted Barb (*Enteromius lineomaculatus*)

IUCN.org

Etymology: ***Enteromius***, Greek, ***enteron***, intestine + Greek, ***myo***, ***mys***, muscle; ***lineomaculatus***, ***lineo-***, line + ***maculatus***, spotted; referring to series of irregular black spots on sides, more or less connected by a dark lateral streak

DS 0, DR iii, 8; AS 0, AR iii, 5; LL 26-32, CP 12

Summary: The Line-spotted Barb has a fusiform-shaped body, translucent brown dorsally and silver ventrally with a series of 4-7 black spots in a row above the lateral line. Dark chevron markings form a broken line along the lateral line, lending the common name. Spawning males turn bright gold. Two pairs of long barbels flank the terminal mouth. It reaches 86 mm SL.

This barb inhabits small streams, large river channels and shallow marginal swamps. They move upstream or laterally to spawn on recently inundated grassy areas. They feed on insects, snails, plants, seeds and algae. They represent a small component of the subsistence fishery.

The Line-spotted Barb is known in Namibia from the Kunene, Cuvelai, Kavango and Upper Zambezi systems upstream into Angola and Zambia. It has not been confirmed in the Kwando-Linyanti Rivers or the Okavango panhandle in Namibia nor the Okavango Delta. It occurs throughout Central and East Africa; Upper Congo, Buzi and Limpopo Rivers; and in upland streams north of Lake Malawi.

Papermouth (*Enteromius mattozi*)

C. Hay

Etymology: ***Enteromius***, Greek, ***enteron***, intestine + Greek, ***myo***, ***mys***, muscle; ***mattozi***, in honour of Ferdinand dos Santos Mattozo (1849-1921), Portuguese politician, diplomat and zoology professor

DS 4, DR 8; AS 0, AR iii, 5; LL 29-35, CP 14

Summary: The Papermouth has a fusiform body with a large terminal mouth and 1–2 pairs of short barbels. It is coloured silver with orange fins and typically has a black bar behind the opercle. The dorsal and caudal fins are slate coloured with the pelvic and anal fins translucent with orange margins. The anal fin is *falcate* shaped.

It inhabits impoundments and large pools in rivers and feeds on zooplankton, crustaceans, aquatic insects and seeds from aquatic plants. Larger individuals (>175 mm SL) feed on small fishes. It can grow to 400 mm SL. It exhibits potamodromic migrations in spring. It may be caught using light tackle and spinning lures.

The Papermouth occurs in the Kunene and Cuvelai systems in Namibia, but is also reported from the Quanza and Upper Zambezi, Angola and Limpopo, South Africa drainages with the latter possibly representing a different species.

Zigzag Barb (*Enteromius miolepis*)

B. van der Waal

Etymology: ***Enteromius***, Greek, ***enteron***, intestine + Greek, ***myo, mys***, muscle; ***miolepis***, **mio**-, less or small + ***lepis***, scale, presumably referring to fewer scale rows below lateral line

DS 3, DR 8; AS 0, AR iii, 5; LL 23-26, CP 12

Summary: The Zigzag Barb derives its common name from the striking black zigzag band of scales running from behind the opercle to the caudal fin. The mouth has two pairs of long barbels. The deep fusiform body is olive-green dorsally and silver ventrally with yellow fins. It obtains a length of 125 mm SL.

This barb occurs in a variety of habitats, including well-vegetated river margins, lagoons and floodplains, but nowhere is it common. Little is known about its feeding habits.

It is found in the Kavango, Kwando-Linyanti and Upper Zambezi-Chobe Rivers in Namibia as well as the Kafue and Congo systems.

Copperstripe Barb (*Enteromius multilineatus*)

B. van der Waal

Etymology: ***Enteromius***, Greek, ***enteron***, intestine + Greek, ***myo, mys***, muscle; ***multilineatus***, ***multi-***, many + ***lineatus***, lined, referring to 3 series of broken bands above the lateral stripe and 2 below

DS 3, DR 7-8; AS 0, AR iii, 5; LL 25-27, CP 12

Summary: This small barb possesses a stout body coloured olive-brown with a distinct coppery stripe from snout to caudal fin in the males from which it receives its common name. There are multiple parallel stripes along the sides, the most prominent running from the snout through the eye to the caudal fin. The belly is silvery and the fins are yellow to orange in adults. The first dorsal spine is serrated. Two pairs of barbels flank the mouth. It reaches 45 mm SL.

The Copperstripe Barb occurs in a variety of habitats, including well-vegetated river margins, lagoons and floodplains in the Kunene, Cuvelai, Kavango, Kwando-Linyanti and Upper Zambezi-Chobe River systems of Namibia. Elsewhere, it is widespread from the DRC to South Africa, including the Okavango Delta. This barb feeds on *aufwuchs*. It spawns from late spring through the summer coinciding with the annual flooding regime. Females deposit eggs on submerged plants and there is no parental care. It is caught by subsistence fishers in rivers and lagoons using traditional gear.

Orange River Chubbyhead Barb (*Enteromius* cf. *oraniensis*)

C. Hay

(National Museum, Windhoek, SMP 1007)

Etymology: ***Enteromius***, Greek, ***enteron***, intestine + Greek, ***myo***, ***mys***, muscle; ***-ensis***, suffix noting place, Orange River drainage

DS 0, DR iii, 7; AS 0, AR iii, 5; LL 32-33, CP 12-16

Summary: The nominal species has a characteristic round, blunt, head profile from which the common name is derived. They possess a large eye located near the tip of the snout and a small, oblique, terminal mouth with a single pair of small barbels. They are coloured greyish-green dorsally, lighter ventrally. A broad dark lateral band may occur, but without chevron markings. A subtle spot occurs at the base of the caudal fin. Males take on a golden hue during the breeding season. Orange River forms have an incomplete lateral line with 90% usually having 2-15 pored scales, but sometimes numbering up to 33. They reach 120 mm SL.

The Chubbyhead Barb is an adaptable species living in a variety of habitats from clear rocky streams to turbid man-made impoundments. Its life history characteristics of fast maturing, high reproduction rate and ability to colonise temporary waters, are favourable characteristics to inhabit stressful environments. Spawning commences in summer at the onset of seasonal rains with multiple clutches of over 3,000 eggs per spawn for up to six months.

This barb is historically known in Namibia only from the ephemeral Kuiseb River and its main tributary, the Gaub River. Both rivers have eroded through the western boundary of the Great Escarpment, thus this is considered to be a relict population from an ancient stream capture. The fish has not been collected since 1973, but is not considered extirpated. This population is disjunct from other Chubbyhead Barb populations, the closest being in the Orange River drainage above Augrabies Falls, where it is widely distributed throughout the upper basin, including isolated fonteins and rivulets of the Karoo ecoregion. Three other Chubbyhead Barb species related to *E. oraniensis* are endemic to the Olifants River of the Western Cape, various Eastern Cape rivers and Gouritz River, South Africa, while the related Shorthead Barb is known in Namibia only from the Kunene River.

Straightfin Barb (*Enteromius paludinosus*)

F. Jacobs

Etymology: ***Enteromius***, Greek, ***enteron***, intestine + Greek, ***myo, mys***, muscle; ***paludinosus***, Latin, ***palud-***, a swamp, refers to type locality

DS 3, DR 7; AS 0, AR iii, 5; LL 32-36, CP 16

Summary: The Straightfin Barb has a fusiform body, plain olive-grey dorsally and silver ventrally. Its fins are pale olive. The first dorsal fin ray is spine-like and serrated; the dorsal rays form a straight margin when erect giving the species its common name. There are two pairs of short barbels that flank a small terminal mouth.

This hardy barb is tolerant of extreme environmental conditions and occupies a wide range of habitats, including vegetated margins of river channels, floodplains and backwater lagoons. It is omnivorous, feeding on a wide variety of small organisms, including insects, small snails, crustaceans, algae, diatoms, detritus and aquatic plants. It reaches 150 mm SL and is commonly caught by subsistence fishers.

Spawning takes place during the summer rainy season and/or arrival of annual floodwaters. They are multiple spawners and potamodromous with lateral movements onto recently inundated floodplains to lay eggs amongst vegetation.

The Straightfin Barb is ubiquitous and found throughout all perennial waterbodies of Namibia as well as the seasonal *iishana* (see *oshana*) of the Cuvelai River and most National reservoirs. Elsewhere, it is widespread in Africa from Ethiopia in the north, through East and Central Africa, extending westwards into Angola and reaching its southern limits in KwaZulu-Natal, South Africa.

Dashtail Barb (*Enteromius poechii*)

F. Jacobs

Etymology: ***Enteromius***, Greek, ***enteron***, intestine + Greek, ***myo, mys***, muscle; ***poechii***, in honour of Austrian anthropologist and ethnologist Rudolf Pöch (1870–1921), who collected the type specimen

DS 3, DR 8; AS 0, AR iii, 5; LL 31-33, CP 14

Summary: The Dashtail Barb is a robust barb species, coloured olive-brown above and silver below. A distinct black dash is present at the base of the caudal fin, leading to its common name; it sometimes has two faint lateral shadow spots. It possesses fairly large eyes; a mouth flanked by two pairs of barbels; and prominent dorsal fin spine. It can grow to 120 mm SL.

This barb commonly shoals and inhabits margins of river channels, floodplains and well-vegetated backwater lagoons of the Kavango, Kwando-Linyanti and Upper Zambezi-Chobe River systems in Namibia. It is also found in the Okavango Delta and in rivers and lagoons from the DRC southwards to Zimbabwe.

This species feeds on insects and other small aquatic organisms. Spawning occurs in sync with the rainy season and arrival of annual floodwaters. Females lay eggs among aquatic plants. It is commonly caught by traditional fishers.

Beira Barb (*Enteromius radiatus*)

F. Jacobs

Etymology: ***Enteromius***, Greek, ***enteron***, intestine + Greek, ***myo***, ***mys***, muscle; ***radiatus***, rayed, probably referring to radial striations of scales

DS 0, DR iii, 8; AS 0, AR iii, 5; LL 24-29, CP 12

Summary: The Beira Barb has a plump body with the dorsal surface olive and sides fading to silvery white. The salmon- or orange-tinted fins and red portion of the eyes are most conspicuous. A small terminal mouth is flanked by two pairs of very short barbels. Length reaches 120 mm SL.

This species is widely distributed and thrives among vegetation-fringing channels, backwater lagoons and floodplains. It is omnivorous, feeding at night on *aufwuchs*. It undergoes seasonal migrations to floodplains to spawn during the rainy season. It is commonly caught by subsistence fishers using fishing baskets, foldable corral traps and mosquito seine nets.

This species has a wide distribution throughout the Kunene, Cuvelai, Kavango, Kwando-Linyanti and Upper Zambezi-Chobe River systems in Namibia and throughout the Okavango Delta. Elsewhere it is found in river systems and lakes from the DRC to South Africa. The common name presumably is in reference to Beira, Mozambique.

Thamalakane Barb (*Enteromius thamalakanensis)*

B. van der Waal

Etymology: ***Enteromius***, Greek, ***enteron***, intestine + Greek, ***myo, mys***, muscle; ***thamalakanensis***, ***–ensis***, suffix denoting place: Thamalakane River, Botswana, type locality

DS 0, DR iii, 8; AS 0, AR iii, 5; LL 25-29, CP 12

Summary: This slender fish is coloured light brown above and silvery-white ventrally. A black lateral stripe extends from the snout to the caudal fin; a black spot occurs at the base of the anal fin. Two pairs of long barbels flank a subterminal mouth. It grows to 50 mm SL.

This common barb inhabits the vegetated margins of rivers, floodplains, lagoons and backwaters. It feeds on *aufwuchs*. It spawns in late spring and summer along quiet channels and in floodplains, laying its eggs on submerged vegetation.

The Thamalakane Barb is found throughout the Kunene, Kavango, Kwando-Linyanti and Upper Zambezi-Chobe River systems of Namibia and elsewhere in the Upper Zambezi drainage of Angola and Zambia. It has been reported from the canal linking the Kunene to the Cuvelai basin. It is common in the Okavango Delta, Botswana where both the species and common names refer to the Thamalakane River.

Threespot Barb (*Enteromius trimaculatus*)

F. Jacobs

Etymology: ***Enteromius***, Greek, ***enteron***, intestine + Greek, ***myo, mys***, muscle; ***trimaculatus***, ***tri-***, three + ***maculatus***, spotted, referring to three black spots, two on side and one at base of caudal peduncle

DS 3, DR 8; AS 0, AR iii, 5; LL 31-35, CP 14

Summary: This full-bodied barb has a terminal mouth with two pairs of long barbels. It is olive-brown dorsally, silver ventrally with two distinct black spots along its sides and one at the base of the caudal peduncle from which both the scientific and common names are derived; however, these spots can be faint, confusing it with the Dashtail Barb. It has a prominent dorsal fin spine. It reaches 150 mm SL.

The Threespot Barb inhabits the vegetated margins of sluggish rivers, floodplains, lagoons and backwaters. It feeds on *aufwuchs* and spawns in late spring and summer laying eggs on submerged vegetation in channels and floodplains.

It is confined to the Kunene and Cuvelai systems in northern Namibia and the Orange-Fish Rivers in the south. It is reported northward through Angola to the southern Congo drainage. It is an important component of the *iishana* fisheries, caught with nets, traps and hook and line.

Longbeard Barb (*Enteromius unitaeniatus*)

G. Merron

Etymology: ***Enteromius***, Greek, ***enteron***, intestine + Greek, ***myo***, ***mys***, muscle; ***unitaeniatus***, ***uni-***, one + ***taeniata***, banded, referring to greyish or blackish band that runs from eye to caudal fin

DS 0, DR iii, 8; AS 0, AR iii, 5-6; LL 29-35

Summary: This slender-bodied barb is light bronze on its upper body with a silvery-white belly. A prominent dark lateral stripe runs along the length of the body. A small inferior mouth is flanked by two pairs of long barbels from which it obtains its common name. It reaches 140 mm SL.

The Longbeard Barb can be found in a range of habitats, including the well-vegetated fringes of flowing rivers, floodplains and backwater lagoons throughout the Kunene, Cuvelai, Kavango, Kwando-Linyanti and Upper Zambezi-Chobe River systems of Namibia into the Okavango Delta. It is widely distributed and is found from the DRC to South Africa.

It feeds on benthic invertebrates, plants and seeds. Spawning occurs amongst aquatic vegetation and is timed to the onset of the annual rainy season and flooding of littoral areas. It is caught in the subsistence fishery using traditional gear.

Cunene Labeo (*Labeo ansorgii*)

F. Jacobs

Etymology: ***Labeo***, Latin, ***labeo***, one who has large lips; ***ansorgii***, in honour of the explorer William John Ansorge (1850–1913)

DS 0, DR iv, 10; AS 0, AR iii, 5; LL 35-40, CP 16-18

Summary: The body is streamlined with a brown back and sides fading to a white belly. Fins are light brown. Lips are moderately wide and subequal with two small barbels each side. This species reaches

250 mm SL. The Cunene Labeo is a benthopelagic species that occurs principally in rocky habitats with flowing water and feeds near the bottom and midwaters or near the surface. It undergoes potamodromic migrations upstream during flooding to spawn.

This fish is restricted to the Kunene and Cuvelai Rivers in Namibia, but also occurs in the Bengo and Kwanza River catchments in Angola. The common name refers to the type locality for the species.

Orange River Mudfish (*Labeo capensis*)

C. Hay

Etymology: ***Labeo***, Latin, ***labeo***, one who has large lips; ***capensis***, noting area of collection, the Cape, South Africa

DS 0, DR iii, 10-11; AS 0, AR iii, 5; LL 42-50, CP 20-22

Summary: Adults have a robust body, light brown in colour fading into an off-white belly. They reach 500 mm TL and 4 kg. The Orange River Mudfish is endemic to the Orange-Fish River system in Namibia and South Africa (from which its common name is derived) and their tributaries including the Vaal River. This species is an excellent fighting fish and provides good sport for anglers.

It prefers slow-moving, quiet, well-vegetated reaches, but also occurs in rocky habitats in moderate currents. It is a benthopelagic species and feeds on rich organic bottom detritus and grazes *aufwuchs* off rocky surfaces and large woody debris, such as logs. This species spawns in summer, gathering in large numbers in shallow rocky rapids where eggs are laid.

Redeye Labeo (*Labeo cylindricus*)

F. Jacobs

Etymology: ***Labeo,*** Latin, ***labeo***, one who has large lips; ***cylindricus***, Latin ***cylindrus,*** referring to the cylindrical form of the body

DS 0, DR iii, 9-10; AS 0, AR iii, 5; LL 34-37, CP 12-16

Summary: This species has a cylindrical-shaped body and is light olive-brown in colour blending to an off-white belly; a dark lateral band is usually present. The head has a prominent snout (and tubercles on spawning males); a relatively large mouth with fleshy lips flanked by a single pair of small barbels; and distinctive red eyes on the upper side, the latter lending its common name. The dorsal fin is moderately tall and the caudal fin forked. The species reaches 230 mm SL and 0.9 kg.

The Redeye Labeo is the smallest labeo in southern Africa. It is a streamlined fish that favours clear running water over rocky habitats in small and large rivers. Their mouth is adapted for grazing *aufwuchs* on the surface of rocks and logs and ingesting detritus on the river bottom. They are potamodromous and often congregate in large shoals with upstream migration against strong currents to spawn. They can use their sucker-like mouths and broad pectoral fins to move upstream over moist rocks, weirs and waterfalls.

They are found in fast-flowing river channels and lagoons of the Kavango, Kwando-Linyanti and Upper Zambezi-Chobe River systems, but are not recorded from the Kunene River. Elsewhere they occur in the Okavango Delta and from the DRC, Angola, Botswana, South Africa, Zambia and Zimbabwe. They are caught in the subsistence fishery.

Upper Zambezi Labeo (*Labeo lunatus*)

B. van der Waal

Etymology: ***Labeo***, Latin, ***labeo***, one who has large lips; ***lunatus***, referring to crescent shape of the dorsal fin

DS 0, DR iv, 9-11; AS 0, AR iii, 5; LL 37-40, CP 16-20

Summary: This is a deep-bodied species, broadest in the middle and coloured silvery-grey. The scales are dark grey at their base and form parallel lines along the body. The Upper Zambezi Labeo has a distinctive crescent-shaped dorsal fin, large caudal fin and subterminal mouth with fleshy lips. This species reaches 40 cm SL and up to 2.5 kg.

This large labeo is found in a variety of habitats from well-vegetated river margins to soft-bottomed lagoons and rocky rapids located midstream. It consumes *aufwuchs.*

They migrate upstream in shoals to spawn in summer coinciding with the rainy season and/or arrival of annual floodwaters. Spawning males have tubercles on their snouts. Females lay their eggs among aquatic plants in recently inundated floodplain habitats.

The Upper Zambezi Labeo is found in the Kavango, Kwando-Linyanti and Upper Zambezi-Chobe River systems of Namibia. Elsewhere, it occurs northward in these drainages into Angola and Zambia and southward to the Okavango Delta. Although not a sought-after sport fish, this species is important in the subsistence fishery when young fish are caught in large numbers in fine-meshed seine nets and traps set across receding waters from floodplains.

Silver Labeo (*Labeo ruddi*)

F. Jacobs

Etymology: ***Labeo***, Latin, ***labeo***, one who has large lips; ***ruddi***, in honour of De Beers Mining Company co-founder Charles Dunell Rudd (1844–1916), who financed the expedition that collected the type specimen

DS 0, DR iii, 9-10; AS 0, AR iii, 5; LL 40-44, CP 18-22

Summary: This species has a plain slender shiny-silver body from which its common name is derived. The mouth is small and subterminal with thin lips and very short barbels. The paired fins are flesh-coloured, tinged with black. The slate-coloured dorsal and anal fins are falcate shaped, the caudal fin strongly forked. It reaches 300 mm SL, 1 kg.

The Silver Labeo has a disjunct distribution. It occurs in Namibia only in the Kunene River upstream into Angola in deep slow-moving reaches over sand or muddy bottoms. Eastward, they are unknown until the lowveld areas of the Limpopo and Incomati Rivers of Mozambique, Zimbabwe and South Africa. The curious distribution infers there might be two distinct species rather than separated populations of the same species.

Little information is available on its habitat and ecology. It is considered a benthopelagic species feeding on organic sediments. The Silver Labeo spawns in summer coinciding with the rainy season and may undertake upstream potamodromic migrations to well-vegetated margins of the river and floodplains for spawning.

Moggel (*Labeo umbratus*)

D. Voorvelt

Etymology: ***Labeo***, Latin, ***labeo***, one who has large lips; ***umbratus***, etymology not explained; could mean "covered" with spots on back & sides or "shaded" (pectoral, anal and ventral) fins and/or eyes

DS 0, DR iii-iv, 8-10; AS 0, AR iii, 5; LL 53-68, CP 26-34

Summary: The Moggel has a fusiform body, light brown to pinkish mottling fading into a silvery off-white belly. There is a relatively small subterminal mouth with two small barbels. It is similar in appearance to the Orange River Mudfish and reaches up to 500 mm TL, 2.9 kg. Its common name apparently refers to its mottled appearance.

Moggel occur in Namibia only in the Fish and lower Orange River drainages. Otherwise, they are widely distributed in the upper Orange-Senqu basin of South Africa into Lesotho and various coastal waters. It is considered benthopelagic, inhabiting slow- or gently-flowing reaches of the river. It is a bottom-feeding detritivore and its diet consists of algae, small invertebrates, detritus and zooplankton. Moggel exhibit potamodromic spawning migrations after summer rains to suitable sites, e.g., shallow river margins with grassy banks and shallow rocky sections. Females lay thousands of small sticky eggs that attach to grass and rocks. It is valued as a strong angling fish and can be an important subsistence food source.

Vaal-Orange Smallmouth Yellowfish (*Labeobarbus aeneus*)

F. Jacobs

Etymology: ***Labeobarbus***, Latin, ***labeo***, one who has large lips + ***barbus***, with reference to the fairly closely related "carps"; ***aeneus***, refers to golden-bronze colour

DS 4, DR 7-9; AS 0, AR iii, 5; LL 36-44, CP 16

Summary: The Vaal-Orange Smallmouth Yellowfish has a fusiform-shaped body, coloured olive-bronze with a creamy belly. It has a subterminal mouth with fleshy lips that can vary in size from thin to bulbous and two pairs of barbels. It can reach 700 mm FL, 7.8 kg.

This species is endemic to the Orange-Vaal River drainage of Namibia and South Africa, leading to its common name. However, it has been introduced into other South African drainages. It is found over sandy or rocky substrates. They are benthopelagic and largely a bottom feeder of a variety of food types ranging from aquatic insects and invertebrates, vegetation, algae, detritus and small fish. Their subterminal mouth is ideally adapted for feeding on the river substrate.

Spawning occurs from late spring to mid-summer with warmer water temperatures. It is a potamodromic species, migrating upstream to spawn in shallow rocky areas. Eggs are laid in a shallow saucer-like nest and covered with gravel after fertilisation. The long spawning season ensures this species can lay eggs whenever favourable conditions occur. It is a very popular sport fish, especially fly fishing.

Upper Zambezi Yellowfish (*Labeobarbus codringtonii*)

R. Bills

Etymology: ***Labeobarbus***, Latin, ***labeo***, one who has large lips + ***barbus***, with reference to fairly closely related "carps"; ***codringtonii***, in honour of Thomas Codrington (1829–1918) who presented a collection of fishes from (now) Zambia to the British Museum

DS 0, DR iii-iv, 9-10; AS 0, AR iii, 5; LL 29-32, CP 12

Summary: The Upper Zambezi Yellowfish has a robust body and a mouth with thick fleshy lips flanked by two pairs of barbels, being gold-orange to olive dorsally fading ventrally to an off-white cream colouration. It is a powerful swimmer with large fins that enable it to manoeuvre in fast-flowing water; the dorsal fin is falcate shaped. They reach in excess of 500 mm TL, 3.2 kg.

This species has a restricted distribution and is known only from the Kavango and Upper Zambezi River systems in Namibia from fast-flowing reaches over rocky substrate and/or rocky rapids. Its common name is derived from the type locality, Zambezi River above Victoria Falls.

It is omnivorous, feeding on aquatic insect larvae, crustaceans, snails and small fish. It is a potamodromic species, migrating upstream to spawn in shallow rocky areas coinciding with the rainy season and/or arrival of annual floodwaters. The Upper Zambezi Yellowfish is a popular angling fish in the Kavango River and important in the subsistence fishery.

Vaal-Orange Largemouth Yellowfish (*Labeobarbus kimberleyensis*)

F. Jacobs

Etymology: ***Labeobarbus***, Latin, ***labeo***, one who has large lips + ***barbus,*** with reference to fairly closely related "carps"; ***kimberleyensis***, ***-ensis***, suffix denoting Kimberley, South Africa, type locality on Vaal River at Riverton

DS 4, DR 8-9; AS 0, AR iii, 5; LL 37-45, CP 16

Summary: This species has a robust body with muscular shoulders and coloured gold-orange dorsally. It is a powerful swimmer with large fins that enable it to manoeuvre in fast-flowing water, but is more slender and paler than the Vaal-Orange Smallmouth Yellowfish. It is known to reach an impressive 820 mm SL, 22 kg.

This species is the largest scale-bearing indigenous fish in southern Africa. It is endemic to the Orange River drainage of Namibia, South Africa and Lesotho in deep reaches or below rapids and co-exists with the Smallmouth Yellowfish. The scientific and common names are both derived from the type locality along the Vaal River drainage near Kimberley.

The Largemouth Yellowfish is a formidable ambush predator initially eating insects and small crustaceans as juveniles, but turning to a diet predominantly of fish as adults, termed *piscivorous*. This is facilitated by a terminal mouth with thick fleshy lips flanked by two pairs of barbels. They are generally associated with structures such as rocky areas or sunken logs, which they use as cover when waiting to ambush unsuspecting prey. They spawn in summer, depositing eggs along flooded margins of the river over gravel beds. This is an exciting sport fish caught with fish-imitating lures or large streamer flies mimicking their prey.

Namaqua Barb (*Namaquacypris hospes*)

R. Bills

Etymology: ***Namaquacypris***, ***Namaqua-*** named for Namaqua region of South Africa and Namibia + ***cypris***, a small carp, a common suffix for cyprinid genera; ***hospes***, ***hostem***, host or guest, referring to Goodhouse, name of farm on south bank of Orange River at Raman's Drift (Namaqualand), South Africa, type locality, owned by the "hospitable" Mr. C. Weidner

DS 3, DR 7; AS 0, AR iii, 5; LL 37-39, CP 16

Summary: The pale-coloured Namaqua Barb has a slender body with clear fins, a deeply forked caudal fin and two pairs of near equal-length barbels at the corners of its subterminal mouth. The dorsal fin is posterior to origin of the pelvic fins. It obtains 75 mm SL.

This barb has a restricted distribution and is endemic to the Orange River system below Augrabies Falls, the Namaqua region of South Africa and Namibia from which it receives its common and generic name. It inhabits open water in the mainstream channel and backwaters. It feeds mostly on aquatic insects and zooplankton.

Northern Barred Minnow (*Opsaridium zambezense*)

F. Jacobs

Etymology: ***Opsaridium***, Greek, ***opsarion***, ***-ou***, a little portion fish; ***zambezense***, ***-ense***, suffix denoting place: Zambezi River

DS 0, DR iii, 8-10; AS 0, AR iii, 11-13; LL 40-44, CP 14

Summary: The Northern Barred Minnow is a colourful species being silvery olive to yellow-green with 12 striking vertical turquoise-to-blackish (usually double) bars from which it derives its common name. Its mouth is large and barbels are absent. Mature males take on a pink-red body colouration, have pinkish fins and head tubercles. The dorsal fin margin is slightly rounded and situated posterior to the pelvic fin origin. The caudal fin is deeply forked. They reach 180 mm TL in the Upper Zambezi and Kavango Rivers.

This specialised rheophilic species is found in the Kavango, Kwando-Linyanti and Upper Zambezi-Chobe River systems in Namibia. They are widespread, known from river systems in the DRC, Zambia, Zimbabwe, Angola, Malawi and Mozambique. They occur in the Okavango Delta, but are extremely rare. Their narrow physiological tolerance for well-aerated flowing rivers is a useful indicator of environmental conditions in an area and is a limiting factor to their distribution.

This minnow forms shoals usually over sandy bottoms in rivers and pools below rocky rapids, such as Popa Falls. They feed on aquatic larvae and insects floating along the water surface and small crustaceans. They undertake seasonal movements up tributaries and onto floodplains to spawn in spring and summer. They are caught in seine nets over sand banks and with small liftnets in the Zambezi River.

Women crossing the Kavango River with fishing baskets above the town of Rundu in 1984 (Photo: G. Merron).

D. Family Distichodontidae (Citharines)

The distichodontids are endemic to Africa. Like many inland fishes of the continent, their taxonomy is in a state of flux and under review; four species are currently identified from Namibian waters. These are small streamlined fusiform fish coloured olive above and white to silvery below, usually with distinct vertical bars, blotches or stripes on the sides. Their caudal fins are forked. They are related to the family Alestidae that includes the robbers and Tigerfish, but have a different tooth pattern, i.e., *distichus*, *di-* two + *odon*, tooth, referring to two rows of teeth on both jaws.

Namibian citharines typically inhabit the roots of papyrus and under reed mats and other aquatic vegetation lining the rivers channels. They can also be found in large perennial lagoons and are abundant in floodplains. They feed by grazing on *aufwuchs*, invertebrates and algae living on the stems and roots of aquatic plants, and also capture insects floating with the river currents.

They all spawn during the summer rainy season and/or after the arrival of the annual flood, which inundates surrounding floodplains and river margins.

Saddled Citharine (*Nannocharax dageti*)

F. Jacobs

Etymology: ***Nannocharax***, ***nanus***, dwarf + ***charax***, typical genus of the Characiformes, from Greek word meaning "palisade of pointed sticks," referring to densely packed sharp teeth; ***dageti***, in honour of ichthyologist Jacques Daget (1919–2009), formerly of the Muséum National d'Histoire Naturelle, Paris

DS 0, DR iii, 10-12; AS 0, AR iii, 7-9; LL 36-41, CP 12

Summary: This species possesses an elongated body that becomes progressively compressed posteriorly to the caudal peduncle. Its head is pointed in lateral profile. Six to 10 dark saddles occur across the dorsal midline of the body for which the common name is assigned. A midlateral longitudinal stripe is overlapped by 7–12 rounded dark blotches that are largely separate from the dorsal saddles. The black caudal spot has whitish blotches either side of it.

Nannocharax dageti is distinguished from *N.* cf. *lineostriatus* by possessing 40–43 scales along the longitudinal line including the lateral line (vs. 34–35 in *N.* cf. *lineostriatus*) and 36–41 pored scales along the lateral line (vs. 6–9 in *N.* cf. *lineostriatus*).

The species occurs in the faster-flowing tributaries, headwater streams and rapids of the Upper Zambezi River.

Broad-striped Citharine (*Nannocharax* cf. *lineostriatus*)

B. van der Waal

Etymology: ***Nannocharax***, ***nanus***, dwarf + ***charax***, typical genus of the Characiformes, from Greek word meaning "palisade of pointed sticks," referring to densely packed sharp teeth; ***lineo-***, line, ***striatus***, striped, referring to black lateral band running from tip of nose to base of tail, transversed by 9-12 vertical black lines

DS 0, DR iii, 9-11; AS 0, AR iii, 7-8; LL 6-9, CP 12

Summary: *N. lineostriatus,* originally described in 1967 from Angola, is a species complex currently under review. One form, *N.* cf. *lineostriatus* (cf., meaning 'related to') occurs on the floodplains of the Kavango, Kwando-Linyanti and Upper Zambezi-Chobe Rivers south to the Okavango Delta. It lacks bars in the caudal fin. There is a tendency for the lateral bars to join and form a complete midlateral stripe from which the common name is derived. It characteristically displays an orange caudal colouration. The black caudal spot has whitish blotches either side of it and the lateral line is incomplete. An adipose fin is present.

Dwarf Citharine (*Nannocharax machadoi*)

B. van der Waal

Etymology: ***Nannocharax***, ***nanus***, dwarf + ***charax***, typical genus of the Characiformes, from Greek word meaning "palisade of pointed sticks," referring to densely packed sharp teeth; ***machadoi***, in honour of António de Barros Machado (1912–2002), distinguished zoologist of the Musée de Dundo, Angola

DS 0, DR iii, 10-11; AS 0, AR iii, 8; LL 8-9, CP 12

Summary: The Dwarf Citharine possesses 12–16 distinct vertical bars along their sides extending onto the caudal fin with a prominent dark spot at its base, surrounded by a triangular yellowish patch, and 2–3 curved bands in the caudal rays. The other fins are light grey and there is no adipose fin. The lateral line is incomplete with 34–36 scales in the lateral series. The head has a terminal mouth and large eyes. It reaches 40 mm SL.

This citharine occurs in the Kunene, Kavango, Kwando-Linyanti and Upper Zambezi-Chobe River systems in Namibia among dense aquatic vegetation lining the mainstream channels, perennial lagoons and floodplains. It has been collected in the canal linking the Kunene to the Cuvelai basin. Elsewhere it is found in the northern Okavango Delta, Upper and Lower Zambezi River systems. It is caught regularly in the subsistence fishery.

Multibar Citharine (*Nannocharax multifasciatus*)

B. van der Waal

Etymology: ***Nannocharax***, ***nanus***, dwarf + ***charax***, typical genus of the Characiformes, from Greek word meaning "palisade of pointed sticks," referring to densely packed sharp teeth; ***multifasciatus***, ***multi-***, many + ***fasciatus***, banded, referring to 16+ brown vertical bars on sides

DS 0, DR iii, 10-12; AS 0, AR iii, 8-9; 40-44 in lateral series, CP 16

Summary: This species has 16–25 dark vertical bars along the body from which its common name stems. There is a prominent dark spot at the base of the caudal fin surrounded by a yellowish triangular patch, followed by a dusky band. An adipose fin is present. The lateral line is incomplete with 40–44 scales in the series. The head is slightly pointed with a terminal mouth and large eyes. It grows to 45 mm SL.

This rheophilic species inhabits all northern Namibian river systems, excluding the Cuvelai. It is also known from the northern Okavango Delta; Upper Zambezi and Kafue-Middle Zambezi, Zambia; and Kasai-Congo River, Angola/DRC systems.

It occurs in well-vegetated margins of rivers and is also found in large perennial lagoons. It is caught in the subsistence fishery.

E. Family Alestidae (Robbers and Tigerfish)

This family of freshwater fishes is represented in Namibia by four species distributed amongst four genera. All are omnivorous/piscivorous with sharp teeth on their jaws. The dorsal and pelvic fins are situated at mid-body and all species have an adipose fin. The lateral line dips characteristically low on the body. They typically have large scales for their body size and the caudal fin is forked. All are shoaling species. This family is an excellent example of phylogenetically-similar species that differ by an order of magnitude in size.

Striped Robber (*Brycinus lateralis*)

F. Jacobs

Etymology: ***Brycinus***, possibly derived from ***bryco***, to bite, gnash teeth or eat greedily; ***lateralis***, of the side, presumably referring to dark lateral band on sides

DS 0, DR ii, 8; AS 0, AR iii, 15-16; LL 30-33, CP 10-14

Summary: The slender body is coloured silvery-blue above and silvery below with a pronounced black lateral band originating on the posterior caudal peduncle and extending through to the fork of the caudal fin. The adipose fin is yellow while the other fins are transparent with a yellowish tinge along their bases. The head is bony with a large eye and terminal mouth lined with sharp multi-cuspid teeth. This species reaches 140 mm SL and females are larger than males and have a straight-edged anal fin, compared to falcate in males.

They inhabit both moderate- and slow-flowing well-vegetated margins of river channels and open water lagoons. This species can be found in the Kunene, Cuvelai, Kavango, Kwando-Linyanti and Upper Zambezi-Chobe systems in Namibia to the Okavango Delta. It has a wide distribution from the DRC to South Africa.

The Striped Robber is an opportunistic predator feeding on a variety of small aquatic organisms, including young of other species. They also consume plant material, including seeds. They are an important dispersal pathway for aquatic plants to recently inundated floodplains of the Okavango Delta as seeds are passed through the intestines and out with faeces. This species can often be seen taking prey from the water surface, such as insects. They form shoals that migrate onto the inundated floodplains to spawn.

It is a very important fish caught in subsistence and commercial fisheries. Fishers catch this species using fine-mesh gillnets (24–30 mm), baskets, traps, and hook and line. Due to its abundance, it is an important prey item for larger piscivorous fish species, including Tigerfish, African Pike, Sharptooth Catfish, largemouth bream and also fish-eating birds.

Tigerfish (*Hydrocynus vittatus*)

D. Tweddle

Etymology: ***Hydrocynus***, ***hydro***, water + ***kyon***, dog; ***vittatus***, ribbon or stripes, referring to several dark blue-black parallel stripes along the sides

DS 0, DR ii, 7-8; AS 0, AR iii, 10-13; LL 43-48, CP 15-16

Summary: The fusiform body is overall silvery in colour with thin blue-black parallel stripes running horizontally, lending its common name; its scientific name literally means "striped waterdog". The head is large with bony cheeks and powerful jaws, each with a set of eight large formidable teeth. The fins are bright red with a brownish base. The adipose fin is black. The dorsal, pelvic and anal fins are concave and the caudal fin is deeply forked. Males reach 500 mm FL and females reach 800 mm FL, weighing up to 15 kg. The closely related Goliath Tigerfish (*H. goliath*) from the Congo River attains a length in excess of 1.5 m and 45 kg.

This voracious predator inhabits well-oxygenated river channels and large lagoons of the Kavango, Kwando-Linyanti and Upper Zambezi-Chobe systems in Namibia, but is absent from the Kunene and Cuvelai Rivers. It is widely distributed throughout Africa from the Nile River in the north to Senegal on the west; Congo Basin; and south to the Phongolo system, South Africa, possibly representing a species complex. They have limited environmental tolerances and avoid anoxic waters.

Tigerfish dominate aquatic fish communities in open waters and form packs of similar-sized fish that will attack any other fish, especially slender shoaling species similar to the Striped and Silver Robbers. They are able to catch prey up to 40% of their length and have been dissected with snakes and small mammals in their stomachs. *Avivorous* behaviour has also been documented for this species, known to catch birds in flight. Whole sets of teeth are regularly replaced throughout their lives with the new teeth developing below the functional teeth in the jaws.

Spawning takes place between October and November prior to the arrival of the annual flood. Fish congregate in sheltered oxbow lagoons and coves of the mainstream with the largest females depositing over 500,000 eggs among the stems of papyrus, reeds and other submerged aquatic plants and rocks. It may also school in eddies below mainstream rapids. When the annual floodwaters arrive, the juveniles move with the floodwaters onto floodplains and feed on aquatic insects, larval fish and other small organisms. Growth is rapid and they move into deeper main channels with the receding seasonal floodplain waters where their diet is piscivorous upon the abundance of juvenile cyprinids, cichlids and other species that are retreating from the floodplains. Tigerfish can roam over vast lengths of river and individual movement has been recorded up to 397 km in the Kavango River.

Tigerfish are a prized sport fish in northern Namibia and one of the greatest acrobatic freshwater fish in the world. They are caught using baited hooks, spoons, spinners or flies. Annual sport fishing competitions historically have been organised in the Kavango and Upper Zambezi Rivers that attracted fishermen from many countries. Tigerfish are also important components of the subsistence and commercial fisheries.

The razor sharp teeth of the Tigerfish (Photos: C. Hocutt (left); C. Hay (right)).

Silver Robber (*Micralestes acutidens*)

P. Bragança

Etymology: ***Micralestes***, ***micro-***, small + ***-alestes***, Greek for miller or grinder, presumably referring to inner row of premaxillary molariform teeth; ***acutidens***, ***acutus***, sharp or pointed + ***dens***, teeth

DS 0, DR ii, 8; AS 0, AR iii, 14-16; LL 23-28, CP 9-11

Summary: The Silver Robber has a slender fusiform body, silvery coloured, often with a greenish sheen and distinctive black tip on the dorsal fin. All fins are pale yellow to white. The mouth is terminal and jaws contain sharp multi-cuspid teeth. The eyes are large and characteristic of a riverine predator. Its name can be interpreted as 'small robber with sharp teeth'. Females grow to 80 mm SL and are larger than males.

This species is a topwater (pelagic) surface predator that is found in all northern Namibian river systems including the Kunene, Cuvelai, Kavango, Kwando-Linyanti and Upper Zambezi-Chobe Rivers. They are also known to occur in the Congo system eastward to Tanzania and Indian Ocean coastal rivers as far south as the Phongolo system in KwaZulu-Natal, South Africa.

They form shoals in flowing open water lined with papyrus and reed beds. They are omnivorous predators feeding on fish eggs, fry, winged insects, crustaceans, zooplankton and seeds. Spawning shoals migrate upstream after the first summer rains and females deposit adhesive eggs on plant stems. They are caught in traditional fish traps, reed weirs and seine nets in floodplains.

Slender Robber (*Rhabdalestes maunensis*)

G. Merron

Etymology: ***Rhabdalestes***, ***rhabdo-***, rod, stick or staff; allusion not explained + ***-alestes***, Greek for miller or grinder, presumably referring to inner row of premaxillary molariform teeth; ***maunensis***, ***–ensis***, suffix denoting place: type locality, Thamalakane River, Maun, Botswana

DS 0, DR ii, 7-9; AS 0, AR iii, 16-19; LL 33-36, CP 12-14

Summary: The Slender Robber is light yellow to light brown above with a silvery-white head and belly. An iridescent blue band runs along the body. The anal fin is long and edged in black along its entire base. All fins are translucent in colour. The caudal fin is forked and an adipose fin is present. The head is small with a slightly upturned mouth and large eye. This species reaches 60 mm SL.

The Slender Robber feeds on small aquatic insects and other invertebrates. Spawning occurs during the summer rainy season in newly inundated floodplains and backwater channels. It is taken in traditional fish traps, reed weirs and seine nets in floodplains.

This pelagic species inhabits well vegetated margins and adjacent coves of the Kunene, Cuvelai, Kavango, Kwando-Linyanti and Upper Zambezi-Chobe Rivers and their adjacent lagoons and vegetated floodplain habitats. They are also known from the Middle Zambezi-Kafue systems in Zambia and the Okavango Delta.

F. Family Hepsetidae (African Pike)

The freshwater Family Hepsetidae is unique to Africa with six species, one of which inhabits northern Namibian rivers, floodplains and lagoons. This family of fish is widespread throughout south, central and west Africa, and is easily recognised by their sharply pointed head, crocodile-like jaws and distinctive colouration. Their dorsal fin sits posterior to the base of the pelvic fins; an adipose fin is present. The caudal fin is broad and forked.

Southern African Pike (*Hepsetus cuvieri*)

F.Jacobs

> **Etymology:** ***Hepsetus***, allusion not explained; ***cuvieri***, in honour of French naturalist Georges Cuvier (1769–1832)
>
> DS 0, DR ii, 7; AS 0, AR ii, 9; LL 48-55, CP 24

Summary: This torpedo-shaped fish is dark brown dorsally and silver ventrally with distinct dark brown blotches along the body. The head is brownish with 3–4 dark stripes radiating from the eye. The snout has crocodile-like jaws and formidable sharp-pointed canine teeth, resembling Northern Hemisphere pikes (Esocidae) that may well have influenced the common name. In adults, both jaws have two pairs of dermal flaps that serve a vital function during spawning and care of the young. The dorsal and anal fins are set far back on the body. All fins have black spots and the adipose fin is orange. The caudal fin is forked. It reaches 470 mm FL and 2 kg with females growing larger than males.

Adult Southern African Pike, like Tigerfish, feed mainly on fish, but will prey on anything swimming, such as aquatic invertebrates, small rodents and *herptiles*. Juveniles grow rapidly reaching maturity after 2 years at a length of about 200 mm (males) and 250 mm FL (females). They have a life span of 4–5 years.

Spawning begins in late spring and continues throughout the summer with each female spawning several times. Adults form pairs and build large bubble nests among the aquatic vegetation where the eggs are deposited. The foam nests float on the water surface among the stems of aquatic plants in the fringes of slow-flowing river channels, lagoons and floodplains. Hundreds of nests can be congregated together over the size of a football field. The bubble nests allow the pike to spawn in poorly oxygenated waters.

The eggs are relatively large and provide a generous amount of yolk for the developing embryos to feed on. Upon hatching, the fry move to the base of the nest where they attach themselves using a gland on the top of their heads and hang downward in the water. The nest is aggressively guarded by the parents. While attached to the nest, the young fish obtain nutrients from the foam nest and organic matter trapped in it. The fry become free swimming within about a week, dependent on water temperature, and feed initially on *aufwuchs* and insects at the surface.

The Southern African Pike is an ambush predator that is well camouflaged among aquatic vegetation in a wide variety of habitats, including slow-flowing areas of rivers, lagoons and floodplains of the Kunene, Kavango, Kwando-Linyanti and Upper Zambezi-Chobe River systems in Namibia. It is reported from the canal system linking the Kunene to the Cuvelai basin, thus it is considered to occur in the latter. It is found elsewhere in the tropical regions of west and central Africa from Senegal to the Okavango Delta, the southern limit of its distribution. They are frequently caught in subsistence and commercial fisheries and are excellent sport fish when hooked.

Lateral view of the head and teeth of the Southern African Pike (Photo: B. van der Waal).

Typical bubble nest of the African Pike (Photo: G. Merron).

G. Family Amphiliidae (Mountain Catfish and Sand Catlets)

These small catfishes are endemic to Africa. They have adapted to fast-flowing water and rocky habitats at high elevations or rapids of large lowland river basins. Morphological adaptations for these habitats include expanded pectoral and pelvic fins; depressed body; and dorsally positioned eyes. They have three pairs of simple barbels, a short dorsal fin and a fleshy adipose fin. There are two genera known to Namibia, *Amphilius* (without pectoral or dorsal fin spines) and *Zaireichthys*. The genus *Zaireichthys* contains the smallest catfishes in Africa and they are distinguished in the Amphiliidae by being the only representatives with stout dorsal and pectoral fin spines.

Stargazer Mountain Catfish (*Amphilius uranoscopus*)

R. Bills

Etymology: ***Amphilius***, Greek, ***amphi***, on both sides + Greek, ***leios***, smooth; ***uranoscopus***, Greek, ***ouranos***, sky + ***skopein***, to watch

DS 0, DR i, 6; AS 0, AR iii, 5-7; LL 0, CP 0

Summary: The Stargazer Mountain Catfish's body is compressed and the head slopes downward to the tip of the snout. The common and scientific names are derived from its eyes situated on top of the head. The colour is brown-beige with many dark brown to black patches which often form a reticulated pattern. The dorsal, anal and adipose fins are short; the caudal fin is forked with a dusky band near its posterior. The large pelvic fins in conjunction with the body form a feeble sucking disc to allow the fish to cling onto rocks.

This small bottom-dwelling predator prefers clear fast-flowing rivers in rocky habitats and under waterfalls where the water is rich in oxygen. It feeds on aquatic insects and other small organisms grazed off rock and plant surfaces. They spawn in summer with water temperature a driving cue. The eggs are laid under stones. It grows to about 130 mm SL.

Distribution is limited to the Kavango and the Upper Zambezi Rivers of eastern Namibia, but excludes the Kwando-Linyanti and Chobe Rivers as well as the Okavango Delta where the preferred rocky habitats are absent. They are also known further upstream in the Cubango and Upper Zambezi systems into Angola and Zambia as well as central and east Africa south to the Mkuze system in KwaZulu-Natal.

Sand Catlets

F. Jacobs

There are four described and one undescribed species of sand catlets in northern Namibian rivers. The above photos present lateral (left) and dorsal (right) views demonstrating general morphology and colouration of all five species. These small fishes grow up to 35 mm SL; some are endemic to the rivers after which they are named. These fish inhabit sandy and rocky habitats, often forming shoals on the bottom. Their bodies are fusiform, somewhat flattened and generally coloured beige-brownish with marked dark brown patches. Their heads are relatively large for their body size with their mouths flanked by three pairs of short barbels. The adipose fin is large and the caudal fin slightly triangular. The sand catlets exude mucous giving them a very slimy feeling when handled.

Sand catlets wriggle into the sand with only their eyes protruding, consuming unsuspecting aquatic invertebrates and larvae that pass by. All spawn in late spring and summer during the rainy season with the arrival of annual floodwaters from Angola and Zambia. They lay relatively few large eggs, suggesting parental care. These species are undergoing taxonomic study.

Blotched Sand Catlet (*Zaireichthys conspicuous*)

Etymology: ***Zaireichthys***, Zaire river + Greek, ***ichthys***, fish; ***conspicuous***, refers to the conspicuous bold markings of this species

DS 2, DR 6; AS 0, AR iii-iv, 6-7; LL 0, CP 0

Summary: This catlet occurs in rocky rapid areas of the Kavango, Upper Zambezi and Chobe River systems. It is not yet recorded in the Kwando and Linyanti Rivers, but may occur. It reaches 35 mm SL.

Kavango Sand Catlet (*Zaireichthys kavangoensis*)

Etymology: ***Zaireichthys***, Zaire river + Greek, ***ichthys***, fish; ***kavangoensis***, ***-ensis***, suffix refers to the Kavango River, type locality

DS 2, DR 5; AS 0, AR iv-vi, 5-6; LL 0, CP 0

Summary: This is a very small catlet found on sandy substrates in flowing waters and on the edges of rocky rapids. It is only known from the Kavango River, Namibia into Angola. It obtains 27 mm SL.

Kunene Sand Catlet (*Zaireichthys kunenensis*)

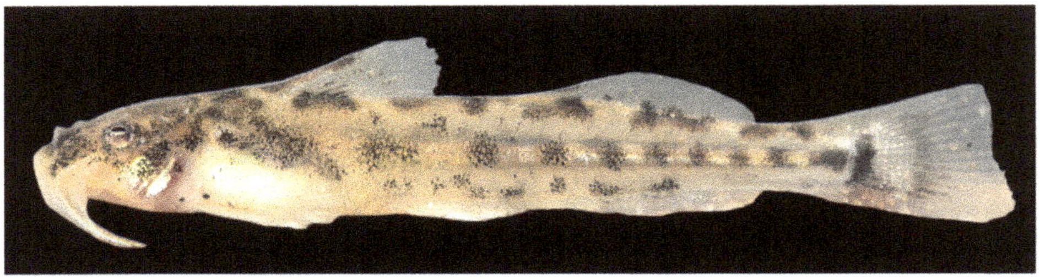

P. Bragança

Etymology: ***Zaireichthys***, Zaire river + Greek, ***ichthys***, fish; ***kunenensis***, ***-ensis***, suffix refers to the Kunene River, type locality

DS 2, DR 5-6; AS 0, AR iv-vi, 5-7; LL 0, CP 0

Summary: The Kunene Sand Catlet is generally more pallid and less conspicuously marked than other species of *Zaireichthys*. It occurs in shallow water with a sandy bottom where it buries in the sand with only its eyes protruding. This species is endemic to the Kunene River, Namibia and Angola. Adults reach 25 mm SL.

Pallid Sand Catlet (*Zaireichthys pallidus*)

Etymology: ***Zaireichthys***, Zaire river + Greek, ***ichthys***, fish; ***pallidus***, pale, colourless

DS 2, DR 4-5; AS 0, AR iv-vii, 5-6; LL 0, CP 0

Summary: The Pallid Sand Catlet inhabits shallow sandy-bottom habitats flushed with gentle currents, usually buried with just the eyes protruding. In life, it is not pallid, that being a misnomer from preserved specimens that are pale yellow to transparent, with a few minute black spots scattered irregularly over the body. It obtains 25 mm SL.

This species is considered endemic to the Kwando-Linyanti system, Namibia and Angola. A closely related, but yet undescribed species, occurs in the Upper Zambezi-Chobe Rivers, upstream into Zambia and the Barotse Floodplain.

Typical habitat of sand catlets in shallow sandy stretches of the rivers (Photo: F. Jacobs).

H. Family Austroglanididae (Rock Catfishes)

Austroglanis is the only genus in the Austroglanididae, an endemic family restricted to southern Africa. There are three species in this genus, one of which inhabits the Orange River system, Namibia and South Africa. These fish have three pairs of barbels and strong dorsal and pectoral fin spines.

Rock Catfish (*Austroglanis sclateri*)

R. Bills

Etymology: ***Austro***, the south + Greek, ***glanis***, a fish that can eat the bait without touching the hook, a catfish; ***sclateri***, in honour of William L. Sclater (1863–1944), Director of the South African Museum

DS 1, DR 7-8; AS 0, AR v–vi, 10-13; LL 0, CP 0

Summary: The body is elongated and scaleless with olive to green colouration. The mouth is inferior and flanked by three pairs of barbels. The prominent pectoral spines are serrated. They grow to 300 mm SL.

This catfish can be easily identified by the odd appearance of its sloping head. It inhabits rocky areas in fast-flowing water or in rapids. It is an opportunistic feeder and preys on crustaceans and other aquatic invertebrates taken from rock surfaces with larger individuals also preying on small fish. It is caught incidentally in the sport fishery.

Gillnets being set by MFMR staff in the Kavango River (Photo: F. Jacobs).

I. Family Clariidae (Air-breathing Catfish)

The family Clariidae includes the air-breathing catfishes. They are recognised by their elongate body shape, long-based dorsal and anal fins, a rounded caudal fin, a flat bony head with small eyes and a broad terminal mouth bordered by four pairs of barbels, often long. The strong-spined pectoral fins are positioned forward on the body and the pelvic fins are far back. Dorsal and anal fin spines are lacking. They are usually grey or brown, but may be mottled, with colour changing to adapt to their environment or as camouflage.

Most adult clariids are distinguished by having a large accessory breathing organ (*labyrinth*) in the back of the large bony head. The labyrinth is composed of modified tree-like extensions *(suprabranchial organ)* from the gill arches that allow them to breathe air directly. Thus, the clariids are hardy fishes that can survive desiccation and anoxia; these adaptations in combination with their use of locked pectoral fin spines to slither across ground has led to some forms being called "walking catfish". Some species attain a large size, but others are small. Catfishes are typically *crepuscular* or *nocturnal*. There are two genera of clariids known from Namibia, *Clariallabes* and *Clarias*, the former being an Africa endemic.

Many *Clarias* species are a major part of the subsistence fishery caught using a variety of traditional gear, such as fishing baskets, traps and modern methods (e.g., gillnets). These fish are a free source of protein to thousands of people. They are also sought after commercially and recreationally, putting up a strenuous fight when hooked.

Broadhead Catfish (*Clariallabes platyprosopos*)

G. Merron

Etymology: ***Clariallabes***, Greek ***clarias***, shining or Greek, ***chlaros***, lively, in reference to the ability of the fish to live for a long time out of water + Greek, ***allabes***, ***-etos***, a fish of the Nile, a kind of lamprey; ***platyprosopos***, ***platy-***, broad or flat + ***prosopos***, face, head

DS 0, DR 73-82; AS 0, AR 56-63; LL 0, CP 0

Summary: This species has a long dark-brown to black body and may be marbled with black spotting. Its head is broadly depressed with bulging cheeks. The mouth is wide and flanked by four pairs of long barbels. The dorsal and anal fins are long and wavy and extend to the caudal fin. There is no adipose fin and the lateral line appears as a whitish line running along the mid-body. The pectoral fins have serrated spines. The air-breathing organ is non-functional or absent. It reaches 283 mm SL.

This uncommon catfish is recorded throughout the Kavango River into Angola living in rocky habitats and crevices. It also occurs in the Upper Zambezi River. It has not been recorded from the Kunene, Kwando-Linyanti or Chobe River systems or the Okavango Delta.

Very little is known of its biology. Adults feed on fish and juveniles probably feed on small invertebrates. Spawning likely peaks during the summer rainy season and/or after the arrival of the floodwaters and inundation of floodplains. It is not commonly caught by subsistence fishers or anglers.

Cunene Broadhead Catfish (*Clariallabes* sp.)

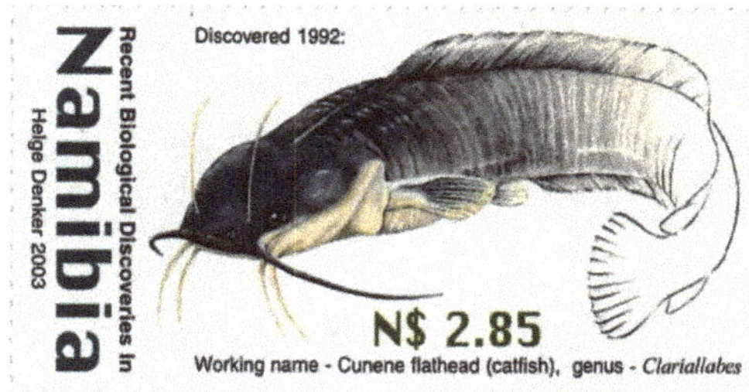

Summary: This undescribed species has been recognised in the literature for a number of years. Its preferred habitat appears to be rapids with fast-flowing water. MFMR data indicate it grows at least to 220 mm SL and 70 g. Small specimens ranging from 50 (0.2) to 90 mm (4 g) were collected in October/November.

This species is considered endemic to the Kunene River, Namibia, extending into Angola. It was featured on a Namibian postal stamp in 2003.

Cave Catfish (*Clarias cavernicola*)

F.Jacobs

> **Etymology:** ***Clarias,*** Greek ***clarias***, shining or Greek, ***chlaros***, lively, in reference to the ability of the fish to live for a long time out of water; ***cavernicola***, cave dwelling
>
> DS 0, DR 64-76; AS 0, AR 51-66; LL 0, CP 0

Summary: The Cave Catfish has a relatively short head, small or no eyes and a long body with creamish, flushed pink or golden-yellow colouration. The pectoral fin spines are barbed on the edges. The closest relative to this unique species is the Snake Catfish (*Clarias theodorae*).

This species feeds on bat faeces, terrestrial insects and animal carcasses that fall into the water. It is an endangered species and found only in the Aigamas Cave system, Namibia, which is more expansive than historically thought. The primary conservation threat it faces is the depletion of ground water.

Sharptooth Catfish (*Clarias gariepinus*)

F.Jacobs

Etymology: ***Clarias***, Greek ***clarias***, shining or Greek, ***chlaros***, lively, in reference to the ability of the fish to live for a long time out of water; ***gariepinus***, ***gariep***, a San name meaning "Great Water", now known as the Orange River + ***-inus***, suffix belonging to

DS 0, DR 61-75; AS 0, AR 45-65; LL 0, CP 0

Summary: The elongate body has a strongly compressed caudal peduncle for fast acceleration to navigate rapids and waterfalls and for active feeding. The body is scaleless with a flattened bony head and four pairs of long highly-receptive barbels around a broad subterminal mouth. The jaws are lined with broad bands of fine pointed teeth. Colour is typically dark grey or black on the back, fading to a white belly. Adults can be mottled with irregular dark blotches. The dorsal and anal fins are long and spineless, but the pectoral spines are strongly serrated anteriorly and can lock in position inflicting a wound if eaten by predators or handled improperly. It can grow to 1.4 m SL and weigh 60 kg, although not obtaining this size in Namibia. It is a long-lived species of up to 15 years and second in size in southern Africa only to the Vundu (*Heterobranchus longifilis*) found in the Lower Zambezi River.

The Sharptooth Catfish is common to all Namibian perennial rivers and National waters, including the seasonal Cuvelai system. This species is adapted to mainstream river channels and tributaries, lagoons, oxbow lakes, backwaters, floodplains and reservoirs. They are the most naturally widespread species of African freshwater fish and possibly have the widest latitudinal range in the world found from the Eastern Cape, Umtamvuna River, South Africa northward throughout most African countries to the Nile and into Asia Minor.

They are very tolerant of extreme environmental conditions and are often the last species to survive in drying floodplains and seasonal river channels. Their air-breathing organs allow them to aestivate and survive in moist mud for long periods of time between rainy seasons and arrival of next annual floodwaters.

Sharptooth Catfish are nocturnal predators and omnivores that detect food using their barbels with their array of sensitive taste buds and electroreception to feel their way around in the dark and find food. Adults feed mainly on fish, but also eat frogs, reptiles (including snakes), small birds, mammals and occasional fruits fallen from riparian trees. They capture their prey by gulping them in with a rapid opening of the mouth that creates a vacuum and prey are swallowed whole. Juveniles feed on small fish, snails, crustaceans and small fruits.

Sharptooth Catfish are highly potamodromous and undertake spawning migrations in response to rising floodwaters, which inundate low-lying areas. They spawn in shallow vegetated margins of rivers, lakes, reservoirs, floodplains and marshes. Courtship and mating take place at night in shallow waters. Eggs and milt are released followed by a vigorous swish of the female's tail to distribute thousands of viable eggs over a wide area. The eggs are extremely sticky when laid allowing them to adhere to submerged vegetation for better oxygenation. Parental care for enhancing the survival of offspring is absent.

On the Chobe River floodplain and tributaries to Lake Liambezi, Sharptooth Catfish migrate in great numbers to spawn each year when the water rises with the annual flood filling shallow depressions on the floodplain. The local fishers call it the *Mulungu wa lindombe*, the catfish run. According to local fishers, it must rain for a day or more on the floodplains before the spawning run begins. The event only occurs at night and fishers harvest thousands of fish. Adults return to the river, lagoon or lake after spawning, whereas the juveniles remain in the inundated area where growth is rapid. As summer passes and *mulapos* drain, juvenile catfish and other species make their way back to deeper waters exposing themselves to a host of predators, including *Clarias* adults, Tigerfish, other piscivorous fish species and fish-eating birds. A feeding frenzy can occur.

Sharptooth Catfish are an important source of protein to rural subsistence communities and are also frequently caught in commercial and sport fisheries. During the spawning migrations into shallow water, traditional fishermen catch large numbers of these fish using spears and other indigenous methods. These catfish form the main catch by weight in the *iishana* (see *oshana*) of the Cuvelai River during the irregular large floods (*efundja*). They are caught by longlines with up to 20 baited hooks and in large funnel nets or traditional valved traps set across flowing *iishana* intercepting their downstream migration. They are smoked, salted or dried by villagers.

Sharptooth Catfish have great economic potential and are cultivated in Namibian aquaculture facilities (Chapter 12). This hardy species easily reproduces in captivity, grows fast and can be densely stocked in low oxygen waters, making it ideal for culture in areas with a limited water supply.

Introduced Sharptooth Catfish can become a successful invasive species often causing adverse ecological impacts such as the loss of biodiversity in natural inland waters.

Smoothhead Catfish (*Clarias liocephalus*)

R.Tate

Etymology: ***Clarias***, Greek ***clarias***, shining or Greek, ***chlaros***, lively, in reference to the ability of the fish to live for a long time out of water; ***liocephalus***, Greek ***leios***, soft, smooth + ***cephalus***, head, i.e., "soft head" in reference to the reduced dermal bones behind the orbits

DS 0, DR 67-79; AS 0, AR 53-67; LL 0, CP 0

Summary: The body is elongate with a short rectangular-shaped head and plain coloured or mottled dark brown to black on the back and flanks fading to light brown below. The snout is rounded with four pairs of long barbels. The labyrinth organ is reduced to a simple branched structure. The pectoral spines are short and serrated on the outer edge, less so on the inner surface. There is a dark edge to the dorsal, anal and caudal fins. Lengths reach 282 mm SL.

This small catfish prefers flowing-water habitats and rocky rapids. It also inhabits papyrus- and reed-fringed habitats along river channels. Adults feed on fish and juveniles feed on detritus, plants, small fish and insects. Spawning peaks during the summer rainy season when floodplains become inundated and serve as important nursery areas for young.

It is strictly an upland headwater species of the Kunene, Kavango and Upper Zambezi Rivers in Namibia and Angola. It is known elsewhere in the Zambian Congo and Kafue River systems through to Lake Victoria. The southern limit of its distribution is the Kavango River. It has not been reported from the Kwando-Linyanti or Chobe River systems or Okavango Delta. It is not commonly caught by subsistence fishers or anglers.

Blunttooth Catfish (*Clarias ngamensis*)

D.Tweddle

Etymology: ***Clarias***, Greek ***clarias***, shining or Greek, ***chlaros***, lively, in reference to the ability of the fish to live for a long time out of water; ***ngamensis*** = ***-ensis***, suffix referring to type locality, Lake Ngami, Botswana

DS 0, DR 56-62; AS 0, AR 50-58; LL 0, CP 0

Summary: Similar in shape to the Sharptooth Catfish, this species is distinguished by having a short low adipose fin and a broad vomerine tooth plate for crushing molluscs from which its common name originates. It is typically coloured dark brown, often marbled, with a grey-cream or off-white belly. The mouth is large, but relatively smaller than that of Sharptooth Catfish, and flanked by four pairs of filamentous barbels. The long soft-rayed dorsal and anal fins nearly extend to the caudal fin. Pectoral spines are strongly serrated. It may grow larger than the reported size of 730 mm TL, 4 kg.

The species is fairly common in rivers, lagoons, floodplains, swamps and quieter waters with dense aquatic vegetation. They feed mainly at night on snails, molluscs, crustaceans, aquatic insects and fish, which are crushed using its vomerine tooth pad and powerful jaws before being swallowed.

Spawning is similar to that of the Sharptooth Catfish, being in sync with the annual floodwaters. It is a very hardy air-breather that is one of the last species to survive in drying-up lakes or seasonal floodplain pools.

Blunttooth Catfish are found in the Kunene, Cuvelai, Kavango, Kwando-Linyanti and Upper Zambezi-Chobe River systems of Namibia, extending into the Okavango Delta. It has a disjunct range from the Zambian Congo to the Lower Zambezi and Limpopo Rivers to as far south as the Phongolo River in KwaZulu-Natal, South Africa.

They are an important source of protein for rural communities and feature strongly in subsistence, commercial and sport fisheries. In the *iishana* region of the Cuvelai, they are preferred either fresh or dried over the Sharptooth Catfish, being fattier and tastier. They are caught using a variety of traditional and modern fishing gear.

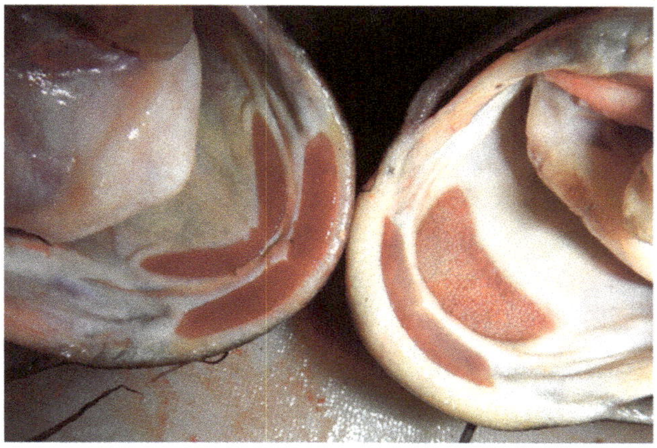

Upper tooth pads of *C.gariepinus* (left) and *C. ngamensis* (right) (Photo: F. Jacobs).

Blotched Catfish (*Clarias stappersii*)

R. Tate

Etymology: ***Clarias***, Greek ***clarias***, shining or Greek, ***chlaros***, lively, in reference to the ability of the fish to live for a long time out of water; ***stappersii***, in honour of physician-biologist Louis Stappers (1883–1916)

DS 0, DR 62-80; AS 0, AR 56-63; LL 0, CP 0

Summary: The elongate body is mottled in dark brown and black blotches (hence the common name) with a light brown or off-white belly. They have a large oblong flattened head with a wide mouth flanked with four pairs of barbels. The soft rayed dorsal and anal fins are long and wavy and extend to the caudal fin. There is no adipose fin. The Blotched Catfish is characterised by its lateral line which appears as a thin white band. The pectoral fins have strong serrated spines and the accessory breathing organ above the gills is well-developed. It can grow to 410 mm SL.

This rare catfish is found in fast-flowing rocky habitats such as Popa Falls rapids in the Kavango River and rapids of the Kunene, Kwando and Upper Zambezi Rivers. It is also known from the Zambian Congo and Kafue Rivers and possibly occurs in other waterbodies, but its habitat specialisation makes it difficult to collect.

This nocturnal predator feeds on small fish, aquatic invertebrates and their larvae, and will consume plant material. As with other *Clarias* catfish, spawning coincides with the annual flood regime when they migrate onto floodplains or into backwater lagoons. Courtship and egg laying take place at night and the eggs adhere to submerged plants. It may be caught in the subsistence fishery, but is not common.

Snake Catfish (*Clarias theodorae*)

D. Tweddle

Etymology: ***Clarias***, Greek ***clarias***, shining or Greek, ***chlaros***, lively, in reference to the ability of the fish to live for a long time out of water; ***theodorae***, species described by German-Dutch zoologist and biogeographer Max Carl Wilhelm Weber (1852–1937) and named in honour of Theodora Jacoba Sleeswijk (1874–1953), the niece of Weber's wife, who accompanied him on his visit to South Africa

DS 0, DR 71-94; AS 0, AR 60-89; LL 0, CP 0

Summary: The slender Snake Catfish is dark brown to black, sometimes mottled in appearance. Its head is very short with a small mouth and long barbels. The dorsal and anal fins are long, extending to the caudal fin with no adipose fin present. The pectoral fin spines are serrated along outer and inner edges. The lateral line appears as a thin white stripe and the air-breathing organ is greatly reduced. Length reaches 350 mm SL.

This small catfish inhabits dense aquatic vegetation, papyrus and reed root balls in moderate- to slow-flowing river channels, backwaters, floodplain lagoons and, occasionally, rocky rapids. They feed on small fish, crustaceans, insects and plant material. Spawning peaks during the summer rainy season with inundated floodplains serving as vital nurseries. During the breeding season, males have enlarged swollen cheeks that may play a role in the pre-spawning stimulation of females. Their air-breathing organs are reduced, but functional, allowing them to migrate across wet ground using their pectoral fins. It is caught in the subsistence fishery in a variety of fish traps, baskets, and hook and line.

This catfish is distributed throughout all the northern rivers of Namibia: Kunene, Kavango, Kwando-Linyanti and Upper Zambezi-Chobe systems. It is also known from one eastern branch of the Cuvelai, but has not been recorded elsewhere in the system. They occur in the Okavango Delta; upper Congo and Kafue River systems; and Lakes Malawi and Tanganyika southwards to the Umfolozi River in KwaZulu-Natal, South Africa. It is the closest relative to the endemic Cave Catfish of the Aigamas Cave system.

J. Family Claroteidae (Claroteid Catfish)

Claroteid catfishes are relatively small with a large mouth, 3-4 pairs of barbels and a well-developed adipose fin. Pelvic fins are posterior to the dorsal fin base; the caudal fin may be rounded, straight or forked. The dorsal and pectoral fins each have a single spine. The family is endemic to Africa.

Zambezi Grunter (*Parauchenoglanis ngamensis*)

F.Jacobs

Etymology: ***Parauchenoglanis***, Greek, ***para***, near + Greek, ***auchen***, neck + Greek, ***glanis***, a fish that can eat the bait without touching the hook, a catfish; ***ngamensis***, ***-ensis***, suffix noting place: Lake Ngami area, Botswana, type locality

DS 1, DR 7; AS 0, AR iv-v, 8-9; LL 0, CP 0

Summary: The Zambezi Grunter has a laterally compressed body behind the head, coloured yellow-brown above and silvery white below with black blotches. The body is with scattered black spots arranged into a series of 5–7 vertical bars. A distinctly flattened large head has a large mouth, fleshy lips flanked by three pairs of barbels with the outer barbel longest and small eyes positioned dorsally. The dorsal and pectoral fins have a leading spine, which is strongly serrated in the pectoral fins. The adipose fin is large and the caudal fin rounded. All fins are spotted and light yellow-brown. It grows to 380 mm SL. The common name 'grunter' is apt as this fish grunts when taken out of water.

This uncommon predatory catfish is the only known member of the family in Namibia. It is found in slow- and moderately-flowing stretches of the Kavango, Kwando-Linyanti and Upper Zambezi-Chobe River systems, often seeking refuge under overhanging trees along river margins and well-vegetated lagoons. It has been recorded from the Okavango Delta and Chobe River near Kasane, Botswana and also from the Kasai-Congo River and Upper Zambezi systems, Zambia.

It feeds on small fish and invertebrates, such as snails, crustaceans and insects. Spawning takes place with the arrival of annual floodwaters on shallow floodplains. Eggs are large and relatively few, evidence of parental care, but few details are known. It is uncommonly caught by anglers, but is a component of the subsistence fishery taken by gillnets and traps.

K. Family Mochokidae (Suckermouths and Squeakers)

The family Mochokidae is the most speciose catfish family on the African continent where it is endemic with 224 recognized species. It is a diverse group of fishes with varying features, characteristics and colouration. In common, however, are complex mouth structures and piercing spines in the dorsal and pectoral fins that can lock into a 90° defensive position, each capable of providing a sharp stinging sensation. The family is represented by two genera in southern Africa, *Chiloglanis* and *Synodontis.* Many are considered 'upside down' catfish due to their foraging habits under submerged logs, boulders and vegetation mats.

Suckermouths (*Chiloglanis*)

The *Chiloglanis* species have modified lips and barbels that form a characteristic inferior suckermouth. Like all other catfish, they have a scaleless body. Sexual dimorphism has been reported with adult males of many of these species having elongate anal and caudal fins. Males may also have an enlarged humeral process, i.e., a distinct bony protrusion lying behind the operculum and originating upward from the pectoral girdle. They possess needle-sharp spines capable of inflicting painful wounds.

Okavango Suckermouth (*Chiloglanis fasciatus*)

F. Jacobs

Etymology: ***Chiloglanis***, ***cheilos***, lip, referring to sucking disc formed by the lips + ***glanis***, general term for catfish; ***fasciatus***, ***fascia***, meaning belt, referring to "traces" of five blackish bands on body

DS 1, DR 6; AS 0, AR iii-iv, 7; LL 0, CP 0

Summary: The Okavango Suckermouth has a fusiform body tapering sharply posteriorly. It is light brown in colour with dark brown or black blotches and a dusky band across a forked caudal fin. The head is curved downward with a pronounced ventrally-situated mouth, which forms a disc that is used to hold onto rocks and plants in river currents. The barbels are moderately long. The dorsal fin has a smooth spine and the small adipose fin is located directly above the anal fin. The pectoral fin spines are serrated. It reaches 50 mm SL.

This small catlet is found in mainstream fast-flowing rocky reaches of large rivers. It feeds on small invertebrates and algae from the surface of rocks and plants. There is distinct pairing during spawning which occurs during the late spring and summer, coinciding with the annual floodwaters.

This species occurs in the Kavango (common at Popa Falls), Kwando-Linyanti and Upper Zambezi-Chobe River systems in Namibia to the Okavango Delta where it reaches the southern limit of its distribution. It is rarely caught in the subsistence fishery due to its habitat specialisation.

Neumann's Suckermouth (*Chiloglanis neumanni*)

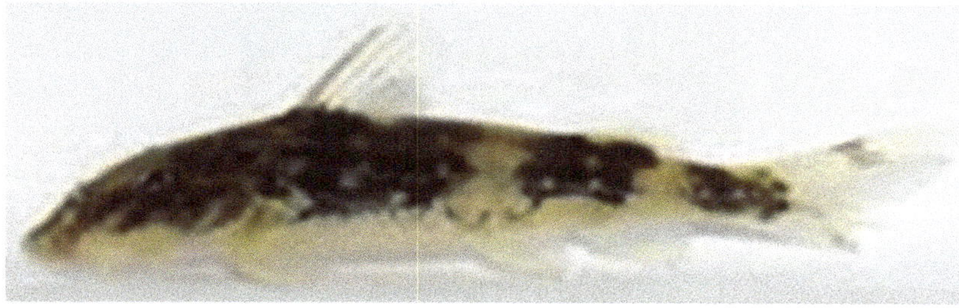

R. Biils

Etymology: ***Chiloglanis***, ***cheilos***, lip, referring to sucking disc formed by the lips + ***glanis***, general term for catfish; ***neumanni***, in honour of German ornithologist Oskar Neumann (1867–1946), who collected the type specimen

DS 1, DR 5-6; AS 0, AR iii-iv, 7; LL 0, CP 0

Summary: This species is slightly dome shaped to the base of the dorsel fin, tapering posteriorly. It is light brown in colour with dark brown or black blotches. The dorsal and pectoral fins are sharp and can cause wounds if not handled carefully. There is sexual dimorphism with the males having a longer anal fin. Distinct pairing occurs during spawning. They may grow to 65 mm SL.

Neumann's Suckermouth inhabits rapids, riffles and pools in moderate- to fast-flowing rivers. The sucking disc helps them to cling onto rocks in rivers with pronounced currents. They feed on the bio-cover (*aufwuchs)* attached to rocks and graze on algae.

This species is restricted in Namibia to the Kunene River, but disjunct populations occur eastward into Zambia, Tanzania, Malawi and Zimbabwe. It is rarely caught due to its habitat specialisation.

Squeakers (*Synodontis*)

This diverse group of mochokids is commonly referred to as 'squeakers' for the distinctive sounds individuals make when removed from the water. The sounds are caused by the base of the pectoral spine rubbing against the pectoral girdle. Their mouths are round and inferior, located ventrally underneath the head. There is one pair of maxillary barbels and two pairs of mandibular barbels that are often branched. The pectoral fins are low on the body and pelvic fins set far back. A large adipose fin is present and the caudal fin is always forked.

The *Synodontis* spp. are difficult to identify based on body markings and colouration alone. All members of the genus have a hardened head cap and a *humeral process*, a bony outgrowth derived from the pectoral girdle, the size and shape of which differs between species. Furthermore, all *Synodontis* species have a premaxillary toothpad on the front of the upper jaw that contains several rows of short chisel-shaped teeth with tooth number and pad shape often used as diagnostic features between species. The dorsal and pectoral fin spines and the structure of their respective barb serrations may also be diagnostic.

Leopard Squeaker (*Synodontis leopardinus*)

F.Jacobs

Etymology: ***Synodontis*** is an "ancient name for an undetermined fish from the Nile" according to Cuvier; ***leopardinus***, ***leoparda***, ***-us***, leopard, referring to its many spots

DS 1, DR 7; AS 0, AR v, 8-10; LL 0, CP 0

Summary: The Leopard Squeaker is a relatively slender fish, tapered posteriorly and coloured light brown with small dark spots over the head and body. The belly is plain or lightly spotted. The dorsal spine is long, equal to head length. They produce audible sounds when disturbed. Length reaches 196 mm SL, weight 0.32 kg.

The Leopard Squeaker inhabits well-vegetated margins of slow-flowing rivers and large open-water lagoons. They are omnivorous, consuming insect larvae, bivalves, crustaceans, algae, plant material and the eggs of other fishes. Spawning takes place during late spring and summer with the arrival of the annual floodwaters.

This squeaker is distributed throughout the Kunene, Cuvelai, Kavango, Kwando-Linyanti and Upper Zambezi-Chobe River systems in Namibia as well as the Okavango Delta.

Largespot Squeaker (*Synodontis macrostigma*)

B. van der Waal

Etymology: ***Synodontis*** is an "ancient name for an undetermined fish from the Nile" according to Cuvier; ***macrostigma***, ***macro-***, large + ***stigma***, spot or mark, referring to large round or oval blackish spots on back and sides

DS 1, DR 7; AS 0, AR iv-vi, 7-9; LL 0, CP 0

Summary: The Largespot Squeaker has a robust fusiform body coloured brown-yellowish with large dark brown or black spots or dashes and belly with smaller spots or plain. The head has small spots and all fins are speckled or spotted. The maxillary barbels have a dark basal membrane and a papillose leading edge, whereas the mandibular barbels have short stubby branches. The humeral process is pointed. The dorsal and pectoral fins have a hardened serrated first spine. It can grow to 170 mm SL.

The Largespot Squeaker is a relatively common upside-down squeaker. It prefers slow-flowing channels, open-water lagoons with sandy substrates and floodplain environments; however, it is occasionally found in rocky stretches of rivers. They feed on detritus, plant material, insects and snails. Spawning takes place during the summer, with a peak coinciding with the summer rains and arrival of the annual flood. There is distinct pairing during spawning.

This squeaker is distributed throughout the Kunene, Cuvelai, Kavango, Kwando-Linyanti and Upper Zambezi-Chobe River systems in Namibia. It is widespread through Angola, Zambia and the Okavango Delta.

Largemouth Squeaker (*Synodontis macrostoma*)

B. van der Waal

Etymology: ***Synodontis*** is an "ancient name for an undetermined fish from the Nile" according to Cuvier; ***macrostoma***, ***macro***, large + ***stoma***, mouth, referring to its distinctive large mouth

DS 1, DR 6; AS 0, AR v–vi, 7–8; LL 0, CP 0

Summary: The Largemouth Squeaker has a triangular-shaped body, creamy-light brown in colour with varied-sized spots and black markings on the body and head. All fins are spotted. The snout is long with a large ventral mouth from which both its common and scientific names are derived. The upper jaw barbels are simple and lower jaw barbels with short stubby branches. The humeral process is narrow, triangular and pointed. The dorsal and pectoral fins have a serrated spine and the caudal fin is forked. Length reaches 92 mm SL.

This is an uncommon nocturnal squeaker that inhabits fast-flowing reaches of rivers and rocky environments. It forages on small invertebrates and algae on rocks and plants. Spawning takes place during the summer with a peak coinciding with the summer rains and arrival of annual floodwaters. It occurs in the Kunene, Kavango, Kwando-Linyanti and Upper Zambezi-Chobe River systems of Namibia; the Kafue River, Zambia; and the Okavango Delta.

Typical subterminal mouth structure of the squeakers (Photo: G. Merron).

Spotted Squeaker (*Synodontis nigromaculatus*)

F.Jacobs

Etymology: ***Synodontis*** is an "ancient name for an undetermined fish from the Nile" according to Cuvier; ***nigromaculatus***, ***nigro-***, black + ***maculatus***, spotted, "covered all over with round black spots, some of which, on the posterior part of the body, are confluent"

DS 1, DR 7; AS 0, AR iv-vi, 7-9; LL 0, CP 0

Summary: The Spotted Squeaker has a rounded body coloured olive-green and entirely covered with small round black spots. The mouth is positioned ventrally and flanked by three pairs of barbels: the maxillary barbels are long and straight without any branches, whereas the two pairs of mandibular barbels are long and slender and have filamentous branches. The humeral process is triangular and ends in a sharp point. The spine of the dorsal fin is straight and smooth in the front and serrated on the back. The spine of the pectoral fin is serrated on both sides. The caudal fin is deeply forked with rounded lobes. When alarmed these fish erect and lock the spiny dorsal and pectoral fins to form a triangle that makes it impossible for predators to swallow and those that try usually perish. It reaches 300 mm SL, 0.5 kg.

This "upside-down" catfish is found in rocky habitats and within vegetation occurring at the edge of rivers and lagoons. It feeds on detritus, algae, plant material, insects, snails and small fish and scavenges readily. Spawning coincides with the rainy season and arrival of the annual flood when males and females pair off. Juveniles retreat from the floodplains with the receding water.

The Spotted Squeaker occurs throughout the Kavango, Kwando-Linyanti and Upper Zambezi-Chobe Rivers of Namibia. It is reported from the DRC to the Okavango Delta. It is a commonly caught species by subsistence and commercial fishermen.

Bubblebarb Squeaker (*Synodontis thamalakanensis*)

G. Merron

Etymology: ***Synodontis*** is an "ancient name for an undetermined fish from the Nile" according to Cuvier; ***thamalakanensis***, ***–ensis***, suffix denoting place, Thamalakane River at Maun, Botswana, type locality

DS 1, DR 7; AS 0, AR v-vi, 8-9; LL 0, CP 0

Summary: The fusiform body tapers sharply from the base of the dorsal fin to the tip of the snout, less so posteriorly and is coloured off-white with numerous dark brown and black spots and dashes. The ventrally-positioned mouth is flanked by a pair of highly papillose maxillary barbels from which the common name is derived. There are two pairs of mandibular barbels with short stubby branches. The humeral process is large and rounded. The dorsal and pectoral fins have a hardened serrated spine. The caudal fin is forked. Length reaches 175 mm SL.

This omnivorous squeaker feeds on detritus, seeds, plant material, algae, insects, crustaceans and snails. Spawning takes place coinciding with the summer rains and arrival of the annual flood, although it has been observed as early as July in the Kavango River of Namibia where it is relatively common. It is also found throughout the Kwando-Linyanti and Upper Zambezi-Chobe River systems into Angola and Zambia. The southern limit of its distribution is the Okavango Delta.

Finetooth Squeaker (*Synodontis vanderwaali*)

F.Jacobs

Etymology: ***Synodontis*** is an "ancient name for an undetermined fish from the Nile" according to Cuvier; ***vanderwaali***, in honour of Ben van der Waal, a pioneer fish biologist working in the Zambezi Region, who collected the type specimen

DS 1, DR 7; AS 0, AR v-vi, 7-9; LL 0, CP 0

Summary: The Finetooth Squeaker has a fusiform body coloured olive-brown, lighter below and covered with numerous dark small dots and dashes that may merge to form a reticulated pattern. The head and fins are lightly speckled. The upper jaw barbels are long and the lower jaw barbels have short stubby branches. The humeral process is broad with a blunt tip and rounded upper margin. The common name refers to the large number of small teeth on the mandible. It obtains 160 mm SL.

This squeaker prefers mainstream river environments and river fringes feeding on detritus, plant material, insects and snails. The Finetooth Squeaker is known from the Kunene, Kavango and Upper Zambezi-Chobe Rivers, but has not yet been collected from the Kwando-Linyanti River system, although it occurs in the Okavango Delta.

Sunset over the Chobe River, a time when the squeakers emerge (Photo: F. Jacobs).

Upper Zambezi Squeaker (*Synodontis woosnami*)

F.Jacobs

Etymology: ***Synodontis*** is an "ancient name for an undetermined fish from the Nile" according to Cuvier; ***woosnami***, in honour of Richard Bowen Woosnam (1880–1915), naturalist and Game Warden of the British East African Protectorate, who collected the type specimen

DS 1, DR 7; AS 0, AR v-vi, 7-9; LL 0, CP 0

Summary: The body of the Upper Zambezi Squeaker is elongated and triangular from a frontal view. It is dark brown on the back, lighter on the underside. The upper body, sides and fins are covered with numerous small black spots, diminishing ventrally; Kunene specimens may have larger spots to blotches. It has a strong bony head and the humeral process is without a distinct point on the end. There is a pair of maxillary barbels and two pairs of mandibular barbels. The maxillary barbels are long and straight without branches. The mandibular barbels have long slender branches. The dorsal and pectoral fins each have a strong serrated spine. The caudal fin is deeply forked. It can grow to 205 mm SL.

This squeaker inhabits perennial river channels and open-water lagoons. They are omnivores consuming aquatic insect larvae, algae, gastropods, bivalves, sponges, crustaceans and the eggs of other fishes. Spawning takes place during late-spring and summer with a peak coinciding with the rainy season and arrival of annual floodwaters. Pairs swim in unison during spawning.

The Upper Zambezi Squeaker is widely distributed, although patchily, in the Kunene, Cuvelai, Kavango, Kwando-Linyanti and Upper Zambezi-Chobe River systems of Namibia. It is found elsewhere in Angola, Zambia, Zimbabwe and the Okavango Delta.

Dried catfish sold along the road (Photo: B. van der Waal).

L. Family Schilbeidae (Butter Catfishes)

Schilbeidae is a family of catfishes native to Africa and Asia. These fish tend to swim in open water. They usually have dorsal and pectoral fins with a sharp spine and long anal fin. However, dorsal, adipose and pelvic fins may or may not be present. There are usually four pairs of barbels flanking the mouth. They are known to pair during spawning.

Silver Catfish (*Schilbe intermedius*)

B. van der Waal

Etymology: ***Schilbe***, ***schilbe***: schérifié, noble, perhaps for the superiority of its taste; ***intermedius***, in between

DS 2, DR 5-6; AS 0, AR iv-v, 47-62; LL 0, CP 0

Summary: This is a distinctive catfish with a deeply-compressed tapered body with a brownish back and head. Its flanks are silver to creamy white, from which its common name is derived. The Silver Catfish has a relatively small depressed head for its body size with large eyes positioned laterally. Four pairs of short barbels flank a large mouth. Dorsal and pectoral fins have sharply-pointed spines. The anal fin is long and the caudal fin is forked, drooping downwards. No adipose fin is present in Namibian populations, but occurs elsewhere.

This common and widespread shoaling fish swims in mid-water unlike other catfishes that live on or near the bottom in lagoons and slow-flowing river channels. It is a nocturnal voracious predator that feeds on fish including mormyrids, cyprinids and small cichlids, along with a wide range of insects, crustaceans, seeds and fruit.

Spawning occurs in late spring and summer coincident with the rainy season. Females lay their eggs on plants and do not guard them. Lifespan is estimated at 6 to 7 years. Size can reach 300 mm SL and 1.3 kg.

The Silver Catfish is widely distributed throughout the Kunene, Cuvelai, Kavango, Kwando-Linyanti and Upper Zambezi-Chobe River systems in Namibia and into the Okavango Delta. It is a near pan-African species found from the Nile River to the Phongola River in KwaZulu-Natal, South Africa.

This is an important species in subsistence and commercial fisheries with abundant high quality flesh of good flavour. It is frequently caught in small-mesh gillnets (24–60 mm), but not especially targeted by sport fishermen. It is also sold in the aquarium trade.

M. Family Anabantidae (Labyrinth fishes)

This family is found in tropical Africa and Asia and has accessory air-breathing organs (*labyrinths*) in chambers above the gills that allow them to extract oxygen from the air and live in water with low oxygen concentrations. Their air-breathing 'lungs' consist of a thin, folded plate that is well supplied with blood vessels where oxygen and carbon dioxide are exchanged.

These fishes have a long series of spines in the dorsal and, especially, the anal fins. The pelvic fins are well developed and far forward under the pectoral fins, which are high on the body. The caudal fin is rounded, not forked. The lateral line is split into two sections. All have ctenoid scales, i.e., scales with minute tooth-like projections (*ctenii*). Two described species of anabantids are known from Namibia (Manyspined Climbing Perch and Blackspot Climbing Perch), while an undescribed form exists in the Kavango and Kwando-Linyanti systems. Anabantids are typically ambush predators hiding among papyrus and reed mats, sunken wood and other bottom structures.

Manyspined Climbing Perch (*Ctenopoma multispine*)

D. Tweddle

Etymology: ***Ctenopoma***, ***ctenos***, comb + ***poma***, operculum, referring to serrated operculum; ***multispine***, ***multi-***, many + ***spinis***, spiny, presumably referring to numerous prickly rays on dorsal and anal fin

DS 16-19, DR 8-9; AS 8-9, AR 8-10; LL 30-32

Summary: This species has an elongate cryptically-coloured body, generally light brown with a series of zigzag dark brown or black bars along the body. The dorsal fin is long and brown. A unique feature of this species is the hind edge of the gill cover, which has numerous strong spines used for locomotion on rain-soaked ground and from which both the common and scientific names are derived. Mature males also have spines on the free edges of the cheek scales. The eyes are large and positioned far forward. They attain 140 mm SL.

The Manyspined Climbing Perch occurs in vegetated floodplains and backwater lagoons, although they also occur along the fringes of the river channels at low water levels when floodplains recede. Strangely, this fish is known to leave the water and move overland to other areas in rain-soaked weather possibly in search of food or mates. Local fishers along the Linyanti and Chobe Rivers call this "*Mbundu*"; its the "*Mbondu*" in Thimbukushu and "*Nvundo*" in Siyeyei as it is told that this fish falls from the sky during heavy rainfall events. The serrated gill cover is bent outward and used like a pulley for traction over muddy and soaked landscapes.

This well-camouflaged Climbing Perch preys on small insects, crustaceans, snails and fish. Adults gather in large groups to spawn at night in floodplains and backwater lagoons where the eggs are scattered in suitable places with no further parental care.

This uncommon species is found in the Kavango, Kwando-Linyanti and Upper Zambezi-Chobe River systems of Namibia. It has not been recorded from the Kunene River, Namibia, but is known upstream in Angola and from the canal leading to the Cuvelai. It is also known from Angola, the DCR, Malawi, Mozambique, Zambia and KwaZulu-Natal, South Africa.

Blackspot Climbing Perch (*Microctenopoma intermedium*)

B. van der Waal

Etymology: ***Microctenopoma***, ***micro-***, small, referring to small size compared to other anabantids; ***intermedium***, described as intermediate in form between two other ***Microctenopoma*** spp.

DS 15-16, DR 8-9; AS 8-9, AR 8-9; LL 25-30

Summary: The body of this species is slightly compressed, dark brown in colour with 7 to 13 dark vertical bars along the body and radiating from the eyes that are relatively large and positioned far forward. A black spot is clearly visible on the caudal peduncle from which its common name originates. The dorsal and anal fins are long and black with whitish margins. The head is pointed with a small terminal oblique mouth. It grows to 62 mm SL.

It occurs in dense aquatic vegetation along the shorelines of rivers, lakes, lagoons and channels of swamps and floodplains. Its body colouration blends with the surrounding vegetation and it preys on aquatic insects, other small organisms and *aufwuchs*. Spawning takes place in late-spring to summer coinciding with the arrival of annual floodwaters. Males build a bubble nest on the water surface under which spawning takes place; the eggs float and are stowed in the nest, to be guarded by the male.

The Blackspot Climbing Perch occurs in northern Namibia in the Kavango, Kwando-Linyanti and Upper Zambezi-Chobe River systems. It is also known from the northern Okavango Delta; Upper Zambezi and Kafue-Middle Zambezi Rivers, Zambia; the southern tributaries of the Congo River; and the St Lucia basin in KwaZulu-Natal.

N. Family Mastacembelidae (Spiny Eels)

The spiny eels are long slender fishes from tropical Africa and Asia. Their head is sharply pointed with a fleshy snout resembling an appendage and mouth positioned subterminally. They have tiny scales difficult to count; there is no lateral line. The dorsal, anal and caudal fins are all joined together around the body. They have no pelvic fins and the pectoral fins are set high on the body. The spiny eels are represented by two species in Namibia: the Longtail Spiny Eel and Ocellated Spiny Eel. Both are rare habitat specialists. Although they may be collected in the traditional fisheries, they in fact seem to be repulsive to fishers (C. Hocutt, pers. observation).

Longtail Spiny Eel (*Mastacembelus frenatus*)

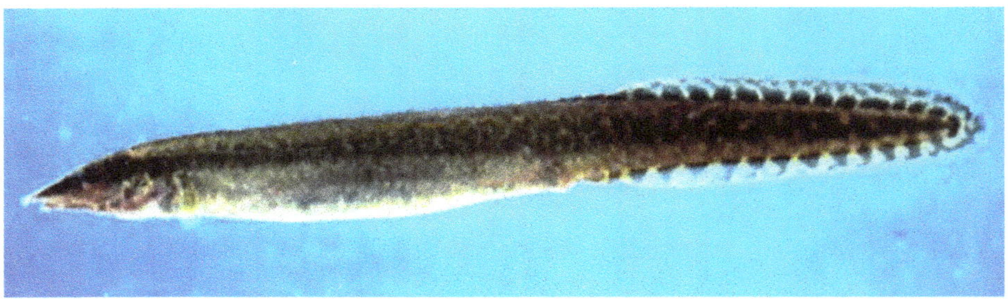

G. Merron

Etymology: ***Mastacembelus***, etymology unknown, but possibly Greek, ***mastax***, bite, mouthful + Greek, ***emballo***, to throw oneself into, on; ***frenatus***, bridled, referring to dark streak on each side of head, passing through eyes

DS 27-34, DR 69-81; AS 2, AR 66-83; LL 0, CP 0

Summary: The body shape is anguiliform (*eel like*) and coloured light to dark brown with a zigzag blotch pattern. The head and snout are elongated and the front nostril is shaped in the form of a fleshy snout appendage; the mouth is positioned subterminal under the snout. The dorsal fin is long with a series of separate short spines in front of it. The anal fin has two spines and numerous soft rays with spots merging to form *eyespots* (*ocelli*) on the caudal fin. They grow up 400 mm SL.

This habitat specialist inhabits dense papyrus and reed stands lining the Kavango, Kwando-Linyanti and Upper Zambezi-Chobe River systems and rapids, such as Popa Falls. It is also reported from the Okavango Delta; Upper Zambezi and Kafue-Middle Zambezi Rivers in Zambia; Congo River; and catchments of Lakes Victoria and Tanganyika.

They feed by waiting for prey items, including aquatic insects and small fish, to swim or drift within range before striking. Spawning is during the flood season in inundated areas among aquatic plants.

Ocellated Spiny Eel (*Mastacembelus vanderwaali*)

F. Jacobs

Etymology: ***Mastacembelus***, etymology unknown, but possibly Greek, ***mastax***, bite, mouthful + Greek, ***emballo***, to throw oneself into, on; ***vanderwaali***, in honour of zoologist Prof. Ben van der Waal, South Africa

DS 22-26, DR 64-75; AS 2, AR 64-79; LL 0, CP 0

Summary: The body shape of the Ocellated Spiny Eel is morphologically similar to that of the Longtail Spiny Eel, but is distinctively bolder in colouration appearing light brown to tan with short vertical stripes and mottling. The head and snout are elongated and the front nostril is shaped in the form of a fleshy snout appendage. The mouth is subterminal. It reaches 180 mm SL.

The type specimen of the Ocellated Spiny Eel was collected in rapids at Katima Mulilo in 1975 when PH Skelton was visiting BCW van der Waal; the species was later named by Skelton for van der Waal. It lives in the Upper Zambezi and Kavango River rapids of Namibia feeding on aquatic insects taken from and within rocky habitats. It is also known from the northern Okavango Delta and Middle Zambezi River in the Batoka Gorge below Victoria Falls where it spawns along the rocky shoreline during the low water period prior to the start of the November rains (J. Minshull, pers. observation). It is unimportant in all fisheries due to its rarity and habitat specialisation.

Rapid areas such as Popa Falls are important habitats for spiny eels and other rheophilic fish species (Photo: F. Jacobs).

O. Family Cichlidae (River Bream, Largemouth Bream and Tilapia)

The Family Cichlidae is both large and diverse. At least 2,000 species have been described, making it one of the largest vertebrate families with new species still being discovered.[20] Cichlids are native to tropical freshwater areas around the world. Between 70% to 80% of all species are African, appearing in great diversity in the major African lakes (e.g., Lake Victoria, Lake Tanganyika, Lake Malawi), where they are particularly well known for having evolved into many closely related, but morphologically diverse and behaviourally distinct species.

Cichlids are typically deep bodied and range from herbivorous to carnivorous representing all trophic levels in the fish community. They are noted for their complex mating and spawning behaviour. This usually involves courtship; preparation, maintenance and defence of the nest; and protection of the offspring. All species perform some form of parental care by either (a) laying eggs in nests and guarding them or (b) spawning in *bowers*, then mouthbrooding the eggs and fry. Mouthbrooding behaviour is common to many Namibian cichlids.

Cichlids support important local subsistence fisheries using a variety of gear, providing a source of protein to tens of thousands of people in Namibia. Certain species are suitable candidates for aquaculture and other larger cichlids, such as the *Serranochromis* genus, are popular sport fish. The cichlids also include many popular freshwater aquarium species kept by hobbyists. Despite these positive attributes, many cichlids introduced into waters outside of their natural range have become nuisances and threaten indigenous fish species through competition, interbreeding and introduction of diseases.

Southern African cichlids are traditionally divided into three main groups: plant- and detritus-eating (e.g., *Oreochromis* and *Tilapia*); mollusc and aquatic invertebrate consumers (e.g., *Sargochromis* spp.); and predators like the Nembwe (*Serranochromis jallae*).

Slender River Bream (*Chetia gracilis*)

B. van der Waal

Etymology: ***Chetia***, from ***Cheti***, local Malawian name for a yellow weaver bird; ***gracilis***, Latin, slender

DS 14-15, DR 10-13; AS 3, AR 7-10; LL 34

Summary: This slender-bodied species has a relatively large head and mouth and is coloured olive-brown dorsally, white ventrally, with 8–9 blackish vertical bars along its sides. The base of scales is dark red. Orange spots occur on the dorsal, caudal and anal fins of males. It grows to 150 mm SL.

This bream prefers vegetated lagoons, backwaters and impoundments. Adults mainly feed on insects and small fish. This species is a mouthbrooder with the female carrying the eggs and larvae. It is found in the Kunene River, Namibia into Angola and the Cubango drainage, Angola.

Angolan River Bream (*Chetia welwitschi*)

R. Tate

> **Etymology:** ***Chetia***, from ***Cheti***, local Malawian name for a yellow weaver-bird; ***welwitschi***, in honour of botanist F. M. J. Welwitsch (1806–1872) who collected the type specimen
>
> DS 14-15, DR 11-12; AS 3, AR 9; LL 31-34

Summary: The Angolan River Bream possess a relatively large mouth and slender body. Males have large eggspots on the anal fin.

Little is known about the habitat and biology of this species. It is found in the Kunene River, Namibia, extending upstream through the Cunene to the Cuanza River, Angola.

Redbreast Tilapia (*Coptodon rendalli*)

F. Jacobs

> **Etymology:** ***Coptodon***, ***copto-***, split or divided + ***odon***, tooth; ***rendalli***, honours British zoologist Percy Rendall (1861–1948), the collector of the type specimen
>
> DS 14-16, DR 12-13; AS 3, AR 9-10; LL 28-32

Summary: The Redbreast Tilapia has a deep compressed body with 5–6 dark vertical bars and a red throat and belly (from which its common name originates) fading to orange-yellow posteriorly onto a predominately grey bicolour caudal fin. The head and body are mid- to dark-olive dorsally. The dorsal fin has a red margin. Juveniles are silver. Lengths reach 400 mm TL and weigh up to 2 kg.

This species prefers quiet well-vegetated lagoons and backwaters and is not normally found in fast-flowing water. Adults mainly feed on plants, algae, insects and crustaceans. Juveniles feed on detritus and plankton.

Spawning occurs along vegetated fringes of river channels and lagoons from spring to late summer, coinciding with the onset of the rainy season and/or arrival of floodwaters. This species is a substrate spawner in shallow water near the fringe of aquatic vegetation. Nests are generally a shallow saucer-shaped depression in the substrate up to 1 m in diameter with usually 5–15 pits in the middle, each up to 30 cm deep in which eggs or larvae are kept and fanned by both adults and regularly moved to adjacent pits (see P. 159). They spawn up to 6–8 times a year. Fish form monogamous pairs and exhibit biparental guarding behaviour.

This species has a wide distribution in the Kunene, Cuvelai, Kavango, Kwando-Linyanti and Upper Zambezi-Chobe River systems and other waterbodies. It is found elsewhere in the Okavango Delta and throughout tropical Africa, including Botswana, Zambia, Malawi, Zimbabwe, Mozambique and South Africa. It is commonly caught in the commercial, subsistence and recreational fisheries with gillnets, traps, fishing baskets and angling.

Banded Jewelfish (*Hemichromis elongatus*)

F. Jacobs

Etymology: ***Hemichromis***, ***hemi-***, partial + ***chromis***, a fish, perhaps a perch; ***elongatus***, presumably referring to oblong/elongate shape of body

DS 13-15, DR 11-13; AS 3, AR 8-10; LL 28-32

Summary: This fish derives its common name from the row of 4–5 distinctive black blotches or spots that run laterally along the length of the body from the gill cover to the base of the caudal fin. The body is elongate, coloured greenish above with the abdomen and underside of the head red in spawning males which are larger and more colourful than females; males also develop elongated dorsal and anal fins. The mouth is terminal with oblique black and red spots on the operculum. Size reaches 190 mm SL.

This relatively rare nocturnal predator is found in moderate-flowing river channels and permanent lagoons with aquatic vegetation, where it preys on small fish, crustaceans and aquatic insects. Spawning occurs during summer coinciding with the onset of the rainy season and arrival of floodwaters. Males and females pair off to spawn on a sand-scraped *bower* that does not provide any protection, unlike a nest. The eggs are mouthbrooded and the parents guard the bower and larvae. As many as 500 eggs can be released by one female.

This species occurs throughout most northern perennial rivers and lagoons of Namibia, including the Kavango, Kwando-Linyanti and Upper Zambezi-Chobe River systems. It is also found in the Okavango Delta and Upper Zambezi, Zambia. It is caught in the subsistence fishery using traps, baskets, and hook and line.

Threespot Tilapia (*Oreochromis andersonii*)

B. van der Waal

Etymology: ***Oreochromis***, ***oreos***, mountains, referring to Mt. Kilimanjaro, locality of type species + Greek, ***chromis***, fish, perhaps a perch; ***andersonii***, in honour of Charles John Anderson (1827–1867), Swedish explorer of Namibia

DS 16-18, DR 11-14; AS 3, AR 11-13; LL 31-35

Summary: This is a deep-bodied cichlid with a rounded snout and small mouth. The body is silvery grey with three distinct mid-lateral black spots usually visible from an early age from which the common name is derived. The subterminal margins of the dorsal and anal fins are light green to brown with bright red lappets. The tops of heads of breeding males are often deep red. This species reaches 500 mm SL, 3.0+ kg.

The Threespot Tilapia is one of the most widely distributed large cichlid species inhabiting main river channels, small tributary channels, permanent lagoons and floodplains with aquatic vegetation cover. It is a *detritivore* feeding primarily on fine particulate matter, including algae, diatoms, detritus and zooplankton. Spawning coincides with the onset of the summer rainy season. Males excavate saucer-shaped bowers up to 150 cm diameter on a sandy substrate in water from 1-3 m deep. They often spawn in an arena setting with multiple bowers in close proximity to each other. Females mouthbrood eggs, larvae and fry. Multiple broods are raised during the warmer months with spawning typically in December and February.

This species is common throughout Namibia's northern rivers and lagoons, including the Kunene, Cuvelai, Kavango, Kwando-Linyanti and Upper Zambezi-Chobe River systems and the Okavango Delta. It is found elsewhere in rivers and swamps throughout these drainages into Angola and Zambia to the Middle Zambezi River below Victoria Falls. This is an important commercial, subsistence and sport fish.

Aerial view of tilapia nests in a Kavango River backwater (Photo: C. Hocutt).

Greenhead Tilapia (*Oreochromis macrochir*)

F. Jacobs

Etymology: ***Oreochromis***, ***oreos***, mountains, referring to Mt. Kilimanjaro, locality of type species + Greek, ***chromis***, fish, perhaps a perch; ***macrochir***, Greek, ***macros***, big + Greek, ***cheir***, hand, or pectoral fins in fishes

DS 15-17, DR 11-14; AS 3, AR 9-12; LL 31-32, CP 18-20

Summary: This is a deep-bodied and overall iridescent fish with a small terminal mouth. Adults are olive to green dorsally with dark green iridescent flecks along the head and on the gill cover giving a marbled appearance. Males have distinctive tasselled urogenital papillae. Juveniles are silvery with up to 10 vertical bars along their flanks. Lengths reach 430 mm SL, weights up to 2.6 kg.

The Greenhead Tilapia inhabits quiet waters along river margins and lagoons with aquatic vegetation. Similar to the Threespot Tilapia, it is a detritivore, but with a stronger diet of algae, diatoms and small organisms. Spawning occurs during summer, coinciding with the onset of the rainy season. Males construct and defend a bower in shallow waters, consisting of a central volcano-shaped mound with a slightly concave top which serves as the mating platform. Males court several females in succession and females may mate with more than one male in a season. Several bowers are often grouped into a spawning arena, which is called a *lek*. Females mouthbrood eggs and fry.

This fish is common throughout Namibia's northern perennial rivers and lagoons, including the Kunene, Cuvelai, Kavango, Kwando-Linyanti and Upper Zambezi-Chobe River systems into the Okavango Delta. Elsewhere, it is found in rivers and swamps throughout these drainages into Angola and Zambia, including Lake Kariba and southern tributaries of the Congo. It has been widely introduced in southern Africa. It is an important commercial, subsistence and sport fish.

Cunene Dwarf Bream (*Orthochromis machadoi*)

P. Bragança

Etymology: ***Orthochromis***, ***orthos***, straight + Greek, ***chromis***, a fish, perhaps a perch; ***machadoi***, in honour of António de Barros Machado (1912–2002), zoologist of the Musée de Dundo, Angola, who collected the type specimen

DS 15-17, DR 9-11; AS 3, AR 7-9; LL 29-33

Summary: The Cunene Dwarf Bream has a relatively long body with a small terminal mouth. The body has 10-12 dark vertical bars, with dark bars across the snout and across the gill covers. The caudal fin is rounded. Males turn red in spawning season. They grow to 65 mm SL.

This species probably spawns in summer, but little is known about its biology. It is also a mouthbrooder. It is endemic to the Kunene and Cuvelai River systems in Namibia and Angola.

Zambezi River Bream (*Pharyngochromis acuticeps*)

F. Jacobs

Etymology: ***Pharyngochromis***, ***pharyngo-***, presumably referring to enlarged pharyngeal teeth and/or stouter pharyngeal bone compared to genus ***Chetia*** + Greek, ***chromis***, a fish, perhaps a perch; ***acuticeps***, ***acutus***, sharp + ***-ceps***, head, referring to its "sharply pointed" head

DS 14-16, DR 10-13; AS 3, AR 7-10; LL 31-36

Summary: This is a smallish slender bream with a straight profile, large pointed head and terminal mouth. Its colourful body is bronze along the flanks with dark vertical bands and individual red spots occurring on each scale. The pelvic fins are well forward (below pectoral fins) and the anal fin is with light-orange eggspots. The long dorsal fin is tipped in orange. It obtains 220 mm SL.

This common cichlid occurs in a wide range of habitats, including slow-flowing river channels, lagoons and floodplains. It feeds primarily on snails, bivalves and insect larvae. Spawning occurs during summer coinciding with the onset of the rainy season. It is a mouthbrooding species and a female can carry 800 eggs and fry in her mouth. It is caught in the subsistence fishery using a variety of methods, including fishing baskets, traps and hook and line.

The Zambezi River Bream is widely distributed in northern Namibia from the Kavango River eastward though the Kwando-Linyanti and Upper Zambezi-Chobe Rivers, southward into the Okavango Delta. It is known elsewhere from the Kafue-Middle Zambezi River, Zambia, through the Lower Zambezi system, Mozambique, south to the Save-Runde River.

Southern Mouthbrooder (*Pseudocrenilabrus philander*)

F. Jacobs

Etymology: ***Pseudocrenilabrus***, Greek, ***pseudes***, false + Latin, ***crenulatus***, cut + Latin, ***labrum***, lip; ***philander***, ***philos***, friend or lover + ***andros***, of man or men, i.e., "friend of man"

DS 13-16, DR 9-11; AS 3, AR 7-9; LL 27-30

Summary: The Southern Mouthbrooder is a vibrant fish with a compressed body, small terminal mouth and rounded caudal fin. It is variable light blue and yellow in colour with a dark oblique bar through the eye, an iridescent blue lower jaw and an orange-tipped anal fin. Diffuse vertical bars can occur on the sides with scale margins being an iridescent blue creating faint parallel diagonal stripes. The dorsal, caudal and anal fins have light blue blotches, while the pelvic fins are dusky. The pectoral fins are short and rounded. It reaches 130 mm SL.

This small cichlid has a wide tolerance to dissolved oxygen, pH and other environmental parameters and is found in various habitat types, including slow-flowing rivers, tributaries, backwater lagoons and floodplains with dense vegetation. It feeds on insects, plant material and smaller fishes. Spawning occurs from early spring to late summer, coinciding with the onset of the rainy season. Males construct a simple cleared bower on the substrate and attract ripe females. The eggs are laid in the bower, fertilised by the male and mouthbrooded by the female when approached by predators. They are an important subsistence fish caught in traps and baskets.

They occur throughout all perennial rivers, lagoons and floodplains of Namibia, including the seasonal Cuvelai River. Elsewhere they are widely distributed throughout southern Africa including Angola, DRC, Malawi, Zambia, Zimbabwe, the Okavango Delta and South Africa.

Rainbow Bream (*Sargochromis carlottae*)

B. van der Waal

Etymology: ***Sargochromis***, Greek, ***sargos***, bream + ***chromis***, a fish, perhaps a perch; ***carlottae***, in honour of Mrs. Charlotte Sclater, wife of zoologist William Lutley Sclater (1863–1944), Director, South African Museum, who presented the type specimen to the British Museum

DS 14-16, DR 11-13; AS 3, AR 9-11; LL 28-32

Summary: The Rainbow Bream is the first species in the genus *Sargochromis* to be discussed in this book. This genus is sometimes commonly called the 'smallmouths' due to that feature, which influences their feeding habits. The *Sargochromis* are closely related to the genus *Serranochromis* that are highly predacious, commonly known as the 'largemouths'. All are mouthbrooders.

This species has a rounded body with a concave head profile and a rather uniform olive-green colour with the scales having dark borders yielding a speckled appearance. All fins are dark with the anal having light orange eggspots, the pelvics being relatively long and the caudal rounded. It grows to 340 mm SL, 1 kg.

The Rainbow Bream inhabits slow-flowing river channels, floodplains and lagoons with aquatic vegetation. They feed on molluscs, crustaceans, snails and aquatic insects. Spawning may occur multiple times from early spring to late summer in sync with the rainy season and flooding regime. Males excavate a small sand nest where the eggs are laid. Females mouthbrood eggs and fry. They are caught occasionally in the commercial, subsistence and sport fisheries.

This bream is distributed throughout the Kavango, Kwando-Linyanti and Upper Zambezi-Chobe Rivers of Namibia, but is absent from the Kunene and Cuvelai systems. It also occurs in the Okavango Delta and Middle Zambezi-Kafue system through Lake Kariba, Zambia and Zimbabwe.

Dusky Bream (*Sargochromis codringtoni*)

F. Jacobs

Etymology: ***Sargochromis***, Greek, ***sargos***, bream + ***chromis***, a fish, perhaps a perch; ***codringtoni***, in honour of Thomas Codrington (1829–1918) who presented a collection of fishes from (now) Zambia to the British Museum

DS 13-16, DR 12-15; AS 3, AR 8-10; LL 34-35

Summary: This uncommon bream has a rounded body, steep straight profile head, moderately upturned mouth and olive-green body, darker dorsally, from which its common name is derived. It obtains 290 mm SL, 2.2 kg.

The Dusky Bream inhabits slow-flowing river channels and floodplain lagoons in vegetated areas. Aquatic plants and seeds, insects, molluscs and small fish are consumed. Multiple spawning can occur during the rainy season. Females mouthbrood eggs and fry. It is a component of the subsistence fishery and a good sport fish.

The Dusky Bream has a limited distribution in the Kavango, Kwando-Linyanti and Middle Zambezi-Chobe River systems of Namibia, extending to the Okavango Delta and the Upper Zambezi-Kafue system, Zambia.

Cunene Bream (*Sargochromis coulteri*)

P. Bragança

Etymology: ***Sargochromis***, Greek, ***sargos***, bream + ***chromis***, a fish, perhaps a perch; ***coulteri***, in honour of George W. Coulter, formerly Senior Fisheries Research Officer, Zambia

DS 14-16, DR 11-13; AS 3, AR 8-10; LL 30-31

Summary: This is a deep-bodied bream with a large head and upturned mouth, coloured olive green with red spots at the base of scales and on all fins except the pelvics which are dusky. Male anal fins have distinct eggspots. It reaches 216 mm SL. Little is known about the biology of this species. It is endemic to the Kunene and Cuvelai Rivers, Namibia and Angola.

Pink Bream (*Sargochromis giardi*)

B. van der Waal

Etymology: ***Sargochromis***, Greek, ***sargos***, bream + ***chromis***, a fish, perhaps a perch; ***giardi***, in honour of zoologist Alfred Mathieu Giard (1846–1908)

DS 14-16, DR 12-15; AS 3, AR 9-11; LL 29-34

Summary: The Pink Bream is a robust bodied fish with a large head having a steep convex profile and upturned oblique mouth. It is greyish-green on its back and head with flanks and abdomen a light red blush to creamy yellow. Its fins are dark grey with red margins and numerous dark red spots; the anal fin has rows of bright yellow-orange eggspots. Adults reach 500 mm SL, 3 kg.

This species typically inhabits floodplains where molluscs are abundant, although it may be found in river channels and large lagoons with sandy bottoms. It feeds primarily on snails, bivalves and insect larvae. Multiple spawning can occur from early spring to late summer coinciding with the rainy season. Females mouthbrood eggs and fry.

It is distributed throughout the Kavango, Kwando-Linyanti and Upper Zambezi-Chobe River systems in Namibia. It occurs elsewhere in the Kafue and Middle Zambezi Rivers and their tributaries, Zambia and Zimbabwe. They are caught in the commercial, subsistence and sport fisheries.

Green Bream (*Sargochromis greenwoodi*)

B. van der Waal

Etymology: ***Sargochromis***, Greek, ***sargos***, bream + ***chromis***, a fish, perhaps a perch; ***greenwoodi***, in honour of Peter Humphry Greenwood (1927–1995), Curator of the Fish Section of the British Museum of Natural History

DS 14-16, DR 12-14; AS 3, AR 8-10; LL 29-34

Summary: This cichlid has a rather shallow body with a long straight head profile. The mouth is terminal, moderately upturned. It is coloured green to olive green dorsally, white ventrally with two black lateral bands and about eight faint vertical bars. Length reaches 320 mm SL.

The Green Bream is an uncommon species found in slow-flowing water with dense aquatic vegetation. It feeds primarily on insects, small snails, other invertebrates and occasionally small fish. Spawning occurs from early spring to late summer, coinciding with the onset of the rainy season and arrival of floodwaters. It may spawn multiple times in a season. Females mouthbrood the eggs and fry. It is caught in the subsistence fishery, although not common.

This is a rare species in the Kavango, Kwando-Linyanti and Upper Zambezi-Chobe River systems of Namibia. It is known elsewhere from the Okavango Delta and Upper Zambezi and Kafue-Middle Zambezi Rivers, Zambia.

Humpback Largemouth (*Serranochromis altus*)

F. Jacobs

Etymology: ***Serranochromis***, Latin, ***serra***, saw + ***chromis***, a fish, perhaps a perch; ***altus***, high, referring to its deep body and high dorsal profile

DS 16-17, DR 15-18; AS 3, AR 11-13; LL 37-39

Summary: This deep bodied olive-brown cichlid has a distinct humpback with a straight profile from the snout to above the eye, then a sharp bend from the eye to the dorsal fin. Its common name is derived from this feature. There is a large upturned protrusible mouth. Scales are flecked in black giving a speckled appearance and a black spot occurs at the tip of the opercle. The belly is white. The pectoral and pelvic fins are dusky, the latter more so. Dorsal and caudal fins have numerous dark spots, while the anal fin is with distinct orange eggspots. It grows to 560 mm TL and 3.9 kg.

This is a relatively large, crepuscular, predatory fish that inhabits river channels and lagoons with good water clarity. It grows larger and seems to prefer heavily vegetated habitats near moderate currents as opposed to the closely related Thinface Largemouth most often found in quiet backwaters and lagoons. It is widespread, but not abundant. It feeds primarily on fishes such as mormyrids, Silver Catfish and certain barbs active at dusk and dawn. Spawning precedes the rainy season. Females mouthbrood the eggs and fry. It is an important commercial, subsistence and sport fish.

It is distributed in the Kunene, Kavango, Kwando-Linyanti and Upper Zambezi-Chobe River systems of Namibia. It is known from the headwaters of these drainages into Angola and Zambia to the Okavango Delta.

Thinface Largemouth (*Serranochromis angusticeps*)

D. Tweddle

G. Merron

Etymology: ***Serranochromis***, Latin, ***serra,*** saw + ***chromis***, a fish, perhaps a perch; ***angusticeps***, ***angustus***, narrow + ***–ceps***, head, referring to its "very strongly compressed" head (i.e., narrow between the eyes)

DS 13-16, DR 14-17; AS 3, AR 11-13; LL 36-39

Summary: This cichlid has a compressed body and head, the latter lending its common name. The head is large with upturned protrusible jaws and is coloured yellow in breeding males and covered with red spots as is the body, which is distinctive from the Humpback Largemouth. The dorsal and caudal fins are speckled with dark spots and the anal with light orange-yellow eggspots. Size reaches 410 mm TL, 2.5 kg.

The Thinface Largemouth is one of many large and widely distributed predatory fish in well vegetated habitats in the Kunene, Kavango, Kwando-Linyanti and Upper Zambezi-Chobe River systems of Namibia; it is recorded from the Calueque-Oshakati Canal, thus is considered present in the Cuvelai drainage. It is also found throughout the Okavango Delta, Middle Zambezi-Kafue Rivers and Luapula-Mweru system, Zambia and the DRC. It is an ambush predator feeding primarily on small fish, crustaceans and insects.

Multiple spawning may occur from spring to late summer throughout the rainy season. Females mouthbrood the eggs and fry after which juveniles move to shallow grassy areas in floodplains and backwaters seeking refuge until floodwaters recede and force them back to the main river channels and permanent lagoons. It is an important commercial, subsistence and sport fish.

Nembwe (*Serranochromis jallae*)

E. Spangenberg

Etymology: *Serranochromis*, Latin, ***serra***, saw + ***chromis***, a fish, perhaps a perch; ***jallae*** = in honour of Rev. Luigi (Louis) David Jalla (1860-1943), Paris Missionary Society, who collected type

DS 15-17, DR 14-16; AS 3, AR 10-11; LL 36-39

Summary: The Nembwe (an indigenous name) is a deep-bodied robust fish, olive- to bright-green dorsally with a dark-olive lateral band along the body fading to a yellow-green belly. They possess a large head and mouth which is slightly inclined. The anal fin of males has small orange eggspots and the olive-coloured dorsal and caudal fins exhibit a red (Kavango) or yellow (Upper Zambezi) margin. Length reaches 560 mm SL, weight 6.1 kg.

This large predator may be found in sheltered riverine areas such as undercut banks, coves and lagoons as well as rocky rapids. Smaller individuals prefer shallow water, inhabiting vegetation-fringed tributaries, floodplain channels and backwater lagoons.

Spawning occurs prior to annual flooding in bowers, simple saucer-shaped depressions of ca. 30-cm diameter that are constructed on sandy substrates among vegetation. Maturity of Nembwe is estimated at approximately 3 years with the average number of mature oocytes per female being 1,165. Females mouthbrood eggs and fry with juveniles moving onto flooded inundated areas that serve as refugia and foraging areas. They prey on a variety of fish, crabs, molluscs and insects. This is an important subsistence species and popular sport fish.

Nembwe were formerly considered as a subspecies, *S. robustus jallae*, but have recently been reclassified as *S. jallae* separate from *S. robustus*. Nembwe are found throughout the Kavango, Kwando-Linyanti and Upper Zambezi-Chobe Rivers in Namibia to the Okavango Delta. They also occur in the Upper and Middle Zambezi River, including the Kafue River, Lake Kariba, Zambia/ Zimbabwe and the Zambian Congo system. The sister species, *S. robustus,* is found from the vicinity of Lake Malawi, Shire River, through Mozambique; it was introduced into the Sand River Dam in Swaziland and has subsequently spread into the Komati River and surrounding coastal streams, South Africa and Mozambique.

Longfin Largemouth (*Serranochromis longimanus*)

B. van der Waal

Etymology: ***Serranochromis***, Latin, ***serra***, saw + ***chromis***, a fish, perhaps a perch; ***longimanus***, ***longus***, long + ***manus***, hand, referring to pectoral fins

DS 14-15, DR 13-14; AS 3, AR 9-10; LL 35

Summary: This compressed largemouth has dark to black blotches on the head and sides, coalescing on the latter to form up to 5–6 bars. The profile from the dorsal fin to the snout is straight and pointed. Both the scientific and common names refer to the noticeably long pectoral fins that are black in colour as are the pelvic fins. Males have dark red eggspots on the anal fin. Size reaches 350 mm TL, 1.5 kg.

This is a relatively rare, nocturnal, predatory fish found only in habitats with little or no current such as backwater lagoons. It feeds on small fishes and aquatic insects. Spawning coincides with the rainy season and arrival of floodwaters. It may spawn multiple times in a season. Females mouthbrood the eggs and fry after which juveniles seek refuge in shallow grassy areas in permanent floodplains, backwaters and lagoons. They are rarely caught in the commercial, subsistence or sport fisheries.

The Longfin Largemouth occurs Namibia in floodplains and lagoons of the Kwando-Linyanti and Upper Zambezi-Chobe Rivers. It is also known from the Okavango Delta and Upper Zambezi floodplain, Zambia west of Mongu.

Purpleface Largemouth (*Serranochromis macrocephalus*)

P. Bragança

Etymology: ***Serranochromis***, Latin, ***serra***, saw + ***chromis***, a fish, perhaps a perch; ***macrocephalus***, ***macro-***, long or large + ***cephalus***, referring to large head

DS 14-16, DR 13-15; AS 3, AR 9-11; LL 34-37

Summary: The Purpleface Largemouth has a deeply compressed body with a large straight-profile head and slightly upturned terminal mouth. It is one of the most colourful largemouths, especially during the spawning season. Colour is generally olive-brown with grey-black vertical body bars with males having distinct eggspots on the anal fin and a deep purple hue over their head and cheeks during breeding from which the common name originates. There is a red margin to the dorsal and caudal fins. It can be recognised by the long sharp teeth in both jaws, more prominent than any other largemouth species. It reaches 350 mm SL, 1.5 kg.

This predatory cichlid is common in habitats from papyrus-lined river margins to open water lagoons with abundant aquatic vegetation. They prey on small fishes and insects. Spawning occurs prior to annual flooding in early spring. Females mouthbrood the eggs and fry after which juveniles move off to shallow grassy areas in floodplains and backwaters seeking refuge until floodwaters recede and force them back to the main river channels and permanent lagoons. It is an important commercial, subsistence and sport fish.

They are distributed in all of Namibia's northern rivers, including the seasonal Cuvelai. They may occur in mainstream habitats and/or adjacent coves and lagoons. They are also found throughout the Okavango Delta, Middle Zambezi-Kafue systems, Zambia and Lower Zambezi to Lake Kariba.

Brownspot Largemouth (*Serranochromis thumbergi*)

B. van der Waal

Etymology: ***Serranochromis***, Latin, ***serra***, saw + ***chromis***, a fish, perhaps a perch; ***thumbergi***, in honour of Swedish naturalist Carl Peter Thumberg (1743–1828), regarded as the founder of South African botany

DS 17-18, DR 13-16; AS 3, AR 9-12; LL 39-41

Summary: The Brownspot Largemouth has a comparatively long slender body with reference to other largemouths. Its dorsum is green-brown above, brown laterally with an off-white belly. Its mouth is large and terminal, slightly upturned indicative of its predacious habits. A black lateral band occurs along the body; scales have brown-red flecks giving it a speckled appearance and its common name. The dorsal and caudal fins have a light-yellow margin, the former more distinct. Males have orange eggspots on the dusky anal fin. Size reaches 350 mm TL, weight 1.2 kg.

This piscivore inhabits slow-flowing river channels and lagoons with abundant aquatic vegetation. It preys mainly on small fishes, but large insect nymphs such as dragonflies and other aquatic insects also form its diet. Spawning occurs along vegetated fringes of river channels and lagoons from spring to late summer, coinciding with the onset of the rainy season and/or arrival of floodwaters. Males construct a sand-scraped bower and females mouthbrood eggs and fry. It is uncommon in the commercial, subsistence and sport fisheries.

The Brownspot Largemouth is patchily distributed throughout the Kunene, Kavango, Kwando-Linyanti and Upper Zambezi-Chobe River systems and floodplains of Namibia. It is recorded from the Calueque-Oshakati Canal, thus is considered present in the Cuvelai drainage. Elsewhere, it is found in the Okavango Delta; the Kafue-Middle Zambezi and Luangwa systems, Zambia; upper Congo, Angola and DRC.

Thicklipped River Bream (*Thoracochromis albolabris*)

P. Bragança

Etymology: ***Thoracochromis***, ***thorax***, breastplate, referring to "small and clearly size-demarcated scales" on thoracic region + ***chromis***, a fish, perhaps a perch; ***albolabris***, ***albus***, white + ***labrum***, lip, allusion not explained or evident

DS 14-16, DR 10-12; AS 3, AR 7-9; LL 32-35

Summary: This is a relatively slender fish with pointed head and a terminal horizontal mouth with thick lips from which its common name stems. It is coloured light yellow with orange on the scales; males have blue cheeks. There is a light-blue lateral band along the mid-body. The lateral line is divided into two; moderately-ctenoid scales occur below the upper lateral line, and cycloid to weakly-ctenoid scales occur on the caudal peduncle and above the upper lateral line. The anal fin is yellowish with orange spots. Dark orange tips occur on the dorsal fin spines and the fin has a bluish submarginal band. The pelvic fins are grey. It grows to 200 mm TL.

This species prefers rocky habitats. It is endemic to the Kunene and Cuvelai Rivers, Namibia and Angola.

Namib River Bream (*Thoracochromis buysi*)

P. Bragança

Etymology: ***Thoracochromis***, ***thorax***, breastplate, referring to "small and clearly size-demarcated scales" on thoracic region + ***chromis***, a fish, perhaps a perch; ***buysi***, in honour of Peter J. Buys, Curator of the National Museum of Namibia, who collected the type specimen

DS 14-17, DR 10-13; AS 3, AR 7-9; LL 32-36

Summary: This relatively deep-bodied bream has large eyes and a small terminal mouth with thick lips, blue in colour as is the snout. Males are green-yellow along the sides. The lateral line is divided into two; ctenoid scales occur below the level of the lower lateral line, and cycloid above it and on the chest. Dark orange tips occur on the dorsal-fin spines and yellow eggspots are on the anal fin. The dorsal fin has a white sub-marginal band. The pelvic fin is grey. It reaches 150 mm TL.

This species occurs in a wide range of habitats and has a diet consisting mainly of algae and aquatic insects. It is endemic to the Kunene and Cuvelai Rivers, Namibia and Angola, being known from the Calueque-Oshakati Canal.

Otjikoto Tilapia (*Tilapia guinasana*)

F. Jacobs

Etymology: ***Tilapia***, Latin for ***thlapi***, the Tswana word for "fish"; ***guinasana***, ***-ana***, belonging to: Lake Guinas, Namibia

DS 12-14, DR 10-11; AS 3, AR 8-10; LL 27-28

Summary: This small cichlid has variable colour patterns ranging from pure black to mottled black and blue to pink and white. This adaptation in colour patterns is unique among tilapias and termed *polychromism*, i.e., multicoloured (see P. 209). Young fish are olive green with 8–9 vertical dark bars and three faint horizontal bars as well as a darker spot at the base of the rayed portion of the dorsal fin (typical of *Tilapia*). Lengths reach 140 mm SL.

The Otjikoto Tilapia is a critically endangered species originally endemic to Lake Guinas near Tsumeb in north-central Namibia. The lake was formed by a subterranean cave collapsing in on itself and is over 100 m deep. The species has been introduced into nearby Lake Otjikoto and other reservoirs in Namibia in an attempt to improve its chances of survival, which is threatened by (a) the depletion of local groundwater resources for agriculture and mining, and (b) the potential impact of introduced fish species. It has also been spread through the aquarium fish trade. Lakes Guinas and Otjikoto are open sinkholes containing very clear water. The common name founded on Lake Otjikoto is enigmatic given that the scientific name in 1936 was based upon the type locality, Lake Guinas.

These fish inhabit the rocky dolomite walls of Lakes Guinas and Otjikoto. The main food is algae and diatoms grazed from the walls. It is a substratum spawner closely related to the Banded Tilapia. Once the spawning season commences, which extends from November to May, males and females leave the shoals and form a tight pair-bond. They develop dark spawning colouration on the lower portion of the throat, abdomen and anal fin. The spawning site consists of ledges on the vertical dolomite sinkhole walls. Females lay their eggs which are fertilised by the males and both parents tend and aggressively guard the nests, alternating to forage while one remains to protect the young. The young fish grow fast and reach sexual maturity within a year.

Okavango Tilapia (*Tilapia ruweti*)

B. van der Waal

Etymology: ***Tilapia***, Latin for ***thlapi***, the Tswana word for "fish"; ***ruweti***, in honour of ethologist Jean Claude C. Ruwet (d. 2007), University of Liège, who donated the type specimen

DS 14-15, DR 10-12; AS 3, AR 8-9; LL 27-29

Summary: Okavango Tilapia have an olive- or iridescent-green body with 8–9 olive-coloured vertical bars and an off-white belly. Their head is small with a small terminal slightly-inclined mouth. They reach 120 mm TL.

This uncommon species is a habitat specialist that occurs in backwater lagoons, floodplain edges and well vegetated margins of slow-flowing channels. It feeds on detritus, soft plants and aquatic invertebrates. Males establish a territory, construct a saucer-shaped nest and then attract a ripe female for eggs to be laid and fertilised. Nests are tended mainly by the female while the male guards the territory. Eggs hatch in 3–4 days and the fry are free-swimming shortly thereafter.

It is known from the Kavango, Kwando-Linyanti and Upper Zambezi-Chobe River systems of Namibia. It is found elsewhere in the Okavango Delta; the Upper Zambezi River, Zambia; and southern tributaries of the Congo River, DRC.

Banded Tilapia (*Tilapia sparrmanii*)

F. Jacobs

Etymology: ***Tilapia***, Latin for ***thlapi***, the Tswana word for "fish"; ***sparrmanii***, in honour of Swedish naturalist Anders Sparrman (1748–1820), who explored the Cape area of South Africa in 1775

DS 13-15, DR 9-11; AS 3, AR 9-10; LL 27-29

Summary: The Banded Tilapia is a small deep-bodied cichlid, olive-green in colour with 8–9 vertical bars along its flanks and two bars between the eyes. A distinct "tilapia spot" is present at the distal end of the dorsal fin. Dorsal and caudal fins have a thin bright red margin. Scales and dorsal, caudal and anal fins are flecked with light blue spots. Spawning pairs have a dark throat, abdomen and anal fin. Length reaches 240 mm TL with weight about 0.5 kg.

This is a widespread and adaptable omnivore. It prefers shallow sheltered waters where plant cover exists in river channels, lagoons and floodplains. It rarely occurs in open water or large lagoons. Adults feed on filamentous algae, aquatic plants and vegetable matter of terrestrial origin (leaves, plants, etc.). Juveniles feed on small crustaceans and insect larvae.

Spawning occurs along vegetated fringes of river channels and lagoons from spring to late summer, coinciding with the onset of the rainy season and/or arrival of floodwaters. Males construct a nest on the bottom where eggs are laid, fertilised and guarded by both parents. Fry are free-swimming after 7–8 days, but remain in a shoal guarded by parents for several weeks. It is commonly caught in the subsistence fishery with traps, baskets, and hook and line.

This species has a circum-Namibian distribution in all the perennial rivers: Kunene, Kavango, Kwando-Linyanti, Upper Zambezi-Chobe and Orange-Fish, as well as the seasonal Cuvelai River. It is widely distributed elsewhere and found in the Okavango Delta; Middle and Lower Zambezi drainages into Zambia, Zimbabwe and Malawi; and southward to KwaZulu-Natal, South Africa.

Typical nests of Redbreast Tilapia at the edge of a floodplain, Kwando River (Photo: B. van der Waal).

P. Family Nothobranchidae (Killifishes)

The nothobranchid killifish are endemic to Africa with perhaps 300 species occurring continent wide. They live a precarious life cycle of only 3–12 months depending upon the species, e.g., the Turquoise Killifish (*Nothobranchius furzeri*) of Zimbabwe and Mozambique has the shortest lifespan recorded for a captive vertebrate, with a median lifespan of only nine weeks and a maximum of 11 weeks. During their life cycle, eggs hatch that were laid the previous year(s) in pools that are destined to dry up after the rainy season. Rapid growth, sexual maturity and spawning all must occur prior to their habitat desiccating completely and all adults die. The desiccation-resistant eggs remain inactive in the substrate until the next rainy season. Due to the narrow limits of their life cycle and environmental requirements, killifish are very susceptible to habitat alterations and pesticides.

The males, particularly, are very colourful fish with red, white, turquoise, orange and black being dominant pigmentation. Males have elongated fins used in display and are larger than females. These fish are very popular with tropical fish hobbyists.

Caprivi Killifish (*Nothobranchius capriviensis*)

B. van der Waal

Etymology: ***Nothobranchius***, Greek, ***nothos***, false + ***branchia***, gill; ***capriviensis***, ***-ensis***, suffix noting place of type locality, Caprivi (now Zambezi) Region, named after the German Count Leo von Caprivi (1831–1899)

DS 0, DR 16-17; AS 0, AR 16-18; LL 29-31, CP 14-16

Summary: The Caprivi Killifish is a small annual fish originally recorded from the Chobe River floodplain by Dr. Ben van der Waal. It is endemic to Namibia and listed as an endangered species. It inhabits the seasonal pans and temporary pools in the Mopane (*Colophospermum mopane*) forest of the eastern Zambezi Region, Namibia, between the Upper Zambezi and Chobe Rivers, and just opposite the range of the Kafue Killifish (*N. kafuensis*) in Zambia to which it is closely related.

The body is robust and iridescent blue in colour with irregular red vertical bars. The mouth is superior with the lower jaw projecting beyond the upper jaw. The dorsal and anal fins are set far back, and almost equal in size. The caudal fin is large and rounded. The dorsal, anal and caudal fins feature narrow blue and red bands. This species reaches 50 mm SL.

This species appears to be a rapid coloniser dispersing during periods of heavy rainfall and flooding, seeking new habitats that are then isolated as pans when water levels subside. The preferred habitats typically have clay bottoms, which allow these pans and temporary pools to retain water for extended periods. They feed on aquatic invertebrates and surface insects.

Males display to attract females and pair off. Spawning occurs daily for an extended period, a few eggs being laid at a time. The fish grow to maturity within a few weeks. When the habitats dry up, the adult fish die and the eggs survive encased in muddy clay during the dry season.

Local boy with day's catch (F. Jacobs).

Thinface Largemouth caught in gillnets (G. Merron).

Q. Family Procatopodidae (African Lampeyes)

The family Procatopodidae is endemic to Africa with five species found in Namibia. Fishes of this family are popularly known as African Lampeyes due to the presence of a characteristic iridescent reflective colouration in the dorsal region of the eye. These species are typically small and elongate or plump with the dorsal and anal fins set far back on the body. The caudal fin is rounded. They lack lateral line scales *per se*, with pored scales occurring only on the head region. Their mouths are upturned with a projecting lower jaw; they feed on small invertebrates, such as mosquito larvae and daphnia. They are non-migratory serial (*multiple*) spawners laying eggs that attach to aquatic vegetation. Eggs hatch within 2–3 weeks and individuals mature within six months. Most recently, the topminnows were reclassified from the genus *Micropanchax* to *Lacustricola.*

Meshscaled Topminnow (*Lacustricola chobensis*)

B. van der Waal

Etymology: Latin, ***lacus-***, lake + ***cola***, inhabit; ***-ensis***, suffix denoting place, Chobe River type locality

DS 0, DR 6-8; AS 0, AR 10-13; 22-24 in lateral series, CP 10

Summary: The short plump body is brownish-yellow in colouration with light iridescent blue reflections from scales. Bars occur on the dorsal, anal and caudal fins. The sooty pigment on edges of the scales gives the appearance of a reticulated mesh effect, hence the common name. Chromatophores form a conspicuous line along the whole mid-body flank. The bottom (basal) sections of the dorsal, anal and caudal fins have dark brown spots. This species reaches 35 mm TL.

The Meshscaled Topminnow is a fairly common species found in floodplains, marshes and slow-flowing reaches of the Kavango, Kwando-Linyanti and Upper Zambezi-Chobe River systems in Namibia, southward to the Okavango Delta. *Disjunct* populations are found in the Congo and Lower Zambezi drainages. It is caught throughout its range by subsistence fishers.

Johnston's Topminnow (*Lacustricola johnstoni*)

F. Jacobs

Etymology: Latin, ***lacus-***, lake + ***cola***, inhabit; ***johnstoni***, original description, Mangochi (former Fort Johnston), Malawi

DS 0, DR 7-9; AS 0, AR 12-15; 27-29 in lateral series, CP 10

Summary: Johnston's Topminnow has a slender elongate body, coloured yellow-olive with silvery blue iridescence on the scales that appear as a mid-lateral band from the head to the rounded caudal fin. The dorsal fin is set far back on the body. The mouth is superior with a projecting lower jaw. Johnston's Topminnow lacks (a) a zigzag pattern along its flanks and (b) bars in the caudal, dorsal and anal fins, distinguishing it from other topminnows. Females have orange-coloured dorsal, anal and caudal fins, such pigmentation absent in other Namibian topminnows. Males have a long rounded dorsal fin that when extended nearly reaches the caudal fin base. This species grows to 85 mm TL.

Johnston's Topminnow is common throughout the Kavango, Kwando-Linyanti and Upper Zambezi-Chobe River systems in Namibia, but uncommon in the Kunene and Cuvelai Rivers. It is also found in the Okavango Delta and in various lakes and river drainages from the DRC to the Lower Zambezi drainage of Malawi and Mozambique.

It occupies a variety of vegetated habitats including floodplains, swamps, lagoons and slow-flowing river channels. It often schools among aquatic vegetation feeding on small invertebrates, such as mosquito larvae and daphnia. This species is a serial *(multiple)* spawner over an extended late-spring to summer period; the females lay eggs on aquatic vegetation. The eggs hatch after 15–20 days and mature within six months when food is abundant. It is caught by subsistence fishers.

Striped Topminnow (*Lacustricola katangae*)

F. Jacobs

Etymology: Latin, ***lacus-***, lake + ***cola***, inhabit; ***katangae***, Katanga, Zaire (now Democratic Republic of the Congo), type locality

DS 0, DSR 8-10; AS 0, ASR 14-15; 25-28 in lateral series

Summary: The body is plump and light brown above with a silvery belly. A distinct blackish zigzag stripe runs along the mid-body from the snout to the rounded caudal fin. This species grows to 50 mm SL.

The Striped Topminnow is found in Namibia along the margins of the Kunene, Kavango, Kwando-Linyanti and Upper Zambezi-Chobe River systems in dense aquatic vegetation and root balls. It is not a floodplain or backwater lagoon species. It is also found in the northern Okavango Delta and in various lakes and rivers from the DRC to KwaZulu-Natal, South Africa. It is uncommon in the subsistence fishery.

Pygmy Topminnow *(Lacustricola pygmaeus)*

R. Bills

Etymology: Latin, ***lacus-***, lake + ***cola***, inhabit; from Latin ***pygmaei***, an allusion to its miniature size, the smallest species within the "***L. hutereaui*** complex"

DS 0, DSR 10-12; AS 0, ASR 12-15; 20-22 scales in lateral series, CP 10

Summary: The body is plump and overall translucent, coloured light-brown dorsally and white ventrally with iridescent blue-green scales on the back of the head and front flanks of the body to the caudal peduncle. The side of the head is predominantly bluish silver, dorsally dark brown. Males possess banded dorsal, anal and caudal fins, the latter having 3–5 bands and coloured light yellow. It reaches 20.5 mm SL.

It occurs in grassy patches in floodplains, shallow swampy areas and permanent *dambos*, the latter being drainage channels in areas of clay soils that are impermeable resulting in wetland conditions. It does not occur in riverine habitats.

The Pygmy Topminnow has been recorded from the Kavango, Kwando-Linyanti and Upper Zambezi-Chobe basins in Namibia. It has been found in the Bukalo Channel in Namibia that connects the Upper Zambezi-Chobe River to Lake Liambezi in high flood years and is reduced to isolated swampy pools at low water levels. It is broadly distributed in the upper Cubango, Cuito and Cuando Rivers, Angola; Upper Zambezi River, Zambia; and south into the Okavango Delta.

III. References

[1]Jubb, R.A. 1967. The freshwater fishes of southern Africa. AA Balkema, Cape Town: 248 pp.

[2]Poll, M. 1967. Contribution à la Faune Ichthyologique de l'Angola. Publicações Culturais 75. Companhia dos Diamentes de Angola (DIAMANG), Lisbon: 381 pp.

[3]Skelton, P.H. 2001. A complete guide to the freshwater fishes of southern Africa. Struik Publ., Cape Town: 395 pp.

[4]Skelton, P.H. 2019. The freshwater fishes of Angola, pp. 207–242. In: B.J. Huntley, V. Russo, F. Lages and N. Ferrand (Eds.). Biodiversity of Angola, Science & Conservation: A Modern Synthesis. Springer Nature, Switzerland.

[5]Bruton, M.N., G.S. Merron and P. H. Skelton. 2018. Fishes of the Okavango Delta and Chobe River. Penguin Random House, South Africa: 120 pp.

[6]Eccles, D.H., D. Tweddle and P.H. Skelton. 2011. Eight new species in the dwarf catfish genus *Zaireichthys* (Siluriformes: Amphiliidae). Smithiana Bull. 13: 3–28.

[7]Kramer, B., R. Bills, P. Skelton and M. Wink. 2012. A critical revision of the Churchill Snoutfish, genus *Petrocephalus* Marcusen, 1854 (Actinopterygii: Teleostei: Mormyridae), from southern and eastern Africa, with the recognition of *Petrocephalus tanensis,* and the description of five new species. J. Nat. Hist. 46 (35–36): 2179–2258.

[8]Skelton, P.H. 2016. Name changes and additions to the southern African fish fauna. Afr. J. Aquat. Sci. 2016: 1 –7.

[9]Bragança, P.H.N., T.G. Smith, E.J.W.M.N. Vreven and A. Chakona. 2020. Integrative taxonomy reveals hidden diversity in the southern African darters genus *Nannocharax* Günther 1867 (Characiformes: Distichodontidae). J. Fish. Biol. 2020: 1–11.

[10]Bragança, P.H.N., R.M. van Zeeventer, R. Bills, D. Tweddle and A. Chakona. 2020. Diversity of the southern Africa *Lacustricola* Myers, 1924 and redescription of *Lacustricola johnstoni* (Günther, 1894) and *Lacustricola myaposae* (Boulenger, 1908) (Cyprinodontiformes, Procatopodidae). ZooKeys 923: 91–113.

[11]Stauffer, J.R., Jr., R. Bills, P.H. Skelton and O.L.F. Weyl. 2020. Re-elevation to species level and redescription of *Serranochromis jallae* and *Serranochromis robustus* (Teleostei: Cichilidae). Zootaxa 4858(1): 126–134.

[12]Kramer, B. and E.R. Swartz. 2010. A new species of Slender Stonebasher within the *Hippopotamyrus ansorgii* complex from the Cunene River in southern Africa (Teleostei: Mormyriformes). J. Nat. Hist. 44: 2213–2242.

[13]Watters, B.R., R.H. Wildekamp and K.M. Shidlovskiy. 2015. Description and biogeography of *Nothobranchius capriviensis*, a new species of annual killifish from the Zambezi Region of Namibia (Cyprinodontiformes: Nothobranchiidae). J. Am. Killifish Assoc. 47(4–6): 97–133.

[14]Stauffer, J.R., Jr., R. Bills and P.H. Skelton. 2021. Four new species of *Serranochromis* (Teleostei: Cichilidae) from the Cuanza and Okavango river systems in Angola, including a preliminary key for the genus. Zootaxa 4908(1): 66–84.

[15]Bragança, P.H.N., P.H. Skelton, R. Bills, D. Tweddle and A. Chakona. 2021. Revalidation and re-description of ''*Lacustricola*'' *chobensis* (Fowler, 1935) and description of a new miniature species of ''*Lacustricola*'' from Southern Africa (Cyprinodontiformes: Procatopodidae). Ichthy. Herpe. 109(1): 123–137.

[16]Skelton, P.H. 2021. *pers. comm.*

[17]Ainsworth, R., I.G. Cowx and S. Funge-Smith. 2021. A review of major river basins and large lakes relevant to inland fisheries. FAO Fish. Aquacult. Circ. No. 1170. Rome: 314 pp.

[18]Betancur-R, R., E.O. Wiley, G. Arratia, A. Acero, N. Bailly, M. Miya, G. Lecointre and G. Orti. 2017. Phylogenetic classification of bony fishes. BMC Evol. Biol. 17: 162. https://doi.org/10.1186/s12862-017-0958-3.

[19]Skelton, P.H., J.R. Stauffer, A. Chakona and J.M. Wisor. 2021. Two new species of African bubble-nesting *Microctenopoma* (Teleostei: Anabantidae) from Angola. Ichthyol. Explor. Freshw., IEF-1134: 1 -16. doi.org/10.23788/IEF-1134.

[20]Stauffer, J.R., M. Geertsa, A.F. Konings, K.R. McKaye and K. Black. 2016. Cichlid fish diversity and speciation, pp. 213-225. In: T.R. Hodkinson and J.A.N. Parnell (Eds.). Reconstructing the tree of life: Taxonomy and systematics of species rich taxa. Systematics Association Special Series, Volume 72.

CHAPTER 5
NON-NATIVE FISHES AND OTHER AQUATIC INVASIVE SPECIES IN NAMIBIA

by

Sarah K. Tribelhorn, Glenn S. Merron and Charles H. Hocutt

I. Introduction

Non-native fishes, also known as *invasive, alien, introduced* and *exotic* species, occur outside their natural range and are largely spread by human activity, intended or unintended. These species can also spread from natural events, such as large-scale flooding that connects previously disconnected waterways, one of which harbours a non-native species.

Exotic fishes have been known from the waters of southern Africa since 1857 when Goldfish were recorded from the Cape, but they had been held in captivity since at least 1726.[1] Schrader (1985)[2] listed only four fishes in his account of invasive species in Namibia, but recognised (a) the real prospect of translocation occurring on a wide scale, and (b) the threat of truly foreign species. The introduction of non-native freshwater fish species in Namibia began with the Common Carp in the 1880s. Largemouth Bass and Mozambique Tilapia were introduced during the 1930s and 1940s and established permanent populations (see Chapter 6).

Non-native species can have various effects on ecosystems, e.g., outcompeting native species for spawning areas; predation leading to lower diversity; carrying the risk of introducing parasites and diseases; and potentially hybridising with indigenous species. There can also be indirect economic consequences of new species introduction, such as the impacts to local subsistence, commercial and sport fisheries. However, there can be both positive or negative impacts; e.g., a South America weevil (*Crytobagous salviniae*) was imported into Namibia to combat the problems associated with the invasive Kariba Weed (*Salvinia molesta*) discussed below, and this introduction can be viewed as a success.

Non-native fishes are found in most Namibian state reservoirs, lower Orange River, several ephemeral and seasonal rivers, and the Upper Zambezi River. The state reservoirs harbour the most non-native species, including Common Carp and Largemouth Bass. The ubiquitous Guppy, introduced to Lake Otjikoto, is exemplary of a genuine concern, i.e., aquarium releases by hobbyists.

The Mozambique Tilapia and Nile Tilapia represent potential threats to Namibia's northern rivers from Zambia, where they are cultured. They are closely related to the native Threespot Tilapia and Greenhead Tilapia, thus there is the distinct risk of hybridisation and competition. Another invasive concern from Zambia is the Australian Red Claw Crayfish (*Cherax quadricarinatus*), discussed below.

II. NON-NATIVE FISHES INTRODUCED TO NAMIBIA

The following non-native fishes are found in Namibia:

A. Family Cyprinidae (Goldfish and Carp)

Goldfish (*Carassius auratus*)

J.R. Stauffer,Jr.

Etymology: ***Carassius***, Latinised form of the French vernacular ***carassin*** or German ***karuse***, for the European carp; ***auratus***, meaning gilded or golden

DS 2, DR 14-20; AS 2, AR 5; LL 26-28, CP 14-16

Summary: Goldfish are native to Asia but have been introduced globally, often the result of aquarium releases. There are several distinct breeds that vary greatly in size, body shape, fin configuration and colouration. Typically, adults have a rotund body, brassy gold in colour, with juveniles being metallic brown. The dorsal fin is long with the leading ray spine-like and strongly serrated. The caudal fin is large, wide and forked. Goldfish have distinctly large scales and eyes. The small terminal mouth lacks barbels. They reach 580 mm SL and can be over 3 kg.

Goldfish are found in numerous water features and ponds in Namibia. They are tolerant of poor water conditions, which also enable them to spread. They can be found in slow-moving river channels, lakes and reservoirs with abundant aquatic vegetation. They often display schooling behaviour. They are opportunistic and feed on plant matter, detritus, aquatic insects and crustaceans. Spawning occurs in spring as water temperatures warm. Goldfish, like all cyprinids, are egg layers. Their eggs are adhesive and attach to dense aquatic vegetation. Owing to their bright colouration they are easily preyed upon by numerous predators including fish, birds, mammals and reptiles.

Grass Carp (*Ctenopharyngodon idella*)

F. Jacobs

Etymology: ***Ctenopharyngodon*** = ***cteno***, comb + ***pharynx***, throat + ***odon***, tooth, referring to comb-like pharyngeal teeth; ***idella***, not explained, but presumably derived from the Greek ***idios***, distinctive or peculiar

DS 3, DR 7-8; AS 3, AR 7-11; LL 37-42, CP 16

Summary: The Grass Carp is a large freshwater fish in the family Cyprinidae. They originate from eastern Asia, but have been introduced into most water bodies in the world for weed control and culture. As the common name implies, they are largely herbivorous with a terminal mouth. They spawn on river beds with currents, but are common to backwaters and pools. They have a wide tolerance to temperature, turbidity and other environmental factors. They are known to grow up to 1.5 m SL, 50 kg, and live for over 20 years.

The Grass Carp is recorded from the Orange River, the border between Namibia and South Africa.

Common Carp (*Cyprinus carpio*)

F. Jacobs

Etymology: ***Cyprinus*** = Greek, ***kyprinos***, carp; ***carpio*** = ***carpio***, carp

DS 3, DR 17-22; AS 2-3, AR 4-5; LL 35-39 when present, CP 14

Summary: The Common Carp has a robust body and is generally brassy to yellowish in colour. The long dorsal fin includes a strong serrated spine and the caudal fin is moderately forked. Scales when present are large; scales are missing in some forms (Mirror Carp, see p. 287) or absent (Leather Carp).

The mouth is subterminal with two barbels on each side of the upper jaw, distinguishing them from Goldfish. This species can tolerate water with very low oxygen levels by gulping air at the surface. They reach over 1 m TL and 40 kg.

Common Carp occur in the lower Orange River, Fish River, Omuramba Omatako, state reservoirs and ephemeral rivers, such as the Kuiseb River. They are raised in numerous small-scale aquaculture facilities in south-central Namibia. They are omnivorous, scavenging the bottom for insects, crustaceans, deitritus, benthic worms and aquatic plants. Carp are broadcast spawners, releasing thousands of adhesive eggs in a single spawn over freshly flooded vegetated margins of rivers and reservoirs. Carp are popular sport fish.

Amur Carp (*Cyprinus rubrofuscus*)

F. Jacobs

Etymology: ***Cyprinus***, Greek, ***kyprinos***, carp; ***rubra-***, red + ***fuscus***, dusky or dark, referring to golden-brown colouration

DS 3, DR 18-22; AS 2-3, AR 4-5; LL 29-33

Summary: The domesticated Amur Carp is closely related to the Common Carp. It has a large number of colour variants, collectively known as Koi, with mutations first developed in China over 1,000 years ago and especially perpetuated in Japan during the 1800s. There are over 20 variants with new varieties still being bred, distinguished by colouration, patterning and scalation. Koi are typically mottled in red, yellow, white, black, orange and blue. They have a terminal oblique mouth below a rounded snout. Koi are found in numerous outdoor ponds, water features and aquaria in Namibia. To date, none have been found in the wild.

B. Family Characidae (Largespot Tetra, *Astyanax orthodus*)

G. Merron

Etymology: ***Astyanax*** = Greek, ***Astuánax***, name of the firstborn son of Hector; ***orthodus*** = ***orthos***, straight + ***odon***, tooth, i.e., straight-toothed, referring to denticles on surface of teeth arranged in a line

DS 0, DR iii, 9; AS 0, AR iv-v, 24-31; LL 35-42

Summary: This oval shaped fish has a laterally compressed body that is shiny silver in colour with orange tips on fins and a characteristic dark spot at the caudal fin base from which it derives its common name. Fins are generally translucent and an adipose fin is present. It matures between 70–80 mm TL and can reach up to 139 mm TL with females being larger than males. It feeds on vegetation, seeds and insects.

A spawning population was reported from a *fontein* (spring) located at Wabi Lodge, Otjahewita Farm in the Waterberg District where it was introduced.[3] These fish are popular for home aquaria and an enormous commercial aquarium trade exists globally for them.

C. Family Osphronemidae (Threespot Gourami, *Trichogaster trichopterus*)

G. Merron

Etymology: ***Trichogaster*** = ***trichos***, hair or ray + ***gaster***, belly, referring to their pelvic fins, each one a single, long, thread-like ray; ***trichopterus*** = Latin, ***tricho-***, hair + ***pteron***, wing, i.e., fin

DS 6-8, DR 7-10; AS 9-12, CP 30-38; LL 30-40 (40-52 in lateral series)

Summary: This gourami gets its common name from the two spots along each side of its body in line with the eye, considered the third spot. They are rather an iridescent light-blue hue in colour. Males tend to be larger, but slimmer with females having a rounder girth, reaching 120 mm SL.

They are omnivores feeding on aquatic plants as well as algae, small invertebrates, zooplankton and detritus. Spawning may occur multiple times in a season and the male shepherds the eggs into a nest. The male then tends to the nest until the eggs hatch, usually in around 20–30 hours.

The Threespot Gourami is a popular aquarium fish that has been widely introduced globally. A spawning population was reported from a fontein located at Wabi Lodge, Otjahewita Farm in the Waterberg District.[3]

D. Family Cichlidae (Mozambique Tilapia and Nile Tilapia)

Mozambique Tilapia (*Oreochromis mossambicus*)

F. Jacobs

Etymology: ***Oreochromis***: ***oreos***, mountains, referring to Mt. Kilimanjaro, locality of type species + Greek, ***chromis***, a fish, perhaps a perch; ***mossambicus***, ***–icus***, belonging to: Mozambique, type locality

DS 15-17, DR 10-13; AS 3, AR 9-12; LL 30-32, CP 16

Summary: The Mozambique Tilapia has a deep compressed body. Adults are silver-olive dorsally, lighter ventrally with red margins on the dorsal and caudal fins. Spawning males become dark grey with white on the lower head and throat. They reach 400 mm SL.

They occur in a variety of habitats from fast-flowing to standing water and have wide environmental tolerances. They are omnivores and feed on algae, insects and invertebrates. Spawning occurs in summer after males build saucer-shaped nests on sandy substrates. This species is a mouthbrooder with the female carrying the eggs and larvae. It is caught in the subsistence fishery using a variety of methods and is an important sportfish.

The Hardap fish hatchery spawns this species and supplies fingerlings to all waters in the southern parts of Namibia, though they are prohibited in areas of the north where they can escape into the rivers and hybridise with native tilapia. The Mozambique Tilapia has been transplanted to a number of privately owned impoundments and state reservoirs in Namibia. It occurs in the Orange and Fish Rivers. They historically were reported from the Omuramba Omatako, a southern tributary to the Kavango River, and also occur in the salt pans near Swakopmund with salinities higher than sea water. They were also introduced into reservoirs of the water supply scheme at Ondangwa, it being the sump of the canal bringing water from the Kunene River.

Nile Tilapia (*Oreochromis niloticus*)

F. Jacobs

Etymology: ***Oreochromis**: **oreos***, mountains, referring to Mt. Kilimanjaro, locality of type species + Greek, ***chromis*** = a fish, perhaps a perch; ***niloticus**: **–icus***, belonging to: Nile River

DS 15-18, DR 11-14; AS 3, AR 9-11; LL 30-34, CP 19-20

Summary: The Nile Tilapia is a large deep-bodied cichlid, usually green-brownish in colour with a silvery-white belly. They have up to 10 thin vertical stripes along the body. The caudal fin is truncated with numerous thin vertical stripes. They reach a length of 70 cm SL and weight of 9 kg with a reported age of 9 years. The Nile Tilapia is primarily herbivorous with aquatic macrophytes, algae and diatoms generally comprising its diet.

The Nile Tilapia has established populations far beyond its natural range in Zambia, Zimbabwe, Mozambique and South Africa, mostly through aquaculture escapees. It was introduced for aquaculture in southern Namibia and an intensive fish farm was in operation below Hardap Dam in the early 2000s. MFMR staff collected individuals in a canal draining overflow water to the Fish River and in pools in the river. Biologists expect this species has reached the Orange River and may pose a threat to indigenous fishes. The first free-ranging specimen in northern Namibia was collected in Lake Liambezi in May 2021; the closest populations in the Upper Zambezi River are at a fish farm in Livingstone, Zambia. It is an aggressive species that can survive in all Namibian freshwaters and is likely to hybridise with the Threespot Tilapia and Greenhead Tilapia.

Nile Tilapia being sold at the Katima Mulilo fish market, May 2021 (Photos: B. van der Waal).

E. Family Poeciliidae (Mosquitofish, Guppy and Swordtail)

Mosquitofish (*Gambusia affinis*)

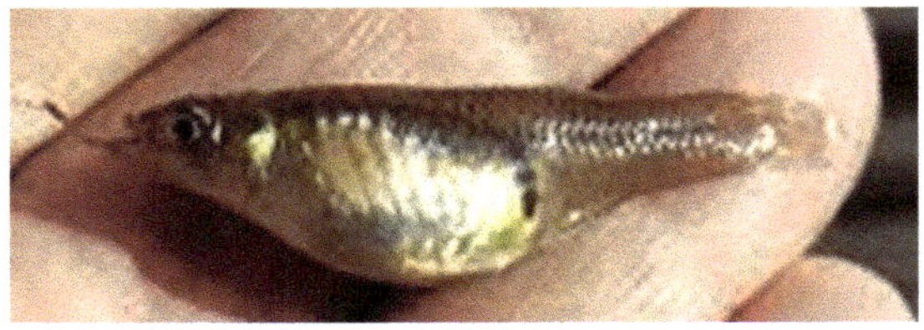
G. Merron

Etymology: ***Gambusia*** = Cuban ***gambusino***, signifying "nothing," referring to their inconsequential size; ***affinis*** = allusion not explained

DS 0, DR 7-9; AS 0, AR 9-10; 26-34 in lateral series

Summary: This is a small robust fish, dull grey to brown with a terminal upturned mouth for feeding at the surface. They exhibit *sexual dimorphism* with mature females being larger than mature males, reaching about 70 and 40 mm SL, respectively. The first few rays of the anal fin are elongated in males.

This aggressive invader has been introduced globally for mosquito control. It is known from sewerage and fish ponds around Namibia. It can threaten native fish as it preys on their larvae. It is also widespread in South Africa, Zambia and Zimbabwe.

Guppy (*Poecilia reticulata*)

F. Jacobs

Etymology: ***Poecilia*** = Greek, ***poikilos***, with a lot of colours; Latin, ***reticulate***, netted

DS 0, DR 6-8; AS 0, AR 8-10; 26-28 in lateral series, CP 14

Summary: The Guppy has been widely introduced worldwide since the early 1900s as a means of mosquito control and via release of aquarium fish into waterways. This species is sexually dimorphic with females being uniform silver-grey and larger with deeper bodies than males, reaching lengths of 60 and 35 mm SL, respectively. Males have conspicuous polymorphic colour patterns consisting of combinations of black, white, red, orange, yellow and green. The rays of the caudal and dorsal fins of males are long, filamentous and colourful in the different varieties of guppies.

The anal fin of males is transformed into an intromittent organ, termed a gonopodium, for internal fertilisation and *ovoviviparous* (live bearing) reproduction. Females give birth to 20-40 live young that are independent from birth. They feed on zooplankton, small insects and detritus.

Interestingly, the Guppy receives its common name from R.J. Lechmere Guppy (1836-1916), a naturalist who sent specimens from Trinidad to the British Museum of Natural History in 1866. This species inhabits Lake Otjikoto and had been introduced into a *fontein* located at Wabi Lodge, Otjahewita Farm in the Waterberg District, although its current status is unknown.[3]

Swordtail (*Xiphophorus helleri*)

F. Jacobs

Etymology: ***Xiphophorus*** = ***xiphos***, sword + ***phorus***, carrier, referring to dagger-like gonopodium and/or extended lower caudal fin; ***helleri*** = in honour of Austrian botanist Karl Bartholomaeus Heller (1824–1880), who collected type while exploring México

DS 0, DR 11-14; AS 0, AR 8-10; 26-30 scales in lateral series

Summary: The Swordtail is a tropical Central American species that was introduced through the aquarium trade and has established itself in Lake Otjikoto. It preys on fish larvae. This species is sexually dimorphic with males having extended rays of the lower caudal fin 'like a sword', hence the common name. The anal fin of males is also modified as an intromittent organ, the gonopodium. Females are typically larger than males.

Like the Guppy, Swordtails are reported to have been introduced into a fontein at Otjahewita Farm in the Waterberg District, where its current status is unknown.

F. Family Centrarchidae (Largemouth Bass, *Micropterus salmoides*)

F. Jacobs

Etymology: ***Micropterus***: Greek, ***mikros*** = small + Greek, ***pteron*** = wing, fin; ***salmoides***, ***salmo-*** = trout/salmon-like

DS 10, DR 11-14; AS 3, AR 10-12; LL 58-69, CP 24-32

Summary: The Largemouth Bass has a large mouth with the maxillary extending beyond the eye. The dorsal fin has a distinct spinous front portion followed by a rayed posterior section. The pelvic fins lie behind the pectoral fins, but are far anterior along the abdomen. It is coloured green to olive dorsally, milk-white to yellow ventrally, with a black lateral band running from the operculum to the base of the forked caudal fin. It reaches lengths of 97 cm TL and weight of 10.1 kg globally, but obtains about 7 kg in southern Africa.

Adults feed on fishes, crustaceans, herptiles, birds and insects; the young feed on crustaceans, insects and small fishes. Spawning takes place in spring to summer. The male builds a nest in shallow water and guards the eggs until they hatch. The Largemouth Bass is a popular game fish in many parts of southern Africa and has been widely introduced elsewhere.

This species is found in Oanob, Von Bach and Fredinau state reservoirs. They have not done well in the Hardap Reservoir with its highly turbid waters. It has been historically reported from the Omuramba Omatako, southern tributary to the Kavango River.

III. Invasive Aquatic Species of Concern

A. Kariba Weed (*Salvinia molesta*)

G. Merron

Etymology: ***Salvinia*** = a genus named in honour of Anton Maria Salvini (1653–1729), an Italian naturalist; Latin, ***molesta***, annoying, troublesome, nuisance

The spread of invasive aquatic species does not respect political boundaries. A visible alien is the floating aquatic plant regionally known as Kariba Weed (*Salvinia molesta*). Kariba Weed was brought to Africa from South America in the 1900s as an attractive ornamental plant to place in garden ponds. This plant has now spread throughout vast areas of the African continent clogging waterways and necessitating control programmes at considerable cost to many governments. It is a floating plant that spreads with the water flow, has a high growth rate and forms huge mats.

Kariba Weed has occurred for over 50 years in northern Namibia along the Kwando, Chobe and Zambezi River systems. Its rapid dense growth literally chokes off waterways, diminishing habitats and diversity. Its mass decomposition can cause anoxic waters. Consequently, fish life is adversely affected. Furthermore, the photosynthesis of native aquatic plants is inhibited, since *Salvinia* diminishes sunlight that would normally reach these species. This further negatively affects *aufwuchs* and the minnows that rely on native macrophytes to sustain their populations, as well as the predators that prey on these minnows.

Kariba Weed can spread through natural waterways and also "hitchhike" on boats or vehicles coming from infiltrated water outside Namibia. Due to the efforts of the Namibian Aquatic Weeds Control Unit, a very successful biological control programme was established using a small weevil (*Crytobagous salviniae*). These weevils feed solely on *Salvinia* plants and pose no threat to any indigenous plants or animals. The weevils can be considered a beneficial introduction that has helped the Republic address the problems associated with Kariba Weed.

B. Australian Red Claw Crayfish (*Cherax quadricarinatus*)

B. van der Waal

Etymology: ***Cherax*** = Thought to be a misspelling of the Greek word 'charax', meaning a pointed stake; ***quadricarinatus***, ***quadri***, four + ***carinatus***, having a keel-shaped ridge, alluding to four ridges along the carapace

The Australian Red Claw Crayfish is an invasive species with a wide environmental tolerance. It was introduced into Zambia as an aquaculture species in 1992. They subsequently escaped the culture facilities invading the Kafue-Middle Zambezi River system. It was recently caught in a fish trap in the Upper Zambezi River near Katima Mulilo, Namibia. They are now spread throughout the Middle and Lower Zambezi River from Lake Kariba into Mozambique. Exacerbating their regional spread was their escape from Swaziland aquaculture facilities into the Komati River in 2002, from which it has rapidly expanded its range in South African waters. It is considered a threat to the Okavango Delta.

Body colouration ranges from dark brown to blue-green and adults have a red patch on the outer margin of their claws. This species feeds on detritus, rotting vegetation and fish eggs, and thus could consume the eggs of native species such as tilapia, cyprinids and other species creating a serious ecological problem. Fertilised eggs are affixed to the underside of the female's tail. Incubation takes approximately six weeks and the newly hatched juveniles rapidly become independent. They can weigh up to 600 g.

IV. References

[1]Raidt, E.H. (Ed.). 1971. Description of the Cape of Good Hope with matters concerning it by F. Valentyn, Amsterdam, 1727.Van Riebeek Society, Cape Town.

[2]Schrader, H.J. 1985. Invasive alien fishes in South West Africa/Namibia, pp. 35–40. In: C.J. Brown, I.A.W. MacDonald and S.E. Brown (Eds.). Invasive Alien Organisms in South West Africa/ Namibia. S. Afr. Natl. Sci. Prog. Rept. No. 119: 35–40.

[3]Hocutt, C.H. and P.N. Johnson. 1993. Fisheries resource assessment of the Kavango and Caprivi Provinces, Namibia. USAID Compl. Rept., Grant No. DAN-5053-G-00-1048-00: 271 pp.

CHAPTER 6
NAMIBIAN WATERBODIES AND THEIR FISHERIES

by

Clinton J. Hay

I. Introduction

Namibia is an arid country with an annual rainfall varying from <50 mm in the southwest to 650 mm in the northeast of the country. Open waterbodies in Namibia are a rarity due to the low rainfall compounded by remarkably high evaporation rates (up to 3,800 mm per annum) leaving a major water deficit with most of the rainfall evaporated, lost through transpiration, river runoff and recharging aquifers. Most rivers in Namibia are *ephemeral* with the only *perennial* rivers found on the northern and southern borders, all forming international borders. The Kunene and Kavango Rivers share their borders with Angola; the Kwando-Linyanti with Angola and Botswana; and the Upper Zambezi and Chobe Rivers with Zambia and Botswana, respectively, while Zimbabwe is only 2 km away from their confluence. The lower Orange River in the south forms the border with South Africa. Catchments of these rivers are predominately outside Namibia in much higher rainfall areas. Major perennial and ephemeral rivers, dams and lakes are shown in Figure 1.

The only perennial rivers that discharge into the Atlantic Ocean are the Orange and Kunene, each with a small *estuary*. The mouths are mainly freshwater except during low river discharge levels or extreme high tides with little variation or mixing with marine saline waters.

The Upper Zambezi-Chobe River passes over Victoria Falls and follows the Zambia-Zimbabwe border before crossing through Mozambique to enter the Indian Ocean (Chapter 3). The Kavango River supplies Botswana's Okavango Delta where water typically evaporates or seeps into the thick Kalahari sands. During exceptional floods, however, these waters may pass through the Selinda Spillway to the Kwando-Linyanti River that drains to Lake Liambezi to join the Upper Zambezi-Chobe River. Traditional fisheries differ between these river systems due to cultural aspects of local ethnic groups, their historical origin and customs as well as the physical conditions of the rivers. The Kunene River fishery is not utilised due to traditional beliefs of the local Himba (plural, OvaHimba) people that do not eat fish. The fish population structure in the Kunene thus reflects a lack of exploitation with the presence of large numbers of large fish, compared to other rivers where fisheries are intense. The lower Orange River is sparsely populated, historically with fish a low priority for the people in the area. This is in contrast with the Kavango and Zambezi where riverine communities heavily depend on healthy fish stocks. In some areas, the fishery has been commer–cialised leading to a decline in many of the larger fish species.

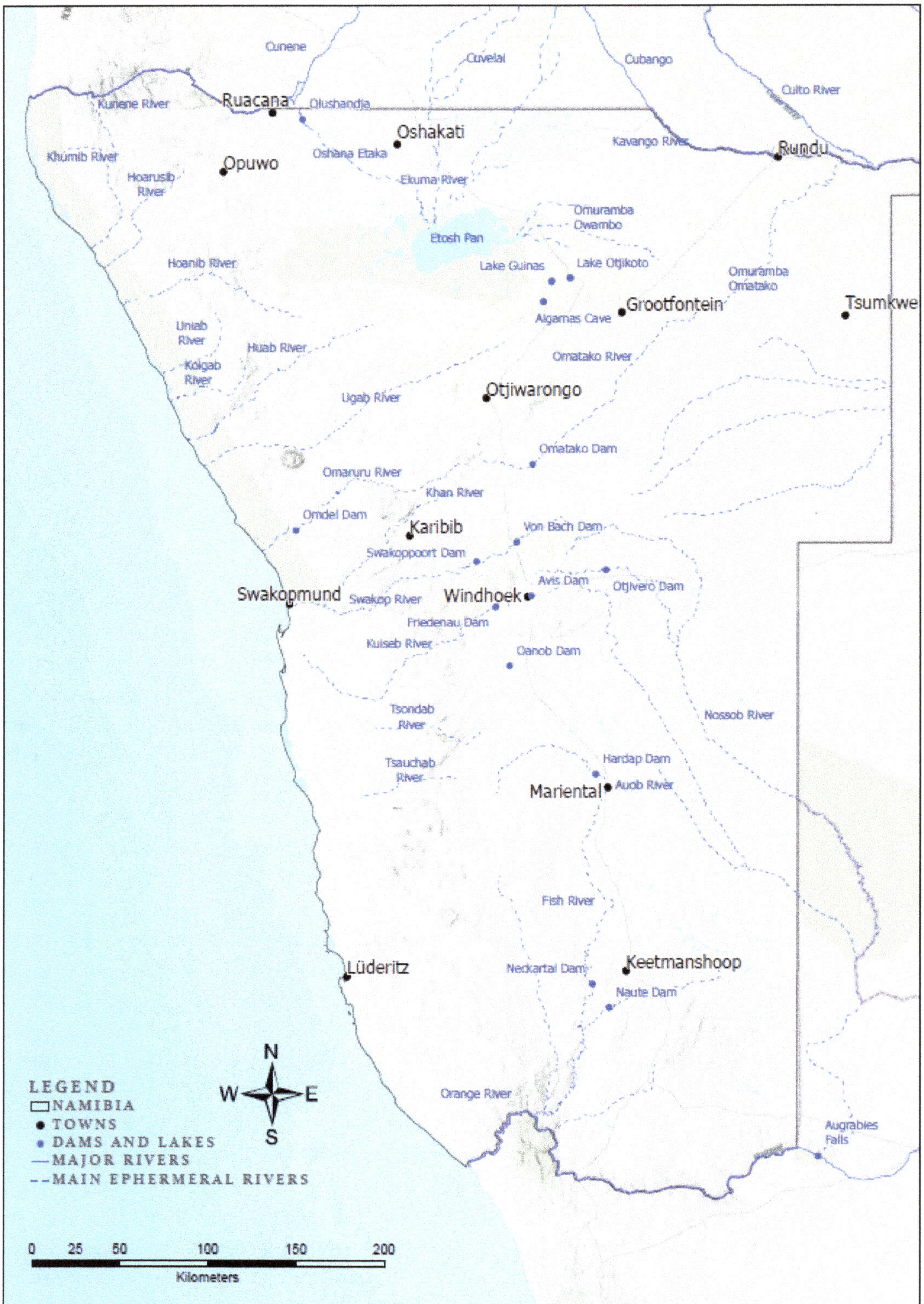
Cunene
Cuvelai
Cubango
Cuito River
Kunene River
Ruacana
Olushandja
Oshakati
Kavango River
Rundu
Khumib River
Opuwo
Oshana Etaka
Hoarusib River
Ekuma River
Omuramba Owambo
Etosh Pan
Lake Guinas
Lake Otjikoto
Hoanib River
Omuramba Omatako
Tsumkwe
Grootfontein
Aigamas Cave
Uniab River
Huab River
Omatako River
Koigab River
Ugab River
Otjiwarongo
Omatako Dam
Omaruru River
Khan River
Omdel Dam
Karibib
Von Bach Dam
Swakoppoort Dam
Avis Dam
Otjivero Dam
Swakopmund
Swakop River
Windhoek
Friedenau Dam
Kuiseb River
Oanob Dam
Tsondab River
Nossob River
Tsauchab River
Hardap Dam
Mariental
Auob River
Fish River
Lüderitz
Neckartal Dam
Keetmanshoop
Naute Dam
N
W
E
S
LEGEND
NAMIBIA
TOWNS
DAMS AND LAKES
MAJOR RIVERS
MAIN EPHERMERAL RIVERS
Orange River
Augrabies Falls
0
25
50
100
150
200
Kilometers

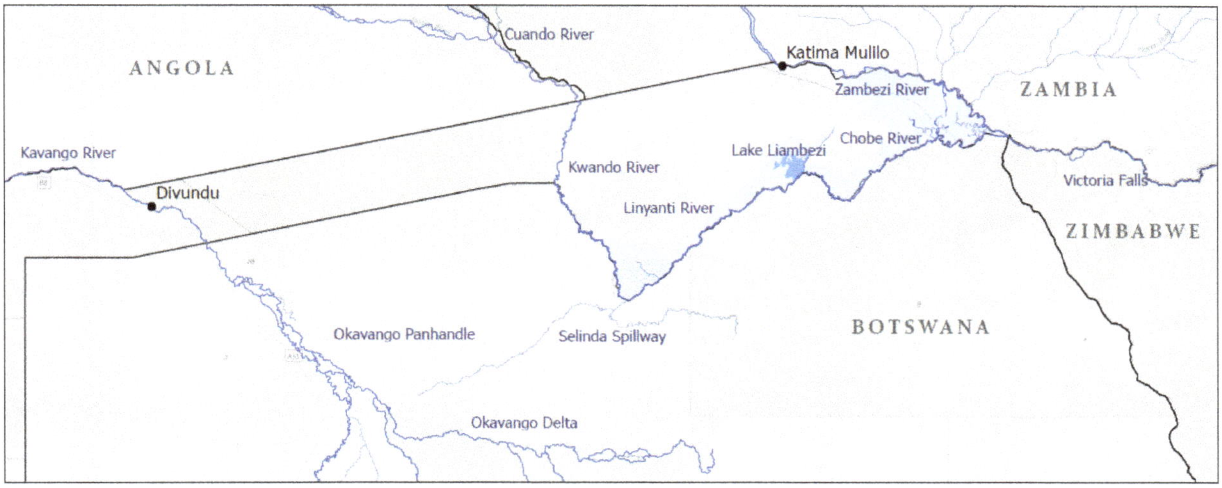

Figure 1. The location of all major rivers, dams and lakes and main ephemeral rivers. Map prepared by G. Jacobs.

The seasonal Cuvelai system (Chapter 7) drains the southern parts of Angola, flows through the Ohangwena, Omusati, Oshikoto and Oshana Regions in shallow depressions called *iishana* discharging into the Etosha Pans during high floods. Although only flooding seasonally, the *iishana* are a lifeline for these rural communities for agriculture, water supply and fish. Other ephemeral rivers in the west discharge into the Atlantic Ocean when in flood, while some drain into the perennial rivers. Seasonal rivers in the east and south disappear into the Namib and Kalahari Deserts.

The only other open waterbodies in the country are manmade reservoirs on the seasonal rivers, impounding water for human and livestock consumption or crop irrigation, supplemented regionally by groundwater supplies. There are no reservoirs on any perennial river in Namibia. Lakes Otjikoto and Guinas in the north are conspicuous *cenote* lakes with collapsed roofs with no other open waterbodies in the area.

The perennial Kavango, Kwando-Linyanti and Upper Zambezi-Chobe Rivers as well as the ephemeral Cuvelai are all *floodplain* rivers in Namibia. The Kunene has floodplains in its upper course in Angola, while the Orange River has very minor floodplains near its mouth. Ecosystem productivity in floodplain rivers is driven by the *flood pulse* (Chapter 1). Reproductive and survival strategies of the fish community are directly linked to this phenomenon and, as seen below, so are the fisheries.

Freshwater fishes play a vital role in the culture, socioeconomics and livelihood of the local communities residing along these floodplain rivers. Fish are considered a safety net for food security that augment diets during droughts when crops fail and livestock die. Inland fisheries also do not have such a delayed turnover as crops or livestock given that all life stages are consumed from young-of-the-year to adults and can be harvested throughout the year, although abundance varies by season.

Namibia's fisheries can be subdivided into three sectors:

(1) **commercial fishery**, where fishers use modern gillnets targeting the high-value tilapia species with the sole aim to make money;
(2) **subsistence fishery**, where fishers use gillnets (usually few and worn out), mosquito nets, hook and line and traditional fishing gear with the catch mainly for own consumption, but surplus fish may be sold or traded to generate income for basic necessities; and
(3) **recreational fishery**, using modern rod and reel to catch game fish (Chapter 10). Organised tournaments are a component of sport fishing, usually with substantial prizes, and associated with the tourist industry, involving lodges, angling guides and local employment opportunities. Recreational anglers usually practice catch and release.

Fish are an important protein source for communities and provide micronutrients essential for a healthy lifestyle. They contain high levels of micronutrients such as vitamins A, D and B and minerals, e.g., calcium, zinc, iron and phosphorus. These micronutrients and minerals are critical for many metabolic processes; early development of children; improving the immune system; and lowering the risk of cancer, other diseases and cardiovascular problems. Smaller fish species contain proportionally higher levels of micronutrients as they are consumed whole, while large fish are usually filleted and carcasses discarded including heads.

Namibia's rich marine fish stocks from the offshore Benguela Current are one of the most important foreign currency earners for the Republic, employing >16,000 Namibians over the past 5 years. However, the freshwater fishing sectors support nearly 50%–90% of the rural communities in the north through nutrition, income and informal employment.

II. PERENNIAL RIVERS

A. Orange River

1. Description

The Orange River catchment originates in the Lesotho highlands less than 200 km from the Indian Ocean and flows westward towards the Atlantic Ocean for 2,300 km. The lower Orange River is formed at Augrabies Falls near Kakamas (South Africa) where it plunges 56 m, descends another 90 m in an 18-km long gorge and flows for ~20 km before becoming the ~550-km long international border between Namibia and South Africa. To the west, the river cuts through the rugged Ai-Ais/ Richtersveld Transfrontier Park characterised by high mountains (~1,250 m), a component of the Great Escarpment (Chapter 3), and the Namib Desert before reaching the Atlantic Ocean. River flow is greatly manipulated by several large impoundments in South Africa, e.g., the Gariep, van der Kloof and Vaal Dams, the former two generating hydroelectric power. The Vaal River, a major tributary of the Orange River, flows through the economic hub of South Africa with large industries in the catchment. These industries are associated with large water extraction and irrigation schemes that hugely impact the water quality and flow regime of the river.

The lower Orange differs from the Kavango and Zambezi with only minor floodplains near the mouth. The main river along the lower section is characterised by large boulders, rapids, vegetated small islands and sandy to silt areas filling up between the rocks. The only major northern tributary is the seasonal Fish River draining the central parts of Namibia. Smaller tributaries such as the Gamkab, Haib, Houms and Gaiab are all ephemeral rivers and seldom reach the lower Orange (Fig. 2).

Figure 2. The lower Orange River near the Gamkab River mouth. The river is characterised by the absence of floodplains and very rocky habitats (left) and the estuary at the mouth where it enters the Atlantic Ocean (right) (Photos: C. Hay).

The regulated nature of the flow and the increased nutrient load due to irrigation schemes in the catchment proliferate the growth of reeds (*Phragmites australis*) along the riverbanks and the many islands (Fig. 3). Alien Mesquite (*Prosopis* spp.) is found along the river and poses a threat to indigenous vegetation. It usually invades areas that were disturbed during floods where it can have major impacts on underground water resources and people's livelihoods. It was originally introduced into Namibia and South Africa in the late 1800s using the pods for animal fodder. Other alien plant species found in the dry riverbeds are Wild Tobacco (*Nicotiana glauca*), Common Apple Thorn (*Datura stramonium*) and White-flowered Mexican Poppy (*Argemone ochroleuca*).

The river's estuary is a declared RAMSAR site, but was placed on the Montreux Record in 1995 due to the degradation of the salt marsh triggered by human activities in the area. The river is sometimes isolated from the Atlantic Ocean due to the build-up of sandbars at the mouth during low flows that are breached during flood conditions or high tides. The important birds visiting the mouth are Great White Pelicans (*Pelecanus onocrotalus*), Greater Flamingo (*Phoenicopterus roseus*), Pied Avocet (*Recurvirostra avosetta*) and Chestnut-banded Plover (*Charadrius pallidus*). Increased water demand upstream is likely to further reduce the amount of freshwater reaching the river mouth endangering this fragile aquatic ecosystem.

Figure 3. Reeds growing on the riverbanks (left) and a small side stream of the Orange River (right) with large boulders and terrestrial vegetation (Photos: C. Hay).

2. Fish and Fisheries

The lower Orange River has no history of a subsistence fishery and the Inland Fisheries Act prohibits the use of any nets. However, the development of irrigation schemes along the river and the subsequent influx of labour force from northern Namibia stimulated harvesting using monofilament gillnets and hook and line. Fish are consumed, sold locally or dried and transported to the north. Residents of the mining towns of Oranjemund (Namibia) and Alexander Bay (South Africa) fish recreationally at the mouth mainly for marine species. Vaal-Orange Largemouth Yellowfish, Vaal-Orange Smallmouth Yellowfish, Mozambique Tilapia and Sharptooth Catfish are the most important freshwater angling species. Largemouth Yellowfish are under pressure from water level manipulation impacting on their breeding. They are slow growing with males maturing at 6 years and females at 8; males reach 17 years with a length of 69 cm.[1] A slow growth/late maturation cycle makes the species sensitive to hydrology changes, pollution and water abstraction. It is also classified as "Near Threatened" according to the IUCN category due to degraded habitats, a decline in numbers, changes in the river hydrology and poor water quality.

The lower Orange River has fewer fish species than Namibia's northern rivers with only 12 indigenous and 9 non-native species (Chapter 4). Mozambique Tilapia, Nile Tilapia and Common Carp were introduced into Namibian waters mainly for aquaculture purposes, while Largemouth Bass were recently sampled from Hardap Dam. Others such as Guppy, Mosquitofish, Koi and Goldfish occur in private and municipal waters. Only one Grass Carp has been recorded from the lower Orange River although it is more common in the upper reaches of the river. Sterile Grass Carp are bred commercially in South Africa and sold for biological control of aquatic weeds in dams and waterways. However, a breeding population was recorded from the Vaal River. Although seen as a very effective biological control agent for aquatic weeds, it can have a significant negative impact on aquatic habitats and should not be distributed indiscriminately. Grass Carp are also carriers of aquatic diseases and parasites that can harm native fish species (Chapter 13). Its longevity (lives up to 50 years in China) makes it cost-effective for weed control compared to the use of herbicides. It is also an excellent angling species that grows to 1 m, weighing up to 45 kg.[1]

The most common fish species are the Orange River Mudfish and the Vaal-Orange Smallmouth Yellowfish. These two species along with the Vaal-Orange Largemouth Yellowfish were considered endemic to the drainage prior to introductions elsewhere in South Africa. The Orange River Sardine, Namaqua Barb and Rock Catfish are endemic to the Orange and Fish rivers. The former two occur only below Augrabies Falls with the falls prohibiting upstream migration. A waterfall in the ephemeral Fish River and the recently completed Neckartal Dam will also prevent upstream migration.

Four species found in the lower Orange River are also present in all the northern rivers: Straightfin Barb, Sharptooth Catfish, Banded Tilapia and Southern Mouthbrooder. The Threespot Barb occurs only in the Orange and Kunene Rivers in Namibia, but is replaced by the Dashtail Barb in the Kavango, Kwando-Linyanti and Upper Zambezi-Chobe Rivers.

B. Kunene River

1. Description

The Kunene River originates at Boas Aguas near Huambo in the Angolan highlands, where it is known as the Cunene. As noted in the text and Species Accounts (Chapter 4), different spellings are used by the two countries for the river and in formal species descriptions. Thus, the Cunene Bulldog is found in the Kunene River.

The Cunene flows in a southerly direction for approximately 650 km until it reaches the Namibian border near Ruacana, turning west and forming the international border between the two countries for approximately 340 km. This dramatic shift in direction is a clear elbow of capture noting a stream piracy of an ancestral inland river by a westward flowing river (Chapter 3).

A weir in Angola on the Kunene above Ruacana Falls blocks the flow of water and diverts it to drive turbines 140 m underground at a power station, generating nearly half of Namibia's electricity. Below the weir, the river plunges over the 120-m high Ruacana Falls, which can be dry when the river flow is low (Fig. 4). All water is diverted through the turbines and released through a manmade tunnel back into the channel below the falls when the river is not in flood. Epupa Falls ("falling waters" in Herero) is located 130 km downstream and comprised of numerous small falls that span the width of the river about 0.5 km and coalesce after a ~40 m drop into the gorge below. This stretch of the river is fast-flowing and cuts through the Baynes Mountains downstream from the falls forming a deep gorge with impressive rapids before passing through the Namib Desert and entering the Atlantic Ocean. This section of the river differs from the Kavango and Upper Zambezi Rivers in Namibia with the absence of floodplains; however, floodplains are found in the upper reaches in Angola on the central plateau.

The Ruacana hydropower plant causes an erratic flow regime where the downstream water level fluctuates daily. This fluctuation impacts negatively on the substrate-spawning fish species such as the Redbreast Tilapia when eggs are exposed to a sudden drop in water level. A second hydropower scheme is contemplated along this section, but the local OvaHimba people have not consented to the dam as it will flood large areas of their ancestral burial grounds. The Kunene flow is erratic from year-to-year mitigating its effectiveness for hydroelectricity. This may worsen in the future with projected climate change impacts and the potential of more dams upstream.

The OvaHimba are the original Herero tribe that crossed the Kunene in the 16th century from Angola and settled in northwestern Namibia. They live a semi-nomadic lifestyle, farming with cattle, goats and sheep and growing maize and millet. Nuts of the palm trees along the riverbanks are also a source of food during dry periods. The OvaHimba are considered a "zero-income culture" as their wealth is solely measured by the number of cattle they own.

Figure 4. Epupa Falls (top left, one of a series of falls) and the exit tunnel (top right) for water released by the hydropower station at Ruacana. The Kunene River mouth (bottom left, Photo: F. Jacobs) blocked by a sandbar with a small opening draining into the Atlantic Ocean. Ruacana Falls (bottom right) during a high flood (Photos: C. Hay).

2. Fish and Fisheries

No subsistence fishery exists along this section of the river as the OvaHimba are traditionally not fish-eating people. MFMR surveys infer the fish stocks are not over utilised with various age classes present for different species. However, there is an increase in localised fishing from people outside the region. Recreational fishing occurs at selected areas, mainly at established lodges along the river. Tigerfish and Nembwe, two highly sought-after fish species by recreational anglers, are absent from the river, perhaps diminishing the Kunene as an inland angling destination.

The Kunene River mouth has a mixture of freshwater and marine fish species and is currently the southern-most location for Nile crocodile and sea turtle concentrations along the Southeast Atlantic Ocean. The native freshwater prawn (*Macrobrachium vollenhoveni*) is common near the river mouth, requiring fresh and brackish water for their lifecycle. It is endemic to West Africa from the Senegal River to the Kunene, demarking their southern limit of distribution.

The Kunene River fish fauna has some interesting anomalies compared to the Kavango and Upper Zambezi Rivers, especially within the cichlid family. Nine cichlids present in the Kavango and

Upper Zambezi Rivers are absent from the Kunene while seven cichlids found in the Kunene are absent from the former two rivers. Five of the latter are shared exclusively with the Cuvelai system as a result of their common drainage history (Chapter 3): Cunene Dwarf Bream, Cunene Bream, Thicklipped River Bream, Namib River Bream and an undescribed *Sargochromis* sp. The Angolan River Bream is endemic to the Kunene, while the closely related Slender River Bream occurs also in neighbouring drainages to the north in Angola.

The Cunene Kneria, an endemic, and Shorthead Barb have only been recorded in fountains and small tributaries flowing from the Baynes Mountains towards the Kunene and from the Otjihipa Mountains 20 km from the main river. Other endemics are the Skeleton Coast Bulldog, Cunene Bulldog, Kunene Sand Catlet and an undescribed topminnow (*Lacustricola* sp.) and Broadhead Catfish (*Clariallabes* sp.). The Cunene River Stonebasher, Cunene Churchill and Kunene River Sardine also occur in the adjacent Cuvelai system, which shared a common drainage history (Chapter 3). Other species appear as *disjunct* populations centred elsewhere: Silver Labeo (Limpopo River), Papermouth (Limpopo River), Threespot Barb (lower Orange River to eastern Africa coastal rivers) and Neumann's Suckermouth (Upper Zambezi eastward through the Lower Zambezi). Five families known in Namibia, but not represented in the Kunene ichthyofauna are Austroglanididae, Claroteidae, Anabantidae, Mastacembelidae and Nothobranchiidae.

The presence/absence of Tigerfish influences species composition and behaviour of fish in Namibia's northern rivers, but it is absent from the Kunene. The African Pike, whose food niche overlaps with Tigerfish, are larger and found in the mainstream of the Kunene compared to pike of smaller sizes with a preference for vegetated and backwater habitats in the Upper Zambezi and Kavango Rivers.

C. Kavango River

1. Description

The river known as Rio Cubango (Angola), Okavango River (Botswana) and Kavango River (Namibia) originates in the Angolan highlands and drains a total catchment area of ~115,000 km^2. The river meanders south-southeast from the Angolan highlands and reaches Namibia at Katwitwi before turning east and forming the 415-km international border with Angola. Two major tributaries enter the Kavango River: the Omatako River from the south, which is an ephemeral river that seldom flows and the perennial Cuito River from the north. The Kavango River then turns south for 53 km passing across Popa Falls and through Bwabwata National Park before entering northwest Botswana and forming the Okavango Panhandle. Thereafter, the Kavango River terminates in the 15,000 km^2 Okavango Delta.

Initially, the Kavango when entering Namibia keeps to the riverbed with a sandy substrate with intermittent rocky areas. Large floodplains develop from Kasivi downstream for more than 200 km. The Cuito River, a major northern tributary from Angola, joins the Kavango River near Katere, contributing 40% to the annual runoff to the Delta. The river then changes character with large rocky areas with numerous rapids flanking many islands until it reaches the well-known Popa Falls. The last 30 km of the river is characterised by large floodplains entering the Bwabwata-Okavango Ramsar Wetland Site at the Mahango and Buffalo Core Areas, the head of the Okavango Delta Panhandle. Papyrus (*Cyperus papyrus*), reeds (*Phragmites* sp.) and bulrushes (*Typha* sp.) dominate the permanent swamp in the Ramsar site. Nearly two-thirds of bird species (more than 400 species) in Namibia are found in the Ramsar site as well as one of the highest diversities of mammal species. Rare birds are Slaty Egret (*Egretta vinaceigula*), Pel's Fishing-owl (*Scotopelia peli*), Wattled Crane (*Bugeranus carunculatus*) and Narina Trogon (*Apaloderma narina*). The "Near Threatened" African Skimmer (*Rynchops flavirostris*) breeds on the sandbanks on the islands in the Ramsar site laying their eggs in shallow hollows as a nest (Fig. 5). The river along the Namibian segment floods annually between February and June, inundating floodplains to supply food and shelter for juvenile fish and other aquatic organisms. Floodplains play an integral part in the livelihoods of the communities, providing food, cultural and social activities.[2]

Figure 5. African Skimmer (left) and a Sable (right) in the Bwabwata-Okavango Ramsar Wetland Site (Photo: T. Naesje).

2. Fish and Fisheries

Different fishing methods and traditional gear are used by the local fishers, subject to water level and fish species they target (Chapter 9). For instance, fishers will construct mud pools that act as traps during low water periods, luring fish into these pools using millet or other bait. Women use fish corral traps (*sintunga*) in the shallower areas on the floodplains also baited with food (Fig. 6). This fishing method is used when people tend to their fields as the traps do not need constant attention. Fish fences are constructed with fish traps when the water retreats from the floodplains back to the river, guiding fish towards the trap. The *tambi* is an oval scoop basket with a handle used mainly by women, dragged through marginal vegetation catching fish taking refuge in these habitats. Mosquito nets and fish funnels (*shikuku*) are used during the receding phase catching fish as they move off the floodplains. Fish funnels are also used in backwaters and rocky habitats (Fig. 7). Hook and line (*erowo*) is a favourite fishing gear used by boys after school or during weekends. Other traditional gears are push baskets (*sididi*), fish spears (*muho*) and bow and arrow (*ngumba*) (see Chapter 9).

Figure 6. A woman collecting her catch from a traditional baited fish corral (***sintunga***) (left) and scientists studying the fish dynamics on the Kamutjonga floodplain (right) (Photo: C. Hay).

Figure 7. A fisher using a traditional basket while trampling the vegetation and chasing fish into it (Photo: C. Hay).

The use of traditional fishing gears is declining, being replaced by modern gears that are much more effective in catching fish. Also, traditional fishing gears are labour intensive to build and use with a limited lifespan. By example, plastic containers are replacing traditional fish traps. The top third of a 5-l plastic container is removed and small holes drilled along the sides. The container is placed in the vegetation on the floodplain, baited with porridge, left for some time and then drained leaving catch behind (Fig. 8). Other modern fishing gears are monofilament gillnets, mosquito nets and seine nets, the latter made from shade cloth. Small-mesh gillnets are also modified and used as seine nets in backwater habitats or even in the mainstream during low water. For seine nets to function effectively in backwaters, fishers remove all aquatic vegetation, destroying important breeding, recruitment and refugia habitats.

Figure 8. A child draining a plastic bottle baited with porridge in the shallow areas along the periphery of the floodplain. Small immature tilapia and barb species, valuable protein sources for subsistence communities, are caught using this method (Photo: C. Hay).

Fishing along the Kavango is still considered an important income source with 60% of the fishers surveyed near Kapako being fulltime fishers/vendors (majority are male) with the rest either fishing seasonally or part-time. The preferred fish species in this area is Greenhead Tilapia which is also the most profitable. Fresh fish are in high demand at restaurants and are profitable food items on the menu.

Gillnets used by local fishers target Sharptooth Catfish, Silver Catfish, Bulldog, and large Redbreast Tilapia, Threespot Tilapia, Banded Tilapia and Largemouth Bream. Fish traps capture the smaller sized barbs, topminnows, mormyrids and juvenile tilapia species.

Historically, fish were plentiful, and fishing was for own consumption or given to family members or relatives. Only traditional fishing gears were used, as nets were not available. The large tilapia species, e.g., Redbreast Tilapia and Greenhead Tilapia, were common and small fish were never taken home. Villagers fished on a rotational basis by so-called experts, i.e., not everyone fished. The use of fishing gear was gender segregated and water level determined the gear and method used. Elders mentioned that earlier fishing was for own consumption, but today fishing is over-commercialised causing catches and size ranges to decline. Also, overgrazing and logging have destroyed riparian vegetation causing increased runoff and erosion.

The Mahango and Buffalo Core Areas located in the Bwabwata-Okavango Ramsar Wetland Site and Bwabwata National Park are by default fisheries reserves. The protected status of the Mahango Core Area plays an important conservation role with higher densities of fish found compared to unprotected sections of the river. Zones are currently being identified along the river to be officially proclaimed as fisheries reserves and cooperatively managed by the local communities *viz a viz* community-based management (CBM). Radiotelemetry studies on Tigerfish demonstrated the value of these reserves by concluding that a 10-km stretch of river could potentially protect 50% of the respective population for at least 75% of the time. A highly mobile species, one individual moved nearly 400 km upstream from the point of tagging while another moved 116 km downstream into the Okavango Delta.[3]

More than 80 fish species have been identified from the Kavango River with the cyprinids and cichlids the most diverse fish families. Most of the species are small and do not grow larger than 15 cm. The species composition differs from the upper catchment in Angola due to different habitat types found in the smaller tributaries, a pattern known as longitudinal zonation. This ecological principle reflects that the physical and chemical characteristics of rivers shift from their headwaters to their mouths, thus influencing alterations in habitats and their species composition, either through addition or replacement.

Common species found in rocky areas of the Kavango are Ocellated Spiny Eel, Stargazer Mountain Catfish, Broadhead Catfish, Blotched Catfish, Okavango Suckermouth, Slender Stonebasher, Upper Zambezi Yellowfish and Redeye Labeo. The Blotched Sand Catlet and Kavango Sand Catlet are species found in the Kavango in sandy areas with clear flowing water. These species tend to wiggle themselves into the sand with only their eyes protruding and prey upon small invertebrates passing by. They are indicator species and significant erosion or siltation may influence their numbers. The Large-bodied Churchill, Okavango Churchill, Kavango Sand Catlet and an undescribed suckermouth catfish (*Chiloglanis* sp.) are considered *endemic* to the drainage. Species found on floodplains, swampy and densely vegetated backwater habitats are Okavango Tilapia, Blackspot Climbing Perch, Manyspined Climbing Perch, Sickle-fin Barb, Upjaw Barb, Copperstripe Barb and Spottail Barb.

No fish species in the Kavango River is considered threatened except Threespot Tilapia and Greenhead Tilapia from potential hybridisation with the closely related Nile Tilapia that was observed (B. van der Waal) in cage culture trials in 2017 in the upper Cubango, Angola, but not yet found in the Namibian section of the river. However, there remains genuine concern over the increased use of modern gear and its impact on the larger *tilapiines* and Tigerfish.[4]

D. Kwando-Linyanti River

1. Description

The Kwando River enters Namibia to the east of the Kavango River from Angola, where it is known as the Cuando River. The Kwando River flows at a low gradient and percolates through the vast Silowana floodplains. The seasonal flood in the Kwando reaches Namibia usually during June, much later than when the Kavango or Zambezi Rivers' flood in Namibia. This is due to the storage effect of the vast floodplains in Angola, lasting for 1–2 months or even for longer periods. The Kwando meanders slowly southeast through Namibia, fringed on both sides by large floodplains, swamps and backwaters of the Bwabwata National Park, then passes through the Mudumu and Nkasa Rupara National Parks where it splits into a number of streams. From here, the Kwando River changes direction and continues northeast as the marshy Linyanti River before entering the ephemeral Lake Liambezi during high flood years. Sections of the Linyanti can dry up during the low water period. During very high floods, the Okavango Delta connects with the Linyanti swamps through the Selinda Spillway. This was the case in 2009 when the two rivers connected for the first time in nearly 30 years.

Numerous isolated backwaters covered with submerged and emerging aquatic plants form when the river is low with large quantities of organic matter on the bottom (Fig. 9). The slow flow of the mainstream allows vegetation to fasten to the substrate, composed mainly of reeds, grasses, and trees. The mainstream substrate is dominantly sand with very few rocky or gravel habitats.

The relative low flood in the Kwando River results in an algal and aquatic macrophyte-based food web. Numerous hippopotami *(Hippopotamus amphibius)* transfer nutrients (carbon, nitrogen, phosphorous and silicon) from the terrestrial environment to the aquatic system, an important function in this ecosystem (Fig. 10). Silicon is an important element for diatoms, a major component of the *aufwuchs* community involved in photosynthesis and nutrient cycling.

Figure 9. Typical isolated pool with a sandy substrate with vegetative detritus visible on the bottom (top left); Mainstream with marginal grasses, with the right foreground showing signs of erosion due to elephant and hippo activities (top right); ***Salvinia molesta*** visible between the reeds, grasses and papyrus (bottom left); and water lilies present in the backwaters (***Nymphaea*** sp.) (bottom right) (Photos: C. Hay; bottom left, B. Hackenberg).

Figure 10. Hippopotami play a critical role in the aquatic ecosystem with the transfer of nutrients from the terrestrial areas to the river and maintaining open channels in dense vegetative habitats allowing for the free flow of water (Photos: C. Hay).

2. Fish and Fisheries

The drainage in Namibia is sparsely populated and little fisheries data are available. Most fishers along the Kwando River are part-time or seasonal with very few fishing fulltime. The main livelihood activities for these communities are farming and piece work. Despite this, most people consider fish an important source of protein and cash or for barter. Women mostly use tin dishes or grass kraals to catch fish, while some use mosquito nets. Men use gillnets and the local wooden canoes (*mikolo*) to set their nets in the evenings. Some commercial fishing takes place, but most is for family consumption. Fish species preferred by the communities are the Sharptooth Catfish, Western Bottlenose, Tigerfish, Threespot Tilapia, Redbreast Tilapia, Greenhead Tilapia and small barb species.

Earlier years' catches were for own consumption using hook and line and fish baskets. Fishers believe that catches declined with the introduction of modern gillnets and illegal fishing methods such as *bashing* and drag netting (Chapter 9) to exclusively sell fish. Historically, no one was allowed to fish in a pool that was drying up without permission, but this changed with the influx of outsiders who weakened traditions in the area.

The Kwando is much smaller than the Kavango and Upper Zambezi Rivers with a much smaller flood. Some tilapia species grow slower in this river compared to the Zambezi and Kavango Rivers and Lake Liambezi. These tilapia breed at a similar size as those from the other rivers, but they grow slower meaning that they breed at a later stage in life, making them much more vulnerable to overexploitation. Tigerfish prefer open water habitats and are not often caught in the Kwando. The low abundance of this dominant predator leads to differences in fish assemblages compared to the Kavango and Upper Zambezi. Catches are dominated by floodplain species: topminnows, Striped Robber, Slender Robber, Upper Zambezi Bulldog, Silver Catfish, Southern African Pike and Southern Mouthbrooder. The Cuando Dwarf Stonebasher and Pallid Sand Catlet are the only known endemic species in the system.

The low flood further ensures the availability and stability of many different microhabitats throughout the year resulting in very little change in the fish assemblages during the different seasons. This is compared to the Zambezi River where large floods 6-8 m over normal stage completely alter the aquatic environment into one homogenous waterbody during high flood. The Kwando River food web is based on a narrow range of resources such as filamentous algae, aquatic macrophytes and dung from the numerous hippopotami (VII.B, below).[5] Low nutrient loads reported in the Kwando may well be an artifact of the vast amount of nutrients absorbed and locked up in the mass of aquatic macrophytes and riparian vegetation.

E. Upper Zambezi-Chobe River System

1. Description

The Zambezi River is the fourth-longest river (2,574 km) in Africa, crossing six countries before reaching the Indian Ocean. It originates in the northwest Mwinilunga District of Zambia, traversing Angola back into Zambia before forming the international border between Zambia and Namibia from Katima Mulilo to Impalila Island, a distance of 120 km. Some 80 km downstream from Impalila Island, the river plummets over the majestic Victoria Falls, demarking the Upper Zambezi River from the Middle/Lower. Rapids are present where the Upper Zambezi enters Namibia, changing into a slow-flowing river widening as the flow decreases. The main river is characterised by a sandy substrate with numerous bends and oxbow lakes. The low topography of the region gives rise to large floodplains of approximately 370,000 hectares with the Chobe River acting as a large backwater of the Upper Zambezi. These floodplains have well-developed perennial channels crisscrossing the floodplains, as well as oxbow lakes, swamps and large backwaters ideal for the local fishery. The Upper Zambezi connects with the Chobe at Impalila Island with major rapids flanking the island. Interestingly, the Chobe changes direction during any particular year depending on the water level of the Upper Zambezi. During high water levels in the Zambezi, the Chobe flows 'upstream' into Lake Liambezi, but reverses direction when the water recedes. Lake Liambezi when flooded links the Kwando-Linyanti system with the Upper Zambezi-Chobe system (Fig. 1).

The floodplains are inundated annually from rainfall received in the catchment. The flood season in Namibia is usually between February and May, depending on the timing and intensity of the rainy season in Zambia and Angola. The annual dry-wet periods yield extremely productive floodplains facilitating the transfer of nutrients from the terrestrial biome to the aquatic system. Marginal vegetation along the open channels on the floodplains are reeds (*Phragmites australis*) and papyrus (*Cyperus papyrus*). Some fishers use plants locally known as "*malindo*" (*Sesbania* spp.) for gillnet floats. Backwaters are covered with a variety of aquatic plants and consist largely of *Sesbania* and *Nymphaea* spp. (water lilies), *Polygonum* spp. (buckwheat family), non-native *Salvinia molesta* (Kariba Weed) and *Ipomoea aquatica* (Water Morning Glory). *Utricularia* spp. (Bladderwort), a carnivorous plant species, is also found in some backwaters (Fig. 11). This plant can survive in poor nutrient environments by obtaining necessary nutrients from consumed organisms. They are not anchored to the bottom, but float in the water column. They have hairs projecting from the tiny bladders that are triggered when a water flea or small organism passes in close proximity. The flattened bladder opens up, sucking water in with the organism, deflating afterwards.

The fertile soils on the floodplains are cultivated during the dry periods, while the floodplains serve as grazing pastures for livestock. During flooding, people and livestock move to the higher areas until the water recedes. Temporary fishing camps rise throughout the floodplains from which fishers go out to the fishing grounds.

Figure 11. Bladderwort (top left) and water lilies on the floodplains (top right) (Photos: C. Hay). A fisherman showing the use of a plant locally known as "***malindo***" as gillnet floats (bottom) (Photos: B. Hackenberg).

2. Fish and Fisheries

The fishery in the Zambezi-Chobe Rivers is the largest in Namibia with an estimated yield of around 5,000 tonnes per annum, including the yield from Lake Liambezi when the lake is inundated (see section V below). The fishery is valued at N$ 100 million considering a conservative market price of N$ 20/kg for fresh fish. Fish are consumed locally, bartered or sold at the Katima Mulilo Fish Market.[6] Many are exported to Zambia, Botswana and the Democratic Republic of the Congo (DRC) to be sold. Sizes of fish range from 50 mm (smaller barb species) to more than 1 m (large catfishes).

This multispecies fishery consists of a gamut of different fishing gear used by the fishers. The modern fishing gears include mono- and multi-filament gillnets, dragnets made of shade netting, mosquito nets, hook and line, and rod and reel, while the traditional gears are spear, fish traps, fish baskets and fish fences. Fishing is gender-based with men mainly using gillnets and large dragnets targeting large tilapia species principally for selling, while women and children fish mostly with mosquito nets and traditional gear targeting small-sized fish species for family consumption (Fig. 12).

Figure 12. Women gathering in the backwaters catching small-sized fish for consumption (Photo: C. Hay).

A commercial fishery exists along the Zambezi-Chobe Rivers and associated floodplains where fishers target the large tilapia species due to their high demand and market price. The top three species contribute 40% of the total gillnet catches in weight of a potential 80+ species that do occur in these waters.[7] These are the Threespot Tilapia, Redbreast Tilapia and Greenhead Tilapia. Fishers use gillnets placed in backwaters and floodplain channels. Some use drag nets, especially during the low water periods when fish are concentrated and tilapia species move to shallower habitats for breeding. *Bashing* is a method where the water surface is hit with a paddle to chase fish into gillnets set near vegetation. Fishers (mainly men) bring their catch to landing sites on the floodplains where vendors (usually women) buy the fish to trade on the Katima Mulilo fish market. Fresh fish are taken to the market on the same day while some vendors accumulate fish over time and dry it before trading on the fish markets. The area where fishing takes place may determine whether fish to be traded are going to Botswana, Zambia or the Katima Mulilo Fish Market. Recent studies show a decline in tilapia catches most likely attributed to the intense unsustainable commercial fishery using monofilament gillnets (Fig. 13). These nets are declared illegal gear under the Inland Fisheries Resources Act, not in accordance with Government's policy of preferring subsistence over commercial fisheries.

There was a drastic change from multifilament to monofilament gillnets by the commercial fishers between 2010 and 2018. Monofilament gillnets are three times more effective than multifilament gillnets, intensifying the fishing pressure. Legislation was amended in 2016 prohibiting the use of monofilament gillnets, an intervention executed by the MFMR to protect the large tilapia species. The species composition did not change between 2010 and 2018, but slightly higher catches of the Sharptooth and Blunttooth catfishes were recorded for unknown reasons.[7] These catfishes are not favoured by the people in this region and large quantities are dried and exported to the northwestern regions of Namibia and to the DRC.

The focus of most women subsistence fishers is to catch fish for household consumption with some surplus sold. Their gillnets are generally in poor condition or they may use mosquito nets and traditional fishing gear when floodwaters recede between May and June and migrating fish are concentrated (Fig. 14). The fish are sun-dried and consumed locally with surplus taken to the Katima Mulilo Fish Market. Similar fishing occurs during June and July at the Impalila Island rapids for mormyrids. Fishers place constricting fishing baskets between the rocks, catching fish as they migrate downstream. Another fish migration that takes place and targeted by the fishers is when catfish migrate from the Chobe to depressions on the floodplains while the water is rising. This only takes place when local rains fall and only during the night. Fishers know when this migration occurs, taking provisions and waiting for the catfish to move from the river into these depressions.

Approximately 85 species are found in the Upper Zambezi-Chobe River floodplains of Namibia with more than 120 species when including the catchment areas in Zambia and Angola. There is both longitudinal distribution and habitat differentiation between the upper catchment area and the Zambezi-Chobe floodplains with some species restricted to the upper reaches and tributaries. Species are distributed based upon their preference for densely vegetated areas; fast-flowing currents with rocky substrate; swamps and backwaters; or mainstream and side channels.

Figure 13. Local fisherman returning from the fishing grounds near Nakabolelwa on the Chobe with gillnet visible (Photo: C. Hay).

Figure 14. Fishing during migrations in the Upper Zambezi River drainage. Men using constriction funnels attached to a cable in a strong current (top; Photo G. Sitengu). Women and children using mosquito-mesh seine nets (bottom left) to catch *Labeo* spp. (bottom right) for household consumption or bartering (Photos: E. Simasiku).

Examples of species preferring fast-flowing rocky conditions include the uncommon Ocellated Spiny Eel and Broadhead Catfish. These habitats usually have higher oxygen levels and fish are adapted to live in crevices. The Okavango Suckermouth has a modified mouth forming a large disc to hold onto rocks or vegetation and not be swept away in the current. The Stargazer Mountain Catfish has a depressed body profile, a reduced swim bladder, dorsally located eyes and large pelvic fins in conjunction with the body forming a sucking disc, all morphological adaptations to live in fast-flowing rocky habitats.

Species found in dense vegetation and on the floodplains are two anabantids, Blackspot Climbing Perch and Manyspined Climbing Perch; some of the small cyprinids such as the Sickle-fin Barb, Upjaw Barb, Copperstripe Barb and Spottail Barb; and the small Okavango Tilapia. The rare Longfin Largemouth also occurs on the floodplains of the Zambezi-Chobe.

The Caprivi Killifish is an endemic species found only in ephemeral pans in mopane forests on the edge of the floodplain of the eastern Zambezi Region. These temporary rain pools in heavy clay soil are seldom connected to any floodplain or river system and dry up after the rainy season. The Caprivi Killifish completes its life cycle in less than one year. It is the only fish species in Namibia whose eggs can withstand desiccation once the pool has dried out, just to hatch the next rainy season when the pool fills again, even after several years. Sexual maturity can be reached within a few weeks, ensuring the propagation of the next generation within a short timeframe. The Caprivi Killifish is considered vulnerable due to its limited and fragmented distribution. Habitat degradation, pressure of land use activities due to increased human population and road construction are potential threats to its survival.[8]

Radiotelemetry studies of the Threespot Tilapia, Nembwe and Greenhead Tilapia indicate these commercially important species are vulnerable to local exploitation due to their residency behaviour and limited movement (Fig. 15).[9,10] These species will benefit from fisheries reserves. A similar study on Tigerfish performed in the Kavango also showed the potential value of fisheries reserves to protect this species known for long movements (Chapter 8).

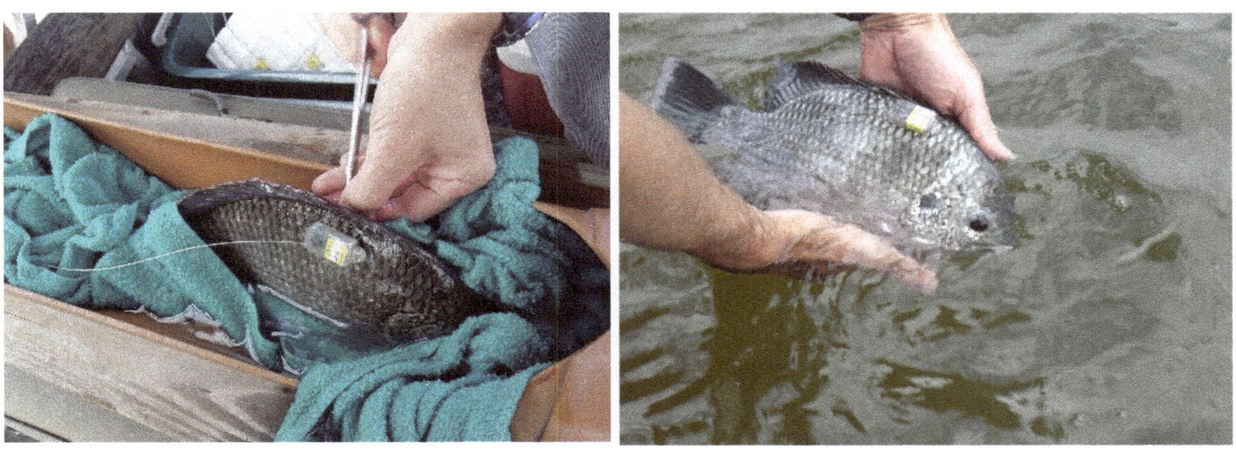

Figure 15. Radio-tagging a Greenhead Tilapia in the Upper Zambezi to study its movement behaviour (Photos: T. Naesje).

3. Katima Mulilo Fish Market

Freshwater fish are sold to the public at the Katima Mulilo Fish Market in the eastern Zambezi Region except between 1 December to the end of February when all fishing activities are closed (Fig. 16). The vendors at the market are mostly women with the majority being head of their households; trading of fish is a particularly important income source. Both dry and fresh fish are sold, but most fish are processed before marketing due to the absence of cooling facilities. Fishers on the floodplains seldom take their catch to the market. Rather, they give their catch to family members to process or to sell to vendors travelling the floodplains.

Figure 16. Weighing of fish at the Katima Mulilo Fish Market entrance. Women negotiating fish prices at the market (Photos: T. Naesje).

The fish market has a retail segment where the general public buys fish and a wholesale segment where fish are sold in bulk for export (Fig. 17) principally via the Wenela border post to the centre and north of Zambia with some going as far as Kasumbalesa, a Zambian town bordering the DRC. The wholesale segment of the fish market only operates when Lake Liambezi is flooded and large volumes of fish are available. During 2015 and 2016 with the tilapia fishery peaking in Lake Liambezi, the average monthly export of fish was 21 tonnes dry fish and 8 tonnes fresh fish valued at N$ 1.9 million. Other fishing hotspots are the Kalimbeza/Lisikili area and the eastern floodplains along the Chobe.[11]

Twenty-four different fish species have been recorded at the fish market (fresh fish) with Threespot Tilapia the most common (32% total number and 35% total weight), followed by Greenhead Tilapia (23% and 20%, respectively) and Redbreast Tilapia (22% and 19%, respectively). The cichlids dominate the market with 14 species, 91% in total number and 88% in total weight. Considering that more than 80 species are present in this system, it is obvious that fishers target the large tilapia species leading to a decline in catches over the years. The price at the market of these three species doubled between 2014 and 2019 indicating either high demand or lack of supply. Fish became a lucrative business stimulated with the flooding of Lake Liambezi in 2009.

Figure 17. Retail section of the Katima Mulilo Fish Market on the left and the wholesale section on the right. Fish sold wholesale are destined for markets in Zambia and the DRC (Photos: T. Naesje).

Nile Tilapia (*Oreochromis niloticus*), a non-native fish farmed at Livingstone, Zambia, have recently appeared in the Katima Mulilo Fish Market. Insufficient supply of local fish was the reason vendors reverted to importing Nile Tilapia. The first wild Nile Tilapia in the Zambezi system in Namibia was collected by fishermen in 2021 from Lake Liambezi.

Price of catch is determined by species, size and condition with fresh tilapia the most popular at about N$ 50/kg, followed by Tigerfish at N$ 40/kg and large catfish around N$ 20/kg. Dried kapenta (*Limnothrissa miodon*) imported from Lake Kariba, Zimbabwe and small barbs from the Zambezi-Chobe floodplain are also sold at the market.

III. EPHEMERAL RIVERS

The interior of Namibia lacks perennial rivers, but several ephemeral river systems exist with the most prominent being the Cuvelai system (Chapter 7) draining from the southern parts of Angola through north-central Namibia into the endorheic Etosha Pan during exceedingly high floods. Several west-flowing rivers have their catchments in the central parts of the country cutting through the Namib Desert and Damaraland further north flowing into the Atlantic Ocean during major floods. A few rivers flow east into the Kalahari Desert in Botswana, but seldom receive water. The only major river flowing towards the northeast is the *Omuramba* Omatako forming a southern tributary of the Kavango River. The Fish River carving through the Namaqua Metamorphic Complex and Nama Group (sandstone and black limestone in the far south) forms the impressive Fish River Canyon and ends in the lower Orange River approximately 120 km from the Atlantic Ocean. Some smaller tributaries in the southern parts of the country also drain into the lower Orange, although these rivers are seldom in flood.

Ephemeral rivers play an important role in recharging underground water such as the Kuiseb (Fig. 18) and Omaruru Rivers where aquifers supply water to the coastal towns of Walvis Bay, Swakopmund and Henties Bay, as well as to the mining town Arandis with most of the water used by the mines. Major dams are also located in these rivers, with Hardap and Neckartal Dams in the Fish River; Naute Dam in the Lowen River; Swakoppoort and Von Bach in the Swakop River; and Omatako Dam in the Omuramba Omatako.

Ephemeral rivers have an impoverished fish diversity due to harsh conditions that exist in these systems. Isolated pools keep water between flood periods, sometimes several years, providing the only refugia for fish to survive. Farm dams located in these systems may burst during heavy rains, releasing fish that may get trapped in these isolated pools once flows cease. The Sharptooth Catfish is found in many ephemeral rivers and is the only species that survives dry periods by aestivating and burying themselves in shallow moist burrows, but will die if the soil dries out. They have accessory air-breathing organs (different from lungs) and can breathe atmospheric air. They can also move across land using their pectoral fins. African lungfishes (*Protopterus* spp.) differ by having true lungs and surviving in totally dry areas by burying themselves in soft sediments while secreting a mucous cocoon around themselves to prevent drying out. They are not found in Namibia, but do occur in the Middle and Lower Zambezi.

A. Cuvelai-Etosha System

The Cuvelai-Etosha drainage is an ephemeral system originating near the Sierra Encoco Mountains in Angola and flanked by the Kunene River on the west and the Cubango River on the east. A maze of smaller tributaries called *iishana* (*oshana* singular) flow southeast through the Oshana, Ohangwena and Omusati Regions in Namibia. They are shallow, grassy, often broad depressions that become inundated during the rainy season and converge towards the Etosha Pan through the Ekuma River from the west and the Omurambas Oshigambo and Owambo from the east. Perennial waterbodies persist in Angola due to higher rainfall, but the smaller tributaries south into Namibia fill only during heavy rains. The *iishana* keep water for 2-5 months forming isolated pools as they dry up. These pools play a vital role in communities' livelihoods through household use, gardening and livestock. They also provide limited refugia for fish during dry periods while deeper pools hold water until the next rainy season.

Fish are an erratic, but an important protein source in this area due to the ephemeral nature of the *iishana*. Fish are imported from the Zambezi and Kavango Regions during dry years, as well as marine fish from Walvis Bay (usually horse mackerel). These are traded on the local markets in the main centres. A more complete description of the Cuvelai basin and its fisheries is presented in Chapter 7.

B. Westward Flowing Rivers

There are 12 major ephemeral rivers flowing west to the Atlantic Ocean which are the Kuiseb, Swakop, Omaruru, Ugab, Huab, Koigab, Uniab, Hoanib, Hoarusib, Khumib, Tsauchab and Tsondab Rivers, the latter two disappearing into the Namib Desert and not reaching the ocean. The remainder traverse the Namib Desert (Fig. 19) that stretches along the southeastern Atlantic coast from the Olifants River in the Western Cape, South Africa to Namibe, Angola. It is only during exceptional floods that water reaches the ocean in these basins, otherwise it disappears into the sandy riverbeds. The frequency of flooding varies between rivers and depends on the size of the catchment, ranging from 2,000 to 30,000 km^2; state and farm dams in the catchment; and volume and distribution of the annual rainfall. Most floods are torrential and short-lived. Subsurface movement of water takes place in most channels throughout the year, although the volume is never large and varies between years.

Groundwater leaches out at fountains and forms pools in erosion resistant bedrock in many of these ephemeral rivers. The Namib Desert is characterised by sparsely covered vegetation with a diverse range of wildlife, some with intriguing lifecycles interwoven with the fragile environment. However, the fish fauna occurs irregularly and is insignificant. Fog from the cold nutrient rich Benguela Current frequently shrouds and dampens the western edge of the Desert, crucial for the survival of many plant and animal species. The remarkable Welwitschia (*Welwitschia mirabilis*) is found here with some specimens as old as 2,000 years (Fig. 19).

Figure 18. Kuiseb River in the catchment area, dry (left) and in flood in 2021 (right) (Photos: C. Deelie).

Figure 19. *Welwitschia mirabilis* (left; Photo: C. Hocutt) and a panoramic of the Namib Desert dunes (right; Photo: C. Hay).

The presence of fish in these rivers is usually a result of an overspill of farm dams in the catchment areas during floods with fish surviving in permanent pools and springs. Common species normally found are Sharptooth Catfish, Mozambique Tilapia, Common Carp and Straightfin Barb. The Chubbyhead Barb was sampled 1968-1973 from isolated pools of the Kuiseb River and the Gaub River, a tributary. The closest populations (*Enteromius oraniensis*) are considered endemic to the Orange River drainage upstream of Augrabies Falls. The status of this species in the Kuiseb drainage is uncertain, but it may represent a relict population similar to *E. breviceps* found in a southern tributary of the Kunene River in Namibia (Chapter 3).[12]

C. Eastward Flowing Rivers

The Central Plateau is east of the Namib Desert and comprised mainly of savanna and scrubland, becoming more wooded towards the north. Central-east Namibia is the core agricultural area in Namibia in terms of both livestock and crop production, although each is localised to specific areas depending upon groundwater and average rainfall.

The Namibian profile slopes gradually west to east, i.e., from the Central Plateau toward the Kalahari Desert and Botswana. The Kalahari is known for its red dunes, covered with drought-tolerant grass and shrubs in the south with more trees to the north where rainfall is slightly higher. Large baobabs are a distinctive feature of the Kalahari terrain in the north (Fig. 20).

Figure 20. Large baobab in northeast Namibia in the Khaudom National Park where the Dorsland (Thirstland) trekkers camped in 1891 on their way to Angola (Photo: F. Jacobs).

The 1,360-km long border between Namibia and Botswana lacks perennial rivers other than the Kavango, Kwando-Linyanti and Chobe Rivers. Permanent fountains support insignificant fish life with the seasonal Nossob River (White and Black Nossob) being the master stream of the area across the western Kalahari Desert (Fig. 21) before joining the Molopo River (Chapter 3). It drains east, then turns southward eventually leading to the Orange River after joining the Auob River in the far southern part of the Kgalagadi Transfrontier Park along the Namibia-Botswana-South Africa border. The Stampriet artesian basin is present in the Nossob-Auob catchment and is important for agriculture in the area. The Nossob seldom flows, but subsurface water movement occurs. Oanob and Nauaspoort Dams are in the catchment of the Auob system. Fish from Hardap Dam were *translocated* to Oanob Dam in the early 1990s and Largemouth Bass were introduced later. Mozambique Tilapia, Common Carp and Sharptooth Catfish are found in farm dams in the drainage. Fish other than the Sharptooth Catfish that enter the Nossob system from these dams are likely to succumb as water sinks into the riverbed or evaporates.

Figure 21. The Kalahari Desert during above-normal rainfall (Photo: B. Liebenberg).

D. North:South Drainages

Between the eastern Kalahari and the western Namib Deserts, there are two other noteworthy drainages of the Central Plateau: the Fish River flowing southward to the Orange River and the Omuramba Omatako channel oriented northeast, tributary to the Kavango River. The seasonal Fish River originates in the Nauchas highlands southwest of the town Rehoboth in the Khomas Region and drains the central and southern parts of Namibia joining the lower Orange River 120 km from the Atlantic Ocean. Portions of the Fish River are deeply incised to form a canyon 85-km long, 27-km wide and up to 550-m deep, often cited as ranking second in size only to the Grand Canyon, USA (Fig. 22). The Orange-Fish system has been highly altered through water extraction and irrigation schemes with numerous agricultural activities concentrated in areas along their banks. The system has a depauperate, but distinct fish fauna. Mozambique Tilapia, Sharptooth Catfish, Straightfin Barb, Common Carp, Orange River Mudfish and Moggel hybrid are found in the river below Hardap Dam. The endemic Namaqua Barb, Orange River Sardine and Threespot Barb are found in the lower Fish River with a waterfall south of Seeheim preventing any upstream movement.[13] Hybrid yellowfish found in Hardap Dam are common in the Upper Fish River, but are replaced by the Smallmouth and Largemouth Yellowfish species further downstream.

Figure 22. Fish River Canyon approximately 500-m deep, often cited as the second largest canyon after the Grand Canyon, USA (Photo: C. Hay).

The Omuramba Omatako is an ephemeral river system originating between Otjiwarongo and Okahandja, seldom flowing, but aligned in a northeastern direction through Bushmanland and reaching the Kavango River 80 km east of the town Rundu. Historically, this omuramba was an important corridor for the transport of goods through the bush to northern Namibia and Angola. The Omatako Dam in this system provides water to Windhoek. Mozambique Tilapia, Common Carp and Sharptooth Catfish are found in intermittent pools throughout the system. Mozambique Tilapia has the potential to interbreed with the Threespot Tilapia should it reach the Kavango River through the Omuramba Omatako

IV. LAKES

Namibia has three waterbodies considered locally as lakes: Lake Liambezi, Lake Guinas and Lake Otjikoto. The former is unique in that some years it is completely dry. Lakes Guinas and Otjikoto are distinguished by being *cenote* lakes.

A. Lake Liambezi

1. Description

Lake Liambezi is a 370 km^2 shallow ephemeral lake in the Zambezi Region seldom deeper than 6 m when flooded. The lake receives water from three major sources: the Kwando-Linyanti River, Chobe River and the Bukalo channel originating from the Zambezi River near Kalimbeza.

The Linyanti River is a 100-km long myriad of swamps and channels between the Kwando River and Lake Liambezi that drain through the Chobe River to its confluence with the Zambezi River 100+/- km away at Namibia's most eastern point. The Linyanti-Chobe system has a general southwest to northeast alignment, but the Chobe may flow in either direction dependent on the flood stage of the Zambezi and floodplains, the Zambezi being seasonally dominant with flooding typically 5 to 6 m above normal stages, but occasionally reaching up to 8 m.

Water may enter Lake Liambezi from the north during local rainfall although this contributes little compared to the other major sources. High floods from the Zambezi or Kwando are needed for water to reach the lake. The lake flooded in the 1930s and again in 1958, before drying up in 1985. Partial flooding occurred in 2002 and 2004 with water reaching the lake intermittently, filling it completely in 2009 when above average rains fell in the Zambezi catchment. It subsequently dried up in 2017 after several years of drought, a natural phenomenon.

During the dry spell, the lake bed is utilised by the local communities for agriculture mainly to grow maize, but also millet, sorghum, spinach, tomatoes and sugarcane in some of the zones. Some community members revert to piecework and collection and sale of natural products such as poles, grass and reeds. The lake supports a major fishery when flooded (Fig. 23). Benefits derived from a full lake during above-average rainfall periods are readily available fish protein and cash income compared to inconsistent crop harvests. The inundated lake, in general, leads to a better standard of living for lakeside communities with income from fish spent on food, clothing, school fees, livestock, electricity and farming materials.

Figure 23. A fisherman returning to the landing site with his catch on Lake Liambezi (left). Weighing the catch from a returning fisherman at Shamahuka landing site at Lake Liambezi (right). Women vendors in the background cleaning fish for processing to be sold at Katima Mulilo Fish Market or exported to neighbouring Zambia (Photos: C. Hay).

Lake Liambezi has moderate nutrient levels compared to other African lakes due to (a) infrequent flooding and (b) limited nutrients released from nutrient-poor Kalahari sand that covers the area.[14] However, the build-up of organic material between floods (the previous dry period lasted 15 years!) do release substantial amounts of nutrients with initial flooding stimulating productivity in the lake followed by a thriving fishery. The initial flood and release of nutrients first promote the production of phytoplankton followed by zooplankton feeding on phytoplankton. Fish that enter the lake via the Chobe/Kwando or the Bukalo channel readily populate the lake abundant with food. The flooded lake goes through different stages of succession with a boom-and-bust cycle of different groups of fish species. The fishers on the lake adapt their fishing methods to maximise daily catch.

Kariba Weed (*Salvinia molesta)* was first reported from the Zambezi in 1948 and by 1967 it was 10 km east of Liambezi. In 1973, 8.4% of the open water of the lake was covered and this increased to 26.1% by 1975 after which mat surface decreased substantially, partly the result of chemical treatment of mats and later the successful release of bioagents (Chapter 5).[14] The filling of the lake in 1976 with accompanying dilution of nutrients, may also have played a role. Kariba Weed never established as a problem after the 1980s. It was not present during the recent filling of the lake between 2009 and 2017.

2. Fish and Fisheries

The lake supports a major fishery when flooded, yielding 600 tonnes in 1974[15] and more than 2,000 tonnes annually between 2011 and 2015.[16] The fishery was mainly focused on tilapia, especially between 2011 and 2015 when large numbers were exported to neighbouring Zambia and the DRC.

A commercial fishery developed around the lake during the most recent flooding (2009) driven by foreign markets, mainly Zambia and the DRC. Fish were also sold at local markets around the lakeside and in Katima Mulilo; nationally in the Kavango, Omusati, Oshana, Oshikoto and Ohangwena Regions; and as far south as Windhoek and Otjiwarongo. As the fishery developed, so did the value chain resulting in secondary business spin-offs. Villages appeared next to the lake where communities processed fish and set up shops selling basic necessities. Fish were dried and salted, according to demand and distance to the market. Communities organised fishing committees, instructed by the Traditional Authorities, to manage the fishery in response to the rapid commercialisation of fish stocks and the influx of foreign traders and fishers. A levy was introduced on fishing and trading to ensure local communities benefited from the open-source fish stocks in the lake. Foreign traders and fishers paid higher levies to support the local management regime.

Gillnet fishing was predominantly by men with little fishing by women using traditional fishing gear. Greenhead Tilapia, Threespot Tilapia, Redbreast Tilapia and Purpleface Largemouth predominated the monofilament gillnet catches with catch rates fluctuating between 15 and 53 kg/canoe/net per night during 2011 and 2012 with 343 canoes (*mikolo*) along the lake from which nets were set.[16] Fishers offloaded their catch early morning at specific landing sites along the lake where vendors, mainly women, negotiated prices before taking the catch to the Katima Mulilo Fish Market. Between 5–6 tonnes/day (maximum of 12 tonnes/day) entered the fish market from the lake during the peak fishing period.[11]

The report of Nile Tilapia in 2021 from Lake Liambezi is the first record of an alien fish species in the Zambezi Region in Namibia. The real concern of it crossbreeding with the Threespot and Greenhead Tilapia now warrants that these two native species be classified as threatened.

B. Lakes Guinas and Otjikoto and Aigamas Cave

The north-central parts of Namibia near Otavi, Tsumeb and Grootfontein area have strong underground aquifers in an area commonly known as the Karstveld. Two famous cenote lakes, Guinas and Otjikoto (Herero "deep water"), are situated within these aquifers (Fig. 25). Nearby is Dragon's Breath Cave containing the largest known non subglacial *hypogean* lake in the world; Aigamas Cave, Ghaub Cave and other smaller caves occur in the area. These sites are sensitive to water extraction as recharge of hypogean waters can be a slow process. These isolated lakes and caves allowed for interesting and rare organisms to evolve over thousands of years. Known endemic fauna includes the Cave Catfish (*Clarias cavernicola*) and a cave-dwelling crustacean isopod (*Namibianira aigamasensis)* in Aigamas Cave and the Otjikoto Tilapia (*Tilapia guinasana*) in Lake Guinas. These rare species are potentially endangered by their limited distribution and the threats of water extraction, introduction of aquatic organisms and pollution.

Lake Otjikoto was declared a National Monument in 1972 and is a major tourist attraction. German troops discarded artillery and ammunition in the lake during World War I while retreating from South African troops; some of the retrieved artifacts can be seen in the Alte Feste Museum in Windhoek. The lake has a surface area of 0.5 ha surrounded by vertical cliffs that widen towards the bottom of the lake. It has a maximum depth that exceeds 70 m. Water is used for irrigation at nearby farms. It is thought that the Southern Mouthbrooder was the only fish species present in Lake Otjikoto before the Otjikoto Tilapia, Mozambique Tilapia and Sharptooth Catfish were translocated.

Lake Guinas is slightly bigger than Lake Otjikoto with a surface area of 0.7 ha and a depth greater than 100 m. The lake is on private land and the water is used for irrigation. There is no evidence to suggest that the two lakes are connected as previously believed. The Otjikoto Tilapia is unique in that it is a non-sex limited, multi-coloured species which is rare for tilapias. It has five principal colours that vary between dark blue, light blue, blue stripe, olive stripe and olive and has been featured on a national stamp highlighting the Biodiversity of Namibia (Fig. 24).

Aigamas Cave is filled with hypogean water to a depth greater than 90 m and visibility more than 15 m. The water temperature in the cave is very stable (27.1–27.2 °C) irrespective of season or water depth. Despite the great depth, the Cave Catfish is found only in the upper 17 m of the water column where it forages for food falling from above (guano and bat droppings) and on the free-swimming isopods. The catfish is a member of the air-breathing catfishes. Its lack of pigment gives it a pinkish colouration and its skin appears to be sensitive to direct sunlight. It is endemic to this cave and is considered vulnerable with a small population of around 200 individuals.[18]

Figure 24. Otjikoto Tilapia featured on Biodiversity of Namibia postage stamp.

Figure 25. Cenote lakes Guinas (top) and Otjikoto (middle) with entrance to hypogean Aigamas Cave (bottom) (Photos: Z. Britz, C. Hocutt and F. Jacobs, respectively).

V. RESERVOIRS

A number of man-made reservoirs are located on ephemeral rivers of Namibia and fall under the jurisdiction of the Namibia Water Corporation Ltd (NamWater). Harvesting of fish with nets is not allowed; recreational angling requires a licence from MFMR.

A. Fish River drainage

1. Hardap Dam

Hardap Dam is located 24 km from Mariental. It has a capacity of 295 million cubic meters (Mm^3) and a surface area of 2,500 ha (Fig. 26). The Hardap Recreation Resort at the dam has chalets, dormitories and camping facilities for visiting tourists. Surrounding the dam is the Hardap Game Reserve covering an area of 252 km^2 with Kudu, Oryx, Mountain Zebra, Springbok, Red Hartebeest and Ostrich present. Water from the dam is for irrigation, household consumption and recreational purposes. The dam typically holds sufficient water for irrigation even though the catchment flows for only a few weeks each year. This dam provides important recreational fishing opportunities for freshwater angling species such as Largemouth Yellowfish, Smallmouth Yellowfish, Sharptooth Catfish, Mozambique Tilapia, Moggel, Orange River Mudfish and Common Carp. Recently the non-native Largemouth Bass was caught by anglers; the origin of this species is uncertain, but clearly introduced.

An interesting fact is that the two yellowfish species as well as the Moggel and Orange River Mudfish have respectively hybridised, indicating environmental stress, perhaps due to changes in the habitat due to the building of the dam. Purer strains of each species are found progressively downstream in the Fish River.

The MFMR Hardap Aquaculture Centre at the dam is responsible for all aquaculture activities in the central and southern parts of the country. A private fish farm experimented with tilapia to investigate their culture potential, but has since ceased operations (Chapter 12).

Figure 26. Hardap Dam (left) with Namibia Wildlife Resorts (NWR) restaurant overlooking the dam (right) (Photos: C. Hay).

2. Neckartal Dam

Neckartal Dam was built in one of the least populated areas in the country on the lower Fish River near Keetmanshoop. It was completed in 2018 and has a potential capacity of 857 Mm^3 and surface area of 40 km^2 making it the largest dam in the country in an area with little surface water. Water will be used to irrigate 5,000 Ha. Recent surveys by MFMR show a similar fish species composition to Hardap Dam situated upstream in the Fish River (Kukuri, L.M. *Pers. comm.*). The dam filled completely during an above average rain season in January 2021 (Fig. 27).

Figure 27. Neckartal Dam near Keetmanshoop at 100% capacity, 2021 (Photos: C. Hay).

3. Naute Dam

Naute Dam (84 Mm^3) was completed in 1972 on the Löwen River, a tributary of the Fish River, supplying water to Keetmanshoop and an irrigation scheme for dates and grapes (Fig. 28). The Naute Game Park surrounds the dam and has Oryx, Springbok, Klipspringer, Duiker and Steenbok. Fishes that occur are Straightfin Barb, Orange River Mudfish, Sharptooth Catfish and the non-native Mozambique Tilapia. Anglers visit the dam for recreational fishing.

Figure 28. Naute Dam and the date farm below the dam (Photos: C. Hay).

B. Swakop River drainage

The Swakop River discharges into the Atlantic Ocean sporadically. It has two dams within its basin, Von Bach and Swakoppoort. Local communities harvest fish from the dams for personal consumption with surplus sold or traded along the roads.

1. Von Bach Dam

Von Bach Dam has a capacity of 48 Mm^3 and is located in the Okahandja District providing drinking water to Windhoek. The dam has a resort with overnight facilities and is popular for aquatic sports. Recreational angling is a common pastime and international fishing competitions are held regularly. A game park surrounds the dam with wildlife such as Zebra, Red Hartebeest, Oryx, Giraffe, Impala and Kudu. Native fish species present are Straightfin Barb and Sharptooth Catfish, while three non-native species occur: Common Carp, Largemouth Bass and Mozambique Tilapia.

2. Swakoppoort Dam

The Swakoppoort Dam is located in the Okahandja District with a maximum capacity of 63 Mm^3. This dam provides water to the town of Karibib and nearby mines and to Windhoek. Native fish species present are the Straightfin Barb and Sharptooth Catfish. Non-native species present are Common Carp and Mozambique Tilapia. Other Namibian species have been translocated to Swakoppoort: Greenhead Tilapia, Redbreast Tilapia and Banded Tilapia. No resort or accommodation facilities are present and NamWater has limited access to this dam for subsistence and recreational angling.

C. Nossob/Auob River drainage

1. Oanob Dam

Oanob Dam was constructed in 1990 and is situated on the Oanob River near Rehoboth (Fig. 29); it historically flowed into the Nossob River. Today, it is an endorheic catchment. The dam has a capacity of 35 Mm^3 and a surface area of 3.6 km^2. Lake Oanob Resort is popular especially during the summer months for water activities. It is well known for its Largemouth Bass and is visited by keen anglers (Chapter 10). It received fish from Hardap Dam in the early 1990s.

Figure 29. Oanob Dam on the Oanob River near Rehoboth (Photos: C. Hay).

2. Daan Viljoen Dam

The Daan Viljoen Dam (0.5 Mm^3) is a gravity concrete structure located on the upper Black Nossob River. Water is pumped from the Daan Viljoen Dam into the adjacent Tilda Viljoen Dam. The Otjivero, Daan Viljoen and Tilda Viljoen (1.2 Mm^3) dams are operated as a subsystem, known as the Molopo Subsystem Namibia. The operating rule for this subsystem allows the dams to supply the demands imposed upon them until the dams reach their respective minimum operating levels.

D. Omaruru River drainage

This drainage has a single reservoir, Omdel Dam west of Uis. This dam periodically dries up and has no value in terms of fish or fisheries.

E. Omuramba Omatako drainage

Omatako Dam (44 Mm^3) is an earth-filled embankment dam about 100 km north of Okahandja in the Otjozondjupa Region of Namibia. It is named after the Omatako Mountains, and it impounds the ephemeral Omatako River. Mozambique Tilapia and Sharptooth Catfish are present.

F. Kuiseb River drainage

Friedenau Dam is a gravity concrete dam in the Khomas Region located 38 km southwest of Windhoek. It impounds the Kuiseb River and provides water to nearby Matchless Mine. It has a capacity of 7 Mm^3 and was completed in 1972. Although not currently open to the public, it is known for Largemouth Bass.

VI. NAMIBIAN EXAMPLES OF INLAND ECOSYSTEM FUNCTIONING

The functioning of Namibia's inland waters and fisheries have been little studied; however, Kamutjonga Inland Fisheries Institute (KIFI), Namibia Nature Foundation (NNF), Norwegian Institute for Nature Research (NINA) and University of Namibia (UNAM) are cooperating to expand the available work. Four studies are highlighted here that consider (a) the relationship of flooding on the fish community structure of the Kavango River;[19] (b) fish assemblages and food-web structure of the Kavango, Kwando and Upper Zambezi Rivers;[5] (c) fish community succession in Lake Liambezi;[20] and (d) an evaluation of water level fluctuations and water quality on juvenile fishes in the littoral zones of the Upper Zambezi-Chobe River floodplains.[21]

A. Fish community response to seasonal flooding in the Kavango River[19]

This was the first published research on the structure and functioning of an inland fish community in Namibian waters. This work in 1992–93 differed from previous annual surveys by the MFMR in that routine sampling was conducted on a seasonal basis. Distinct community responses supported the concept that the annual flood pulse drives fish productivity.[22]

In the highly variable environment of the Kavango River flood cycle, fish diversity and catch were least in low-flow months, October–December, and increased with rising water levels through May-June, peak months of flooding. The reproductive strategies of K-selected piscivorous Largemouth

Bream and Tigerfish were in advance of flooding with young appearing in November during low flows, growing to juveniles by January–February (Fig. 30). Among the herbivores, Redbreast Tilapia bred earlier than Banded Tilapia, with young occurring in November and February, respectively. Numbers of Redbreast Tilapia peaked during February, whereas numbers of Banded Tilapia did so in May.

Many *r-selected* invertivores, especially cyprinids, were in relative synchrony with flooding and the stimulation of littoral zone plant growth, peaking in May–September and were prey for the developing piscivores spawned in October-November. Herbivores had their lowest relative abundance during peak flooding of May–June, increasing in September; this numerically inverse relationship with the invertivores was not replacement of the herbivores, but rather the swamping of the system with young *r*-selected invertivores. Scavengers and detritivores lagged behind herbivores and invertivores in the annual flood cycle. Research has demonstrated that invertivores experience high prey availability in newly flooded habitats, while detritivores have access to unlimited detritus during receding water phases.[7]

In short, the cyclical nature of the annual flood inferred that the most stable environmental conditions for the breeding of K-strategists were during low-flow conditions when the river is generally within well defined banks. Conversely, flooding was both a time of an unstable environment as well as rapidly increasing habitat for spawning sites, food availability and refugia favourable to *r*-strategists.

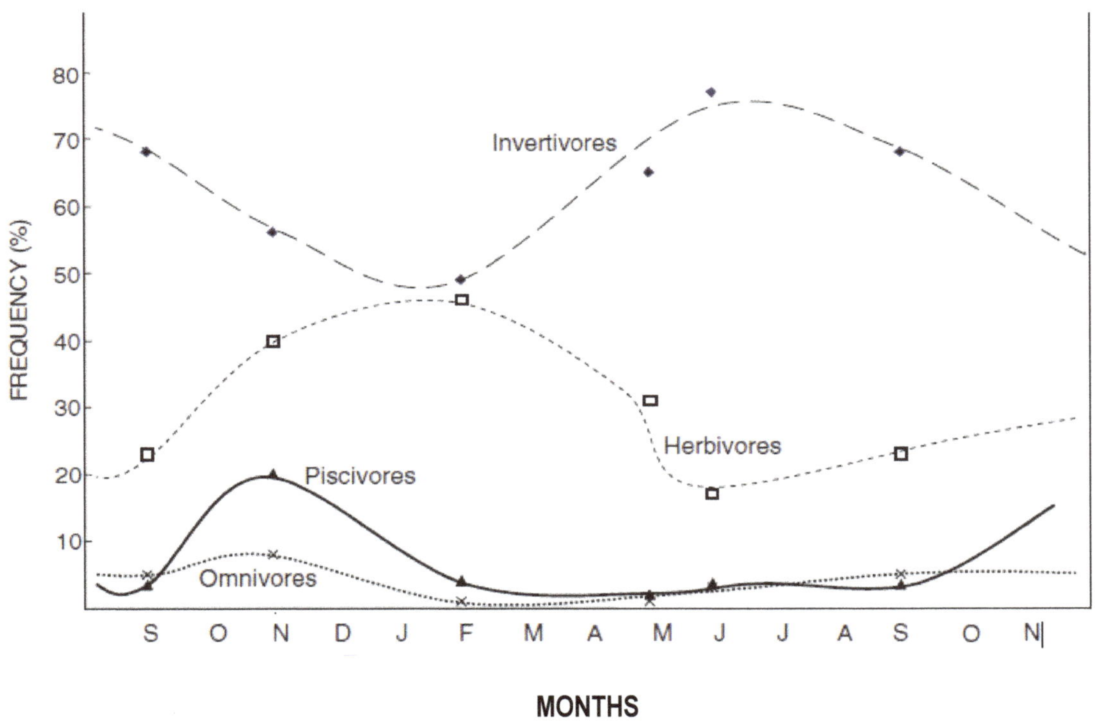

Figure 30. Seasonal trends in trophic levels of fish along the Kavango River floodplain in 1992–1993 (piscivores ×10).[19]

B. Fish assemblages and food-web structure in the Kavango, Kwando and Upper Zambezi Rivers[5]

Ecosystem productivity in Namibia's three perennial floodplain rivers (Kavango, Kwando and Upper Zambezi) was driven by the annual *flood pulse*, being highest in the Upper Zambezi and Kavango, March-June, and lowest in the Kwando, July-August. All three systems are interconnected during very high flood years and have a related drainage and ichthyofaunal history (Chapter 3).

Key findings include:

1. Seasonal assemblages were significantly different in the Upper Zambezi and Kavango Rivers, while seasonal assemblages in the Kwando River during rising and falling water were not significantly different;

2. There was a seasonal difference in fish communities among rivers. Tigerfish and Silver Robber were consistently represented in both the Upper Zambezi and Kavango Rivers in all seasons, while Silver Catfish, Striped Robber and Upper Zambezi Bulldog were present in the Kwando River for most seasons. The Spotted Squeaker occurred in the Kavango River throughout low, rising and falling water seasons;

3. The core habitat area of total fish communities was largest for the Kavango River, followed by the Kwando River and smallest for the Upper Zambezi River. Photosynthesis uptake and detritus data suggested that the Upper Zambezi and Kavango River's food webs were based on a broader range of resources derived from riparian vegetation from the floodplain, while the Kwando River food web was likely based on a narrower range of resources such as filamentous algae, aquatic macrophytes and dung from the numerous hippopotamuses. Piscivores did not occupy elevated trophic positions in the Upper Zambezi River compared to those in the Kavango and Kwando. This was presumably due to high fishing exploitation of tertiary consumers, resulting in a reduced food chain length; and

4. Kavango River fishes had more distinct niche areas with 81% of all possible group interactions resulting in overlap, whereas the niches in the Upper Zambezi and Kwando Rivers overlapped 87% and 95%, respectively (NOTE: overlap is reverse of distinctness). Core niche areas varied with species groups between rivers in a similar pattern; for example, barbs, benthic and herbivorous cichlids had large core niche areas, while citharines, labeos, piscivores and serranochromines had smaller core niche areas.

C. Fish community succession in Lake Liambezi[20]

The most important abiotic factor influencing fish communities in shallow, ephemeral lakes like Lake Liambezi is the hydrological regime. Life history strategies strongly influence the rate at which fishes are able to recover from major habitat disturbances such as desiccation. Succession and recovery of stressed ecosystems will first favour fast maturing opportunistic species with high reproductive rates (*r*-selection) followed by large, late-maturing periodic species that can endure longer periods, perhaps years, without good recruitment when environmental conditions are unfavourable.[24]

Fish succession was studied over an 8-year period in the ephemeral Lake Liambezi, 2007–2014. The lake had a thriving fish community of up to 35 species, the success of which was attributed to (1) the integration of multiple lake food chains and (2) behavioural adaptations to the changing dietary resources, permitting the lake to flourish despite its ephemeral nature. The lake filled in 2007 and then declined in 2008, but was sustained until major flooding in 2009 before attaining maximum levels in successive floods in 2010–2011, covering ~370 km^2, and reaching maturity in 2014.

Three phases of community succession were identified based on species influx to the lake during flooding. The lake was colonised in 2007 by fishes from the adjacent Upper Zambezi-Chobe River floodplains, predominantly by floodplain specialists: Straightfin Barb, Dashtail Barb, Upper Zambezi Bulldog, Sharptooth Catfish and Silver Catfish. The decline of the floodplain in 2008 resulted in a concentration of fishes in deteriorating environmental conditions, dominated in abundance by the Straightfin Barb (88%), Striped Robber (5.6%) and Silver Catfish (3.2%). The second successional flood phase in 2009 witnessed (1) a decline in the two barbs; (2) an explosive growth of the Slender Robber populations 2009–2010; and then (3) their rapid decline and replacement by the Striped Robber and Silver Catfish that dominated the fish community 2011–2014. The third flood phase in 2011 was characterised by larger, slower growing cichlids that increased steadily in abundance and became the dominant component of a 2,700 tonnes/annum artisanal fishery that developed on the lake as it approached a more stable state. The lake received no inflow after 2014 and by mid-2016 the lake recessed to shallow isolated lagoons accompanied by a collapse of the cichlid fishery with the Sharptooth Catfish dominating catch.

Data over the 8-year period implied that the fish community of Lake Liambezi underwent significant changes over the period of inundation from a colonising assemblage dominated by *r*-selected floodplain specialists to a *lacustrine* K-selected assemblage dominated by species adapted to more stable environments. It was hypothesised that predation by Sharptooth and Blunttooth Catfish may have contributed to lower numbers of certain prey species such as mormyrids. Sight-feeders like the Largemouth Bream would have been constrained by turbid waters.

D. Effects of water level and water quality on juvenile fishes in the littoral zones of the Upper Zambezi-Chobe River floodplains[21]

The behaviour of juvenile fishes in the *littoral* zones of the Upper Zambezi-Chobe River floodplain was assessed between March 2017 and February 2018 during rising, peak, recession and low flooding phases. Littoral zones represent the land:water interface along the shorelines of rivers and lakes, extending onto floodplains during high water. The Slender Robber (15.5%), Straightfin Barb (15.2%) and Hyphen Barb (14.2%) were the most numerous species during the rising phase, with all cyprinids combined representing 40.7% of the catch. Cyprinids also were dominant during receding (45.3%) and low flooding phases (40.9%); however, the dominant species varied by flood stage. Species richness was lowest on the rising flood and highest during flood recession.

The Zambezi River Bream (20.5%) and Redbreast Tilapia (19.1%) along with other cichlids (63.1%) dominated the peak flooding phase, characterised by high dissolved oxygen content, neutral pH and moderate water temperatures. Juveniles of the commercially important Threespot Tilapia and Redbreast Tilapia were important during all flood phases. They reached highest relative abundance during peak flooding, mimicking the pattern observed[19] in the Kavango River (VI.A, above). Both utilise plant material as a food source during their early life history with the latter being a true herbivore. Their populations in the littoral zone crashed as waters receded and they migrated into the mainstream, having reached 70-80 mm SL.

The Straightfin Barb (23.6%), Southern Mouthbrooder (11%) and Okavango Tilapia (16.8%) dominated catch during the receding phase, when species richness was significantly higher. Straightfin Barb (19.8%), Slender Robber (19.7%) and Okavango Tilapia (17.6%) were most prevalent during the low flood phase. Tigerfish, Purpleface Largemouth and Sharptooth Catfish, all predators, were poorly represented in the littoral zone throughout the flood phases.

VII. REFERENCES

[1]Skelton, P.H. 2001. A complete guide to the freshwater fishes of Southern Africa. Struik Publ., Cape Town: 395 pp.

[2]Mendelsohn, J. and S. el Obeid. 2004. Okavango River: The flow of a lifeline. Struik Publ., Cape Town: 176 pp.

[3]Jacobs, F.J., T.F. Naesje, E.M. Ulvan, O.L.F. Weyl, D. Tiyeho, C.J. Hay, G. O'Brien and C.T. Downs. 2019. Implications of the movement behaviour of African Tigerfish *Hydrocynus vittatus* for the design of freshwater protected areas. J. Fish Biol: 1–9.

[4]Peel, R. A. 2012. The biology and abundance of three cichlid species from the Kavango and Caprivi regions, Namibia. M.Sc. Thesis, University of Namibia, Windhoek: 169 pp.

[5]Taylor, G.C., O.L.F. Weyl, J.M. Hill, R.A. Peel and C.J. Hay. 2017. Comparing the fish assemblages and food-web structures of large floodplain rivers. Freshw. Biol. 62: 1891–1907.

[6]Tweddle, D. and C.J. Hay 2011. Data collection analysis: Report on workshop conducted from 26–27 October 2011. Field Document no. MFMR/NNF/WWF/Phase II/5, Namibia Nature Foundation: 16 pp.

[7]Hay, C.J., T.F. Naesje, F.J. Jacobs, E.K. Simasiku and D. Tweddle. 2020. Results and lessons learned after eight years of monitoring gillnet catches in the Zambezi and Chobe Rivers. Norwegian Institute for Nature Research, Trondheim, NINA Proj. Rept. 1791.

[8]Tweddle, D., B.C.W. van der Waal and R.A. Peel. 2014. Distribution and migration of the Caprivi Killifish, *Nothobranchius capriviensis* Walters, Wildekamp and Shidlovskiy 2015, an assessment of its conservation status, and a note on other killifish in the same area. J. Am. Killifish Assoc. 47 (4–6): 134–151.

[9]Thorstad, E.B., C.J. Hay, T.F. Naesje, B. Chanda and F. Okland. 2005. Movements and habitat utilization of Nembwe, *Serranochromis robustus* (Günther, 1864), in the Upper Zambezi River. Afr. Zool. 40(2): 253 –259.

[10]Thorstad, E.B., C.J. Hay, F. Okland, N. Nickanor and T.F. Naesje. 2007. Spatial behaviour and management of Greenhead Tilapia (*Oreochromis macrochir*) in the Zambezi River, Namibia. Norwegian Institute for Nature Research, Trondheim, NINA Project Report 287: 26 pp.

[11]Simasiku, E.K. 2019. Dynamics of the floodplain fisheries of the Zambezi and Chobe floodplain, Zambezi Region, Namibia. PhD Dissert., University of Namibia, Windhoek: 179 pp.

[12]Skelton, P.H. 2021. *Pers. comm.*

[13]Hay, C.J. 1991. The distribution of fish in the Fish River, Namibia. Madoqua 17: 211–215.

[14]Seaman, M.T., W.E. Scott, R.D. Walmsley, B.C.W. van der Waal and D.F. Toerien. 1978. A limnological investigation of Lake Liambezi, Caprivi. J. Limnol. Soc. South. Afr. 4: 129–144.

[15]van der Waal, B.C.W. 1980. Aspects of the fisheries of Lake Liambezi, Caprivi. J. Limnol. Soc. South Afr. 6 (1): 19–31.

[16]Peel, R.A., D. Tweddle, E.K. Simasiku, G.D. Martin, J. Lubanda, C.J. Hay and O.L.F. Weyl. 2015. Ecology, fish and fishery of Lake Liambezi, a recently refilled floodplain lake in the Zambezi Region, Namibia. J. Aq. Sci. 40: 417–424.

[17]Simasiku, E.K. 2014. Assessment of the Lake Liambezi fishery, Zambezi Region, Namibia. M.Sc. Thesis. University of Namibia, Windhoek: 154 pp.

[18]Jacobs, F.J., P.G. Jacobs, C.J. Hay and T.F. Naesje. 2021. Status update of the endemic and critically

endangered cave catfish *Clarias cavernicola* Trewavas 1936, from the Aigamas Cave system, Namibia. Afr. J. Aq. Sci. 46(1): 18–21.

[19]Hocutt, C.H. and P.N. Johnson. 2001. Fish response to the annual flooding regime in the Kavango River along the Angola/Namibia border, pp. 449–464. In: A.I.L. Payne, S.C. Pillar and R.J.M. Crawford (Eds.). A decade of Namibian fisheries science. S. Afr. J. Mar. Sci. 23.

[20]Peel, R.A., J.M. Hill, G.C. Taylor, D. Tweddle and O.L.F. Weyl. 2019. Species succession and the development of a lacustrine fish community in an ephemeral lake. J. Fish. Biol. 95: 855–869.

[21]Simasiku, E.K., C. Hay and J. Abah. 2021. Effects of water level and water quality on small-sized and juvenile fish assemblages in the littoral zones of the Zambezi/Chobe floodplain. Afr. J. Ecol. 2021: 1–13.

[22]Junk, W.J., P.B. Bayley and R.E. Sparks. 1989. The flood pulse concept in river-floodplain systems, pp.110–127. In: D.P. Dodge (Ed.). 1986. Proceedings of the International Large River Symposium (LARS), Honey Harbour, Ontario, September 1986. Can. Spec. Publ. Fish. Aquat. Sci. 106.

[23]Winemiller, K.O. 1990. Spatial and temporal variation in tropical fish trophic networks. Ecol. Monogr. 60: 331 –367.

[24]Winemiller, K. O. 1996. Factors driving temporal and spatial variation in aquatic floodplain food webs, pp. 298 –312. In: G.A. Polis and K.O. Winemiller (Eds.). Food webs: Integration of patterns and dynamics. Chapman and Hall, New York.

CHAPTER 7

FISH AND FISHERIES OF THE CUVELAI RIVER SYSTEM

by

Ben C.W. van der Waal

I. Introduction

There are no perennial rivers originating in northern Namibia, but that is easily compensated for by the unique and special rivers on its borders. The Cuvelai is a prominent example with most of its catchment in southern Angola, where rainfall is considerably higher than in the Namibian reaches. This river basin hosts a dense rural population along the shared border with Angola, estimated at 1.3 million persons, most of whom live in Namibia in the administrative Omusati, Oshana, Ohangwena and Oshikoto Regions (Chapter 1, Fig. 1). Fish form one of the valuable renewable resources that is used extensively when it is available. The structure and functioning of this unique basin are so interlinked that the Namibian/Angolan border is disregarded in the following discussion and the reference to Kunene/Cunene River is used interchangeably.

This drainage has been underexplored for decades. The climate is harsh with temperatures often over 40°C during the day with little shade or surface water. Namibia's freedom fighters were based in the region from the 1960s until the United Nations Mt. Etjo Accord was signed in December 1988, followed by the Ruacana Accord in April 1989 paving the way for Namibia's independence in 1990. Plus, an Angolan-wide civil war loomed 1975–2002 between the populace. The Angolan frontier between the Kunene and Zambezi River remains remote today and littered with abandoned ammunition and land mines.

A. Description of the Cuvelai River system

The Cuvelai River system lies wedged between the Kunene River in the west and the Kavango River in the east (Fig. 1). It originates almost 300 km north of the border in Angola where the rainfall is nearly 1,000 mm per year. However, only the main channel of the Cuvelai itself is a permanent river with long deep pools and granitic rock bars holding water over dry years and seasons. The eastern tributaries of the catchment run over thick Kalahari sand and floodwater seldom reaches the lower section of the catchment. A different system is found in the west and south where a maze of parallel and anastomosing shallow channels flow over fine-clay soil. This system is 130-km wide at the Angolan-Namibian border, but then converges like an inverted delta into the Omadhiya Wetlands, consisting of a series of large shallow pans that keep water for most years, but dry out during droughts. Although this wetland is more permanent than the *iishana* (see *oshana*) the water becomes very salty, brackish and muddy over time allowing only the most adaptable fish to survive. In flood years, water drains into Etosha Pan and can fill most of the pan up to a depth of 1+ m, but all this water evaporates in a few months. Good flood (*efundja*) years allow flamingos and pelicans to breed successfully as there are sufficient plankton and fishery resources to raise their young.

Etosha has one other source of water and fish coming from the east, the Omuramba Owambo. This catchment lies over calcareous and sandy soils. Wells cut into the water-rich calcareous rock by farmers in the Tsintsabis area offer small refugia for catfish enabling spawning and regular repopulation of the eastern part of Etosha (Fischer's Pan) during major floods.

B. Ephemeral nature of the Cuvelai system

The Cuvelai system has a fascinating geomorphic history that explains its uniqueness today (Chapter 3). Until around 2 million years ago (Ma) in the late Pliocene/early Pleistocene, the *palaeo*-Kunene River drained southwards into a huge inland lake, Lake Kunene, a remnant of the Etosha Owambo Basin covering an area ten plus times the size of present day Etosha Pan.[1,2] The *proto*-Kunene River, an Atlantic coastal river, cut back into the Great Escarpment, eventually diverting the flow of the *palaeo*-Kunene toward the coast. This river capture-took place near Calueque and explains the presence of Ruacana Falls in its middle course (Chapter 3).

Lake Kunene shrank to a smaller size with the loss of the *palaeo*-Kunene flow, but the remaining Cuvelai maintained a semi-permanent wetland environment until ~4,000 years ago (Ka), based upon fossil remains of swamp-adapted Sitatunga (*Tragelaphus spekei*) and other water-dependent vertebrates found at the deltaic mouth of the present Ekuma River in Etosha Pan.[2] Although the *palaeo*-Kunene waters were all diverted westwards down the Ruacana Falls and no longer southwards to Etosha, the Cuvelai and its fish fauna still retain natural connections with the Kunene that are seasonally functional today.

The river capture left the Cuvelai and other disjunct tributaries as the only seasonal sources of water for that ancient lake which over time has dried up leaving the soils salinised in the present day Etosha Pan and other smaller salt pans around it. During flood conditions, the extensive delta becomes an integrated network of shallow grassy channels with very low gradient, draining slowly and converging towards the lowest point in Etosha Pan. The catchment is bordered on the south by uplifted dolomite hills, forming an endorheic system with no connection to an ocean.

C. Frequency of flooding

The flows of the *iishana* are unpredictable as they are dependent on rainfall, both locally and in the upper catchment in Angola. Rains start in November, peak in February or March and diminish by April. The *iishana* fill with local floodwater first, followed by flooding from the main Cuvelai and Mui Rivers, that overflow into the shallower channels and *iishana* that generally drain southward.

This forms the *efundja*, a period of flooding that disrupts and inconveniences local communities, but also a time of bountiful protein in the form of frogs and fish brought by the flood. Such *efundja* are, however, unpredictable and occur about once in six years. Medium floods and major *efundja* occur 46% of the time, smaller floods with no fish 21% and droughts 33% of the years. Serious droughts in the past have led to human famine and deaths. Figure 2 summarises available data over a period of 80 years.[3] The duration of droughts can vary from 1 to 7 years with a mean of 2.7 years. This duration has serious implications for fish survival in the *iishana* region as most waterbodies will become either dry or have very high salinities.

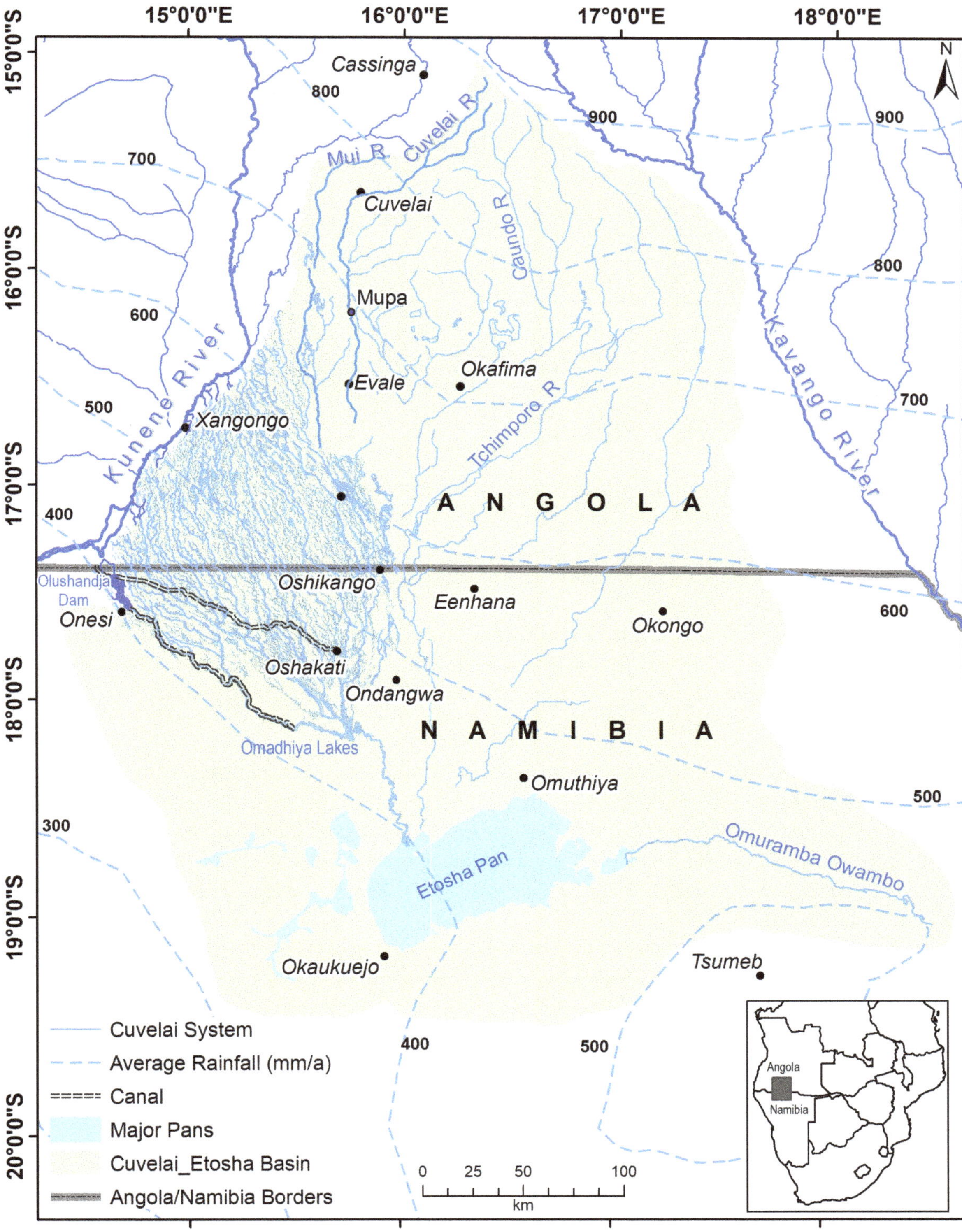

Figure 1. Cuvelai River with its tributaries and neighbouring river systems (Courtesy of M.H.T. Hipondoka, University of Namibia (UNAM)).

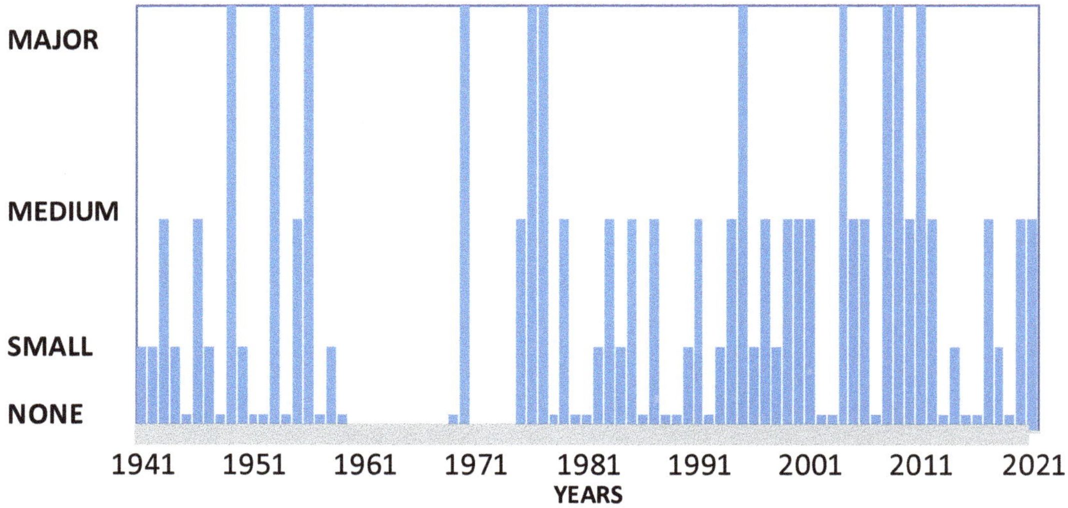

Figure 2. Frequency of occurrence of small to major floods in the Cuvelai basin over an 80-year period (13 years missing). The difference between medium and major floods is hundredfold.[3]

D. Where do all the fishes suddenly come from?

Persons acquainted with the parched *iishana* in the Omusati, Oshana, Oshikoto and Ohangwena Regions (Fig. 1, Chapter 1) in Namibia are amazed at the rich fish life during major *efundja* when fishing is practised at culverts and bridges where *iishana* are crossed (Chapter 8). Different anecdotal explanations were offered for over 100 years for the sudden plentiful presence of fish in the *iishana*:

1. Fish survive in the dried mud in the bottoms of *iishana* and manage to breed successfully and spawn in the multitudes. (Counter: Although it is possible for some fish to survive in liquid mud, it is only Lungfish that can survive in dry mud, but they do not occur here);
2. Fish eggs survive in the dry mud and hatch with fresh flooding. (Counter: Only the specialised Caprivi Killifish, present only in the Zambezi Region, lays eggs that survive in dried mud);
3. Fish survive in the pools of the upper permanent reaches of the Cuvelai and Mui Rivers and spawn there at the beginning of the rainy season to spread into the lower river course. (Counter: It is not feasible for these relative small fish communities to spawn such phenomenal numbers, spread throughout all the *iishana* and grow to mature size in the course of a few weeks);
4. Fish survive in the hundreds of ponds and wells dug by the local people themselves and by the Angolan and Namibian Governments from the 1960s. (Counter: To a degree this is possible, but these ponds and wells provide limited habitat and dry up during longer droughts. Additionally, similar mass fish migrations were reported before any deep dams had been built in the region); and
5. Fish originate from the Kunene and cross the divide during high floods into the Cuvelai basin, then follow the floodwaters downstream to populate the *iishana*. (Counter: This is

impossible as Kunene flood water cannot spill into the Cuvelai catchment, based on a 1926 survey that concluded that the divide between the Kunene and *iishana* is at least 10 m higher than the Kunene during peak flooding).

Explanation 1 and 2 do not apply to the Cuvelai River system as the fish present are not specially adapted for survival in dried mud. Explanation 3 cannot explain the sudden abundant appearance of the fish although recent evidence confirms the presence of small numbers of offspring during small and medium floods in the Cuvelai and *iishana* receiving the flood water. Explanation 4 is plausible, but only to the limit of habitat available for successful spawning and dispersion from one year to the next.

Relative to the fifth alternative above, is it possible that fish actually migrate from the flooding Kunene into the *iishana* with the large floods (*efundja*)? During the 19th and early 20th centuries when Angola and South West Africa were colonised, the sudden occurrence of fishes during large floods was explained by the Kunene bursting its banks and overtopping into the *iishana*. This explanation was refuted, however, after land surveys in 1926 concluded there was no spillway, but a continuous low divide separating the two rivers at least 10 m higher than the highest Kunene flood level.[4]

In response to a survey of fishing activities in the Omadhiya Wetlands in 1999, a third of the 39 local fishers questioned about the origin of the fish in the pans said they came from the Kunene. This suggestion was not taken seriously until 2017 when mass migrations of fish from the Kunene into small tributaries originating on the divide and into the top end of *iishana* were reported to scientists by locals living on the divide (Fig. 3). The divide was visited in February 2020 and possible surface water connections between the two catchments in the form of sheets of standing water were observed, potentially permitting fish migration from the Kunene to the top end of *iishana* (Fig. 3). However, such migrations were not observed in 2020 due to insufficient flooding, reflected in the low number of fish available in *iishana* later in the season.[5]

The upper *iishana* and Kunene tributaries are connected during heavy downpours by sheets of almost standing water offering access for adult and young fish to migrate down the *iishana* into virgin environments with plenty of food and shelter. The divide's headwaters on the Kunene side dry out much earlier than the top end of the *iishana*, preventing fish to return to the Kunene River.

Figure 3. Catch using a small-meshed net during the 2017 *efundja*. The main species during *efundja* are Straightfin Barb and Threespot Barb (left, Photo: M. Hipokdoka). Shallow temporary connections between the Kunene and Cuvelai basin's divide near Xangongo, southern Angola, February 2020 (Photo: B. van der Waal).

II. Biology of the Main Fish Species in the *Iishana* and Functioning of the System

A. Fish species of the Cuvelai system

The main fish species of the *iishana* are well known for their migratory colonising habits and wide adaptability. They are dominated by the cyprinids (Straightfin Barb, Threespot Barb), clariids (Sharptooth Catfish, Blunttooth Catfish), schilbeids (Silver Catfish) and cichlids (Threespot Tilapia, Banded Tilapia, Southern Mouthbrooder and Redbreast Tilapia). Other less common fishes in the *efundja* include mormyrids and alestids. Table 1 lists 46 described species plus 1 undescribed form from the Cuvelai basin, totalling 47 species. All are considered 'Cuvelai fauna' for the purpose of this book even though the table distinguishes between the Cuvelai River itself (some 100–300 km north of the Angolan/Namibian border); *iishana* in Namibia; Omadhiya Wetlands at the bottom of the *iishana* region and near Etosha National Park; and the man-made pipeline and open canal conveying Kunene water to the *iishana* region.

Some 47 species are reported in Table 1 for the Cuvelai basin, Namibia and Angola combined, considerably lower than the adjacent Kunene (82 species) or Okavango system (104 species).[6] The manmade Calueque-Oshakati Canal and Olushandja Dam has a higher diversity than the *iishana* and the Cuvelai River itself, which is not unexpected given (1) a rather permanent supply of water and (2) the dam and canal offer access to at least 14 Kunene species not previously reported from the Cuvelai, although most of these were temporary vagrants unable to establish viable populations. The *iishana* and the sump lakes at Omadhiya Wetlands offer suitable habitat only to those species that have the characteristics of colonising species (Chapter 1). The *iishana* fish community is dominated during *efundja* by two barbs, the Sharptooth Catfish and Silver Catfish, with another ten fish species making up only a small percentage (Fig. 4). The low occurrence of Sharptooth Catfish in 1976 is the result of not having gill nets available to catch these larger fish.

Table 1. Occurrence of fish species in the Cuvelai basin, southern Angola and Namibia (Cuvelai = CV; Calueque-Oshakati Canal and Olushandja Dam = CAN; *iishana* = IIS; Omadhiya Wetlands = OM).

		CV	CAN	IIS	OM
Mormyridae: Snoutfishes					
Slender Stonebasher	*Hippopotamyrus ansorgii*		X		X
Cunene River Stonebasher	*Hippopotamyrus longilateralis*		X		
Cunene Bulldog	*Marcusenius multisquamatus*	X	X	X	
Western Bottlenose	*Mormyrus lacerda*	X	X		
Cunene Churchill	*Petrocephalus magnoculis*	X	X	X	X
Dwarf Stonebasher	*Pollimyrus castelnaui*	X	X	X	
Cyprinidae: Barbs, minnows, yellowfishes and labeos					
Upjaw Barb	*Coptostomabarbus wittei*		X		
River Sardine	*Engraulicypris howesi*	X	X	X	X
Blackback Barb	*Enteromius barnardi*	X	X		
Hyphen Barb	*Enteromius bifrenatus*		X	X	
Redspot Barb	*Enteromius kerstenii*		X		
Line-spotted Barb	*Enteromius lineomaculatus*			X	
Papermouth	*Enteromius mattozi*	X	X	X	
Copperstripe Barb	*Enteromius multilineatus*	X			
Straightfin Barb	*Enteromius paludinosus*	X	X	X	X
Beira Barb	*Enteromius radiatus*	X	X	X	
Thamalakane Barb	*Enteromius thamalakanensis*		X		
Threespot Barb	*Enteromius trimaculatus*	X	X	X	X
Longbeard Barb	*Enteromius unitaeniatus*	X	X		
Cunene Labeo	*Labeo ansorgii*	X	X	X	
Silver Labeo	*Labeo ruddi*		X		
Distichodontidae: Citharines					
Dwarf Citharine	*Nannocharax machadoi*		X		
Alestidae: Robbers and Tigerfish					
Striped Robber	*Brycinus lateralis*	X	X	X	
Silver Robber	*Micralestes acutidens*	X	X	X	
Slender Robber	*Rhabdalestes maunensis*	X	X	X	

Hepsetidae: African Pike		**CV**	**CAN**	**IIS**	**OM**
Southern African Pike	*Hepsetus cuvieri*		X		
Clariidae: Air-breathing Catfishes					
Sharptooth Catfish	*Clarias gariepinus*	X	X	X	X
Blunttooth Catfish	*Clarias ngamensis*	X	X	X	X
Snake Catfish	*Clarias theodorae*			X	
Mochokidae: Squeakers and Suckermouths					
Leopard Squeaker	*Synodontis leopardinus*		X	X	
Largespot Squeaker	*Synodontis macrostigma*		X		
Upper Zambezi Squeaker	*Synodontis woosnami*	X		X	
Schilbeidae: Butter Catfish					
Silver Catfish	*Schilbe intermedius*	X	X	X	X
Anabantidae: Labyrinth Fishes					
Many Spined Climbing Perch	*Ctenopoma multispine*		X		
Cichlidae: River Bream, -Largemouth Bream and Tilapia					
Redbreast Tilapia	*Coptodon rendalli*	X	X	X	X
Threespot Tilapia	*Oreochromis andersonii*	X	X	X	X
Greenhead Tilapia	*Oreochromis macrochir*	X	X	X	X
Cunene Dwarf Bream	*Orthochromis machadoi*	X	X		X
Southern Mouthbrooder	*Pseudocrenilabrus philander*	X	X	X	X
Cunene Bream	*Sargochromis coulteri*	X	X		X
"Bream"	*Sargochromis* sp.		X		
Thinface Largemouth	*Serranochromis angusticeps*		X		
Purpleface Largemouth	*Serranochromis macrocephalus*		X		
Brownspot Largemouth	*Serranochromis thumbergi*		X		
Thicklipped River Bream	*Thoracochromis albolabris*		X		
Namib River Bream	*Thoracochromis buysi*	X	X	X	X
Banded Tilapia	*Tilapia sparrmanii*	X	X	X	
Procatopodidae: African Lampeyes					
Johnston's Topminnow	*Lacustricola johnstoni*		X	X	
Total number		**28**	**43**	**27**	**14**

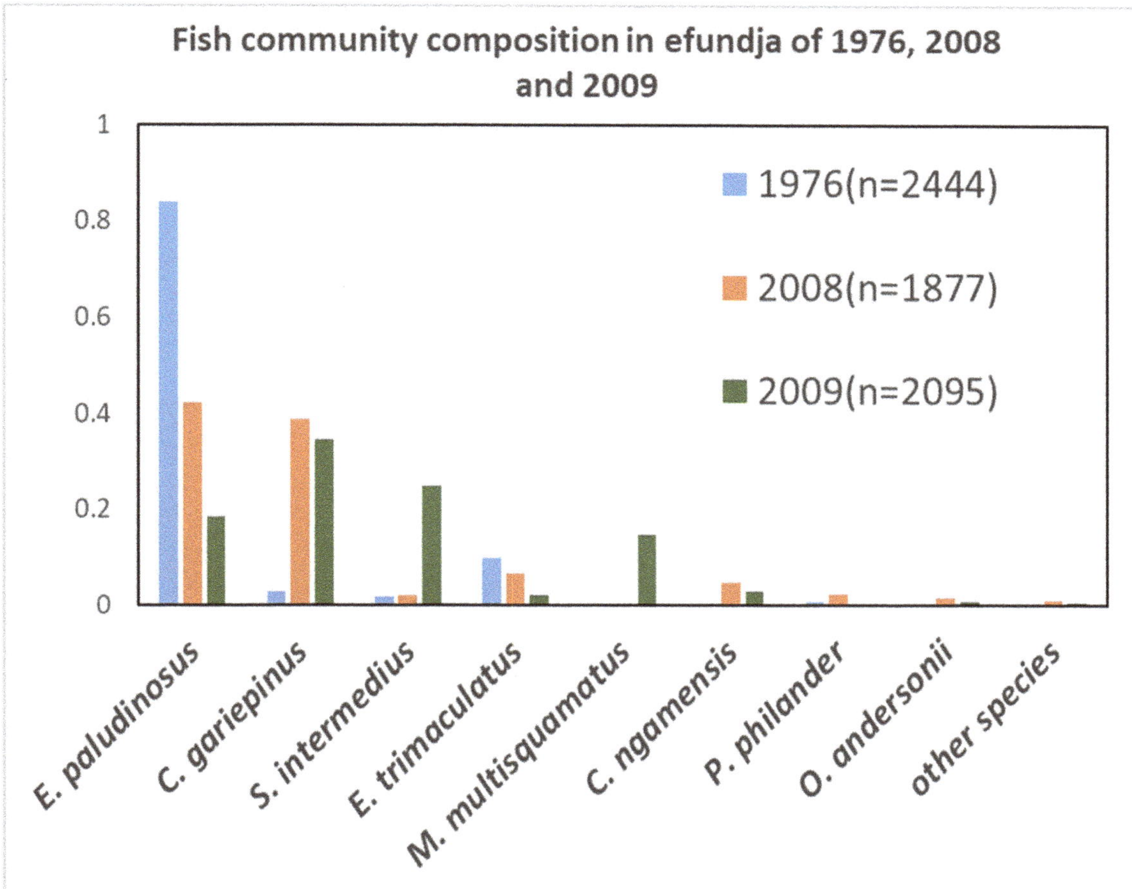

Figure 4. Composition and relative abundance (%) of experimental fish catches during 1976, 2008 and 2009 *efundja* that flooded *iishana* in northern Namibia.[7,9]

B. Fish migration in the *iishana*

Efundja last from February to April in good years. There is a general slow movement of water from north to south with one or more peaks, dependent on floodwater coming from Angola and local downpours. Fish generally move with the current. The most efficient gear to catch migrating fish is the modern funnel net, up to 20 m long and set facing upstream below constrictions in the *iishana*, e.g., road crossings (Chapter 8). Traditional funnels, traps and fishing with hook and line remain common. Once *iishana* stop running and fish migrations end, various types of seine nets and traditional thrust baskets are employed to catch fish stranded in deeper pools. Clashes sometimes develop between fishers and cattle herders as seine netting stirs up mud and renders the water undrinkable for livestock.

Fish seem to be more available in certain *iishana* during medium floods, reflecting abundance of breeders in refuges connected to those *iishana*. Fish size can vary between *iishana* dependent upon the source of the fish, i.e., either from (1) a nucleus of mature fish breeding in the Cuvelai or in deep semi-permanent dams in the *iishana* or (2) arriving from across the Kunene divide facilitated by local rainstorms. This difference between fish communities of *iishana* tended to disappear during major *efundja* and these populations also showed growth in size over time.

All fish migrating downstream may end up in the Omadhiya Wetland pans and even Etosha Pan, the ultimate sump lake of this endorheic system. These fish face annihilation in time as the water quality deteriorates over months with an intolerable high salinity and eventual desiccation. Some fish do, however, occasionally succeed in spawning in these lower reaches resulting in a small upstream migration of young fish. Young barbs and tilapia were observed migrating upstream against the current in March 1976 and February 2020. These migrations of only younger fish demonstrated successful spawning in some of the downstream water bodies with the juveniles dispersing upstream. Some fishers harvested upstream migrants by placing valved traps facing downstream.

C. Fish passage into the *iishana* region via the Calueque-Oshakati canal

The need for drinking water in the densely populated *iishana* region (including the Oshana, Ohangwena, Omusati and Oshikoto Regions) of Namibia necessitated the development of a pipe and canal system to pump Kunene water to those regions. Originally, a pipeline was in operation from Ruacana into an earthen canal on top of the divide from where water was gravity-fed to Ogongo and Ondangwa. This system was upgraded in the early 1970s to a larger pipeline from the newly-built Calueque Dam in southern Angola to the Olushandja balancing dam built right on the divide between the two rivers, necessitating two dam walls! The intake screens of the pumping scheme are coarse, allowing smaller fish to be entrained into the pipeline and then into the canal.

A lined open canal takes water to purification dams in Ogongo and Oshakati from where potable water is distributed by pipelines to communities. The open canal is not well isolated from the *iishana*, especially during major floods, and this allows many Kunene fish direct access from the canal into the *iishana*. Table 1 lists fish species collected in the *iishana,* Omadhiya lakes and canal with Olushandja Dam and smaller balancing dams near the canal. The canal and associated water bodies host 43 fish species (19 more species than recorded for *iishana*) dependent upon flood discharge, but they do not appear to be successful in populating the *iishana* and none have thus far populated the Cuvelai River upstream. Some of the larger cichlids like Purpleface Largemouth were found frequently in balancing dams and Olushandja, indicating successful breeding there, but it is unlikely they can successfully establish in the unstable *iishana* environment.

D. Fish diversity in *iishana*

The *iishana* represent a special habitat type that is advantageous to certain fish species, but not to others. Characteristics of *iishana* that can seriously impact many fishes include an unstable, murky, shallow, warm, oxygen-stressed and saline environment. Only fish adapted to such unstable conditions are successful (Chapter 1).

Data collected of fish community composition of *efundja* floods in 1976, 2008 and 2009 are presented in Figure 5, demonstrating the domination of the fish community by Straightfin Barb, Sharptooth Catfish and Threespot Barb. However, Silver Catfish, Blunttooth Catfish, Threespot Tilapia and Bulldog were also important during certain years. In terms of weight, the biomass in the *iishana* was totally dominated by the larger catfish species, representing >80% weight of fish sampled in 2008 (Fig. 5).

The *iishana* are not suitable for all species as demonstrated by the Calueque-Oshakati Canal having 19 more species. Most fish species entering the canal were unsuccessful in establishing spawning populations in the *iishana*. There is the additional complication that fish living in *iishana* must survive in small refugia when the *iishana* dry out. Straightfin Barb and Sharptooth Catfish completely dominate the fish community in *iishana* in terms of numbers and biomass, respectively. They are able to thrive in the harsh environmental conditions of *iishana* for a number of reasons (Table 2). Principally, the Straightfin Barb has all the characteristics of an *r*-selected species: plasticity to a number of environmental factors, high fecundity and maturing at an early age (Chapter 1). The Sharptooth Catfish also has a short generation time, but its *labyrinth* system gives it an insurmountable advantage over other species in that it is not dependent on dissolved oxygen in the water by utilising aerial oxygen.

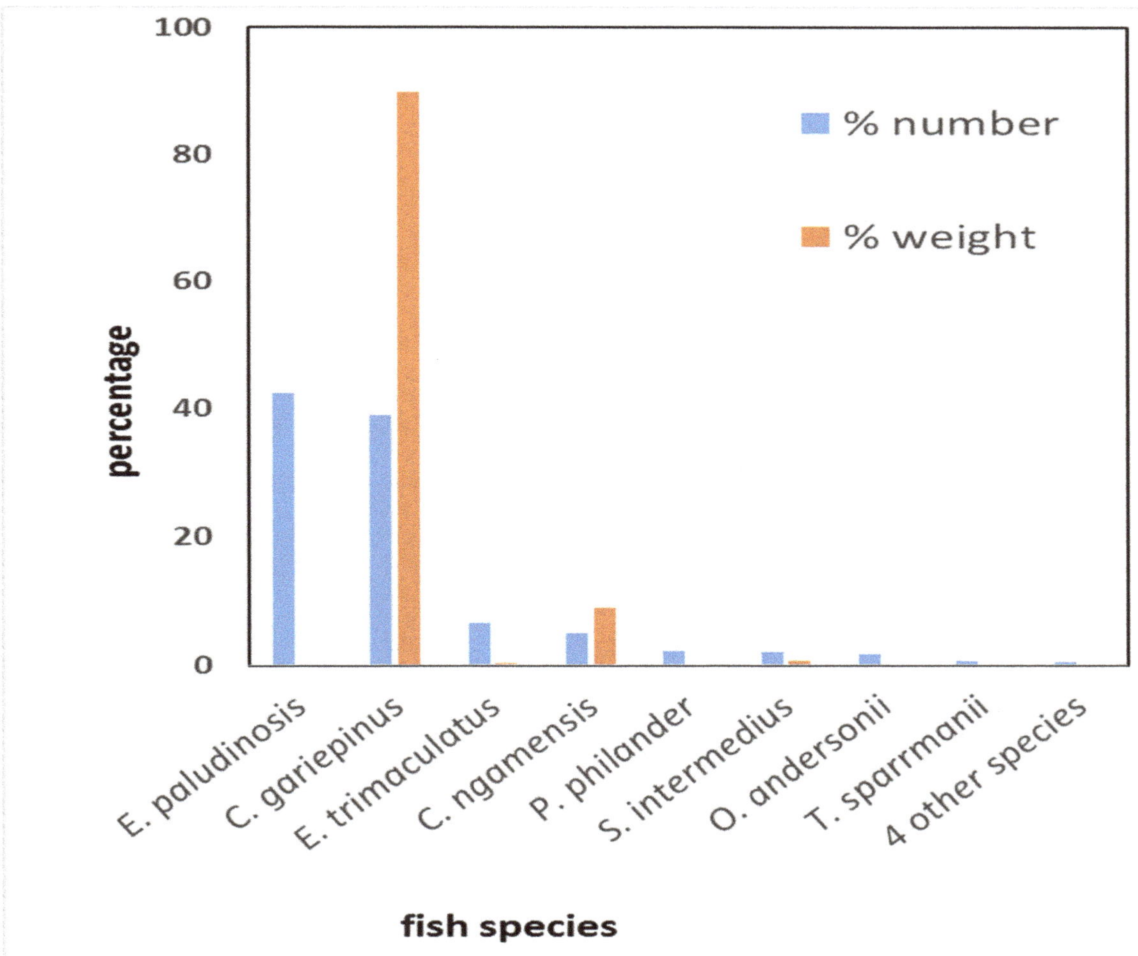

Figure 5. Relative importance of fish species in experimental catches in *iishana* during the 2008 *efundja* expressed as % numbers and % weight.

Table 2. Adaptive features of the Straightfin Barb and Sharptooth Catfish as successful colonisers of the *iishana*, Cuvelai River basin.

Straightfin Barb	Sharptooth Catfish
Can live successfully in warm, turbid and saline water with low oxygen content.	Can survive in warm turbid water and even in deoxygenated water and muddy environments as it breaths air to supplement aquatic respiration.
Short generation cycle, less than one year. Can breed when 5 cm long and only a few months old.	Short generation cycle, one per year. Can spawn when 40 cm long and less than a year old.
Multiple spawner, female repeatedly lays batches of eggs.	Multiple spawner, females can spawn more than once per season.
High fecundity, 1,000 eggs per female (5,000 max).	Very high fecundity, 100,000 up to 500,000+ eggs per female.
High growth rate, reach 5+ cm in a summer season.	Very high growth rate, 40 cm and 400 g per year.
Bottom feeder on aquatic insects, small crustaceans, algae, detritus.	Feed on small fish, mainly *E. paludinosus*, snails, terrestrial and aquatic insects. Exhibits pack hunting (Fig. 6).
Migrate downstream into unpopulated warm waters with little competition for food.	Migrate downstream together with its main prey species.
Migratory species, enters floodplains to spawn and feed.	Migrate onto floodplains to spawn and feed.
	Have a labyrinth structure for direct respiration of air.
	Have strong pectoral spines that can be locked into position to enable it to crawl overland from one pool to the next.

Figure 6. Example of pack hunting by Sharptooth Catfish on barb species, their main prey, at Mpundja Bridge south of Oshakati during the *efundja* 14 April 2008. The catfish swim underneath the small barbs while blowing bubbles and then rise to the surface, loudly clapping their mouths (Photo: B. van der Waal).

The single-most important reason for the success of catfish during an *efundja* is the readily available food. At the beginning of the *efundja*, both the barbs and catfish are small. While they swim downstream, the fish enter a pristine unpopulated environment with abundant planktonic and benthic invertebrate life. This is favourable to barbs and young catfish alike. When catfish have grown beyond 10 cm, their diet switches to barbs (Fig. 7). This enables them to grow to considerable size in a few months, representing the main catch harvested by the local population.

Figure 7. Night scene in a running *oshana* near Ogongo, Namibia, April 2008 with loud clapping and splashing sounds of Sharptooth Catfish juveniles hunting barbs in concerted chase (Photo: B. van der Waal).

E. Omadhiya Wetlands and Etosha Pan

The main *iishana* from the north carrying Cuvelai water converge with *iishana* from the west into a series of extensive shallow pans that are fed during medium to large floods and can hold water for consecutive years. The water quality under the latter conditions becomes so stagnant and saline that it cannot be used by the cattle herds grazing the expansive plain and fish eventually succumb. Up to 14 fish species have been recorded during fresh conditions, but this declines to only the resistant Straightfin Barb, Sharptooth Catfish and Threespot Tilapia as waters evaporate. When water levels drop and become more saline, the barbs die first and catfish develop lesions on their skin.

The Ekuma River draining the pan complex (Fig. 8) is a narrow winding stream with some deeper pools entering the Etosha Pan at the northwest. Sharptooth and Blunttooth Catfish, Straightfin Barb, Threespot Barb and Threespot Tilapia were present in the Ekuma River in both the 2008 and 2009 *efundja*, with four additional fish species in each year. Only Sharptooth Catfish and Threespot Tilapia have been reported in Etosha. Floodwaters stand for only a few months and all fish succumb long before the very saline water evaporates (Fig. 9).

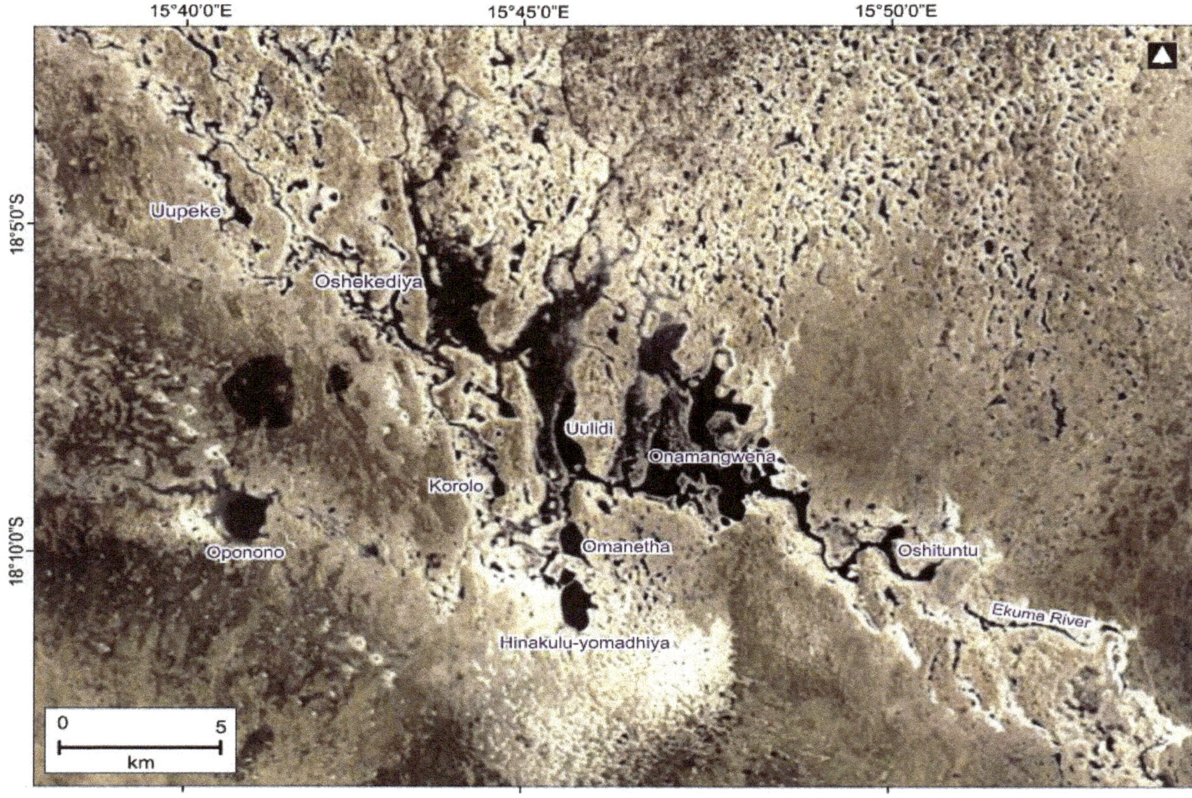

Figure 8. The Omadhiya Wetlands drain the *iishana* and consist of a series of pans connected during floods discharging water into the Ekuma River (lower right) in the northwestern corner of Etosha (GoogleEarth, 31 December 2007, map prepared by M.H.T. Hipokonda, UNAM).

Figure 9. Ekuma River draining the Omadhiya Wetlands during a medium efundja in 2021 (Photo: B. van der Waal).

F. Fisheries in the *iishana* during efundja

The region is characterised by sandy wooded ridges (*ofima*) between the *iishana* that are intensively used for settlement and farming by the highest density of people for rural Namibia. This population concentration is due to the combination of relatively fertile soil, availability of fresh groundwater resources accessible by shallow wells, grazing for stock and natural resources of which fish are a major part. The fishery resources have been utilised for centuries by the local population practising effective traditional methods that promoted conservation of the fishery (i.e., "wise use"). Fishing in *iishana* was controlled by the local chief or king, who only permitted fishing once the migrating fish were considered to be of reasonable size. Local tribesmen were allowed to fish in local waters and visitors had to ask permission and compensate for fishing rights by sharing part of their catch. Traditional fishing methods are still practised today, although the use of modern, easily-obtained, monofilament nylon netting has increased significantly (Chapter 8).

G. Fish productivity of the *iishana*

There is meagre information on the contribution of fisheries to the wellbeing and food security of communities living in the four regions representing the '*oshana'* region (Chapter 1). A 1-day long survey of all fishing activities at *iishana* crossing the road between Oshakati and Ondangwa (35 km) was undertaken in the 1976 *efundja*. Some 123 persons were actively fishing, using funnel nets, rod and line and dip nets. The estimated daily catch per funnel net varied from 10 to 200 kg per haul (mean 150 kg), anglers caught around 3 kg of fish and dip-netters 10 kg during the same time frame. The total 1-day catch was calculated at 4.2 tonnes. With an expected fishing season of two months during an *efundja* year, the section between Oshakati and Ondangwa was estimated to produce some 250 tonnes of fish.[7]

An indirect indication of the wealth of fish present in the Omadhiya pans can be deduced from the dietary needs of 3,000 pair of pelicans raising 1,000 young successfully over a 3-month period in 1971. This required a calculated mass of 975 tonnes of fish.[8] The pelicans were breeding on Etosha Pan, but caught fish in the Ekuma River and Omadhiya Wetland pans.

The Omadhiya Wetlands pans are regularly harvested after the flood during winter by fishermen who camp there, previously with large-wheel ox wagons, now with trucks, to fish and process catch. Fresh fish from Omadhiya were also recorded at the local market at Ondangwa in 1984. At the Open Market at Oshakati, fresh and dried fish from *efundja* are regularly on sale, but on small scale. Most of the fish caught is directly consumed or dried and kept or sold locally. The demand for fresh fish in the northern regions is considerable and since the 1970s horse mackerel (Maasbanker, *Trachurus capensis*) was sold at markets in all the main centres as a replacement for freshwater fish, especially in dry years when no local fish were available. The marine fish were accepted in time, but local freshwater fish are still preferred. The demand for catfish created a market for dried catfish from the Zambezi Region where it is less preferred, e.g., it was freely available on the open market in Oshakati in 2008.

The unreliability of floods prevents the development of a permanent fishery. However, the feasibility of aquaculture in the *iishana* should be explored for possible protein supplementation (*see* Chapter 12) and management of available water supplies.

H. Threats to the fish resource of the *iishana*

The fish life in the *iishana* is completely dependent on regular flooding by local rainfall and floodwaters from southern Angola. Apart from the availability of water, fish have to survive in refugia during the dry season and spawn successfully the next season. Overexploitation (Fig. 10), poisoning, pollution, nutrient enrichment and water quality deterioration have the potential to negatively affect fish populations the following year (s).

The main source of fish in the Cuvelai *iishana* originates from the Kunene River. Any disturbance of the flooding regime of the Kunene River itself may have a detrimental effect on the ability of fish to migrate successfully into the upper sections of western *iishana*. Impoundment and irrigation projects planned for the upper Kunene may detrimentally affect the potential fish migration into *iishana* during high floods. Furthermore, such development activities may change the composition of the natural fish fauna evolved to survive the variable environment of the region. Critical to the fisheries of the *iishana*, and food security of communities dependent upon the fisheries, is the proper construction and repair of roadways along the common divide with the Kunene; this is a case where upgrading of roadways will be detrimental unless precautions are taken for fish passage (Fig. 11).

The shift away from traditional gears and conventional social customs is indeed a challenge. Historically, fish were allowed free access from the north until a stage where the local chief gave permission to harvest the fish. This allowed most migrating fish to populate the *iishana*. The complete blocking of the *iishana* during floods (Fig. 10) is counterproductive, perhaps representing the classical clash between short-term gains and long-term benefits.

Figure 10. Typical situation at bridges during the 2008 *efundja* west of Oshakati. A dozen long and effective funnel nets block the passage of fish migrating downstream in the *oshana* (Photo: B. van der Waal).

Figure 11. Poorly designed culvert preventing upstream fish migration near Xangongo, southern Angola, along the divide between the Kunene River and western Cuvelai *iishana* (Photo: B. van der Waal).

III. References

[1]Hipondoka, M.H.T. 2005. The development and evolution of Etosha Pan, Namibia. Ph.D. Diss., Universität Wuerzburg: 152 pp.

[2]Hipondoka, M.H.T., H.J. Jousseb, J Kempf and D. Busche. 2006. Fossil evidence for perennial lake conditions during the Holocene at Etosha Pan, Namibia. S. Afr. J. Sci.102: 93–95.

[3]Mendelsohn, J.M. In press. Droughts and floods in Angola's Cuvelai Basin. In: A. Gomes (Ed.): Drought in southwestern Angola (Book to be published in Portuguese).

[4]Stengel, H.W. 1963. The Cuvelai. A contribution to the hydrography of South West Africa, pp. 368–380. In: H.W. Stengel (Ed.). Wasserwirtschaft / Waterwese / Water Affairs in SWA, Afrika-Verlag Der Kreis, Windhoek.

[5]Van der Waal, B.C.W., M.H.T. Hipondoka and M.N. Ekandjo. 2021. Towards answers to the secret of the plentiful fish life in running iishana. Namibian J. Environ. 5(B): 7–10.

[6]Skelton, P.H. 2019. The freshwater fishes of Angola, pp. 207–242. In: B.J. Huntley, V. Russo, F. Lages and N. Ferrand (Eds.). Biodiversity of Angola, Science & Conservation: A Modern Synthesis. Springer Nature, Switzerland.

[7]Van der Waal, B.C.W. 1991. Fish life of the Oshana Delta in Owambo, Namibia and the translocation of Cunene species. Madoqua 17(2): 201–209.

[8]Berry H.H., H.P. Stark and A.S. Van Vuuren. 1973. White Pelicans *Pelecanus onocrotalus* breeding on the Etosha Pan, South West Africa during 1971. Madoqua. 1(7): 17–31.

[9]Ekandjo, M.N. 2009. A biological survey on the fish life of iishana of the Cuvelai System in northern Namibia, 2009 Efundja. Report to Ministry of Fisheries and Marine Resources, Directorate Aquaculture, Ongwediva Inland Aquaculture Centre: 23 pp.

CHAPTER 8
MANAGEMENT OF NAMIBIA'S INLAND FISHERIES

by

Clinton J. Hay

I. History of Inland Fisheries Management in Namibia

Community-based resources management (CBM) is not a new concept. Historically, locales were managed by communities who had regulations in place not necessarily to protect natural resources, but to secure access to a particular geographical area for specific social groups. Such CBM measures included fisheries. The low number of people and the use of traditional fishing gear (that has a natural effort limitation) prevented overfishing. Although traditional fishing gear types were the dominant fishing gear used, gillnets were utilised as early as 1893 in the Zambezi Region. Fish were originally caught for food security and trade. Certain areas and seasons were closed for fishing to allow recruitment to take place. The local headmen controlled fishing in these areas. Rules enforced by communities included:

- Banning of certain fishing methods such as drag netting (first used in 1990s in Zambezi Region), poison (1950s in Kavango and 1970s in Zambezi) and explosives;
- Limitations on large fences in Zambezi Region;
- No fishing on the floodplains before the chief had opened the season;
- No use of lamps during the night;
- No *bashing* to drive fish;
- Minimum mesh restrictions for gillnets; and
- Access to mainstream rivers was generally open to fishing, but floodplains were more restricted with outsiders requiring permission. Control was exerted over local embayments and ponds.

Inland fisheries management was under the jurisdiction of Nature Conservation within the Ministry of Agriculture prior to Independence, 1990. The Nature Conservation Ordinance (No. 4 of 1975) included inland fish. Initially, the only rules for inland fish were that:

- A person needed a permit to use a hook and line;
- No live fish were allowed to be released in any water body; and
- No poison, explosives or any intoxicating materials were allowed in any waterbody.

The interests of the local communities were not protected through this legislation and few management studies were conducted in north and northeast Namibia. The emphasis on inland fish was mainly on central Namibia with the focus on recreational fishing in man-made reservoirs (Chapters 6 and 10) and their potential for aquaculture (Chapter 12). Inland fisheries management was integrated into the newly established Ministry of Fisheries and Marine Resources in 1991, transferred from the Ministry of Agriculture, Fisheries, Water and Rural Development. Since Independence, there has been a strong focus on fisheries management in northeast Namibia where fish are an important protein source for rural communities.

The MFMR developed comprehensive legislation on inland fisheries after Independence through a transparent consultative process, visiting fishing regions in the country, and incorporating comments and suggestions from local communities and other stakeholders. Areas were revisited to ensure proposed suggestions were sufficiently incorporated. The emphasis was for communities to play a more prominent role in the management of fish stocks rather than a top-down approach. The Inland Fisheries White Paper was approved in 1995 and the Act and Regulations gazetted in 2003.

II. Inland Fisheries Legislation

The Inland Fisheries Resources Act and regulations came into force on 6 June 2003. These documents are "living documents", amended as the state of fish stocks changed over time. For instance, Gazette Notice No. 296 dated 2016 prohibits the use of monofilament nets. Gazette Notice No. 297 dated 2016 declared that the Zambezi-Chobe River system shared with Zambia and Botswana is a *fisheries reserve* for the period 1 December to the end of February every year and fishing is not allowed during this closed fishing period. During the initial writing of the act, monofilament gillnets were not a threat to the fish stocks with very few monofilament nets used. Since then and despite being illegal, monofilament gillnets became the dominant nets used due to them being three times more effective than multifilament nets.

The Act mandates the Minister of the MFMR to determine the inland fisheries policy in consultation with the regional council and any local or traditional authorities of the areas. The Minister must further take into account all relevant economic, social and environmental factors including the best scientific advice when formulating the general policy. Various parts of the Act stipulate:

- The conditions for fishing licenses and the registration of nets (Part IV);
- Illegal fishing methods (Part V);
- Management, conservation and protection measures, including (a) prohibition of the transfer of live fish between waterbodies, (b) regulations concerning threatened and endangered species and (c) conditions for the establishment of fisheries reserves (Part VI);
- Duties of Inland Fisheries Inspectors (part VII);
- Offences and penalties (Part VIII); and
- Regulations (Part IX) for (a) registration of nets, the fees payable, restriction on number of nets, tagging of nets and cancellation procedures; (b) prohibited fishing methods; (c) recreational angling, including minimum sizes; (d) recreational charter boats and the holding of fishing competitions; and (e) offences and penalties.

III. COMMUNITY-BASED FISHERY MODEL

A. Background of Community-based Management (CBM) in Namibia

Namibia is globally praised for its Community-based Natural Resources Management (CBNRM) programme to alleviate poverty that is based on the conservation and sustainable utilisation of natural resources through empowering local communities to share management responsibilities. CBNRM arose from the need to protect and manage wildlife in areas of human contact and conflict. It stems from the Namibian constitution that stipulates that the state is responsible for *"the maintenance of ecosystems, essential ecological processes and biological diversity and the utilisation of living natural resources on a sustainable basis for the benefit of all Namibians, both present and future"*. Communities that live near wildlife areas pay the highest price for wildlife conservation and should therefore have the right to share in the benefits derived from protecting and the sustainable use of these resources.

The first communal *conservancies* were gazetted by the Ministry of Environment and Tourism in 1998. These communal conservancies are democratic entities, recognised by the Namibian government, but not managed by government. Conservancies have fixed boundaries, managed by community committees elected democratically by community members. Conservancies have management plans, benefit distribution plans, rules regarding community meetings and financial accounting. Land-use plans define areas reserved for wildlife and areas for other activities such as farming. Local members are appointed as game guards to monitor these areas and report suspicious activities. Conservancies also have mechanisms in place to address human-wildlife conflicts.

B. Community-based Management for Fisheries

With this as background, the CBNRM concept was recently expanded whereby communities co-manage their fishery resources. The Inland Fisheries Act of 2003 makes provision for the establishment of *fisheries reserves* (Fig. 1). Importantly, designation of an area as a fisheries reserve (FR) does not imply that fishing will be prohibited. A range of rules and regulations can be implemented within each reserve as long as they are within the framework of the Inland Fisheries Resources Act.

1. Community ownership

Through more than a decade's intensive work by the Namibia Nature Foundation (NNF), University of Namibia and the MFMR, a model was developed to incorporate communities in the management of their fish resources on a sustainable manner benefitting both the resource and the community. Reserves can fall within the borders of an established conservancy or it may be a separate entity managed by the community. Eight milestones were developed to facilitate the establishment of FRs (Table 1).

Table 1. Milestones for the development of community-based management (CBM) fishery reserves.

1	Assess the willingness of communities. The first step is to confirm the willingness of a community in establishing a fisheries reserve. This is done through initial consultations with communities and other stakeholders.
2	Develop a fisheries constitution. A fisheries constitution is developed either as part of the conservancy constitution or as a stand-alone constitution. The fisheries constitution is the document as stipulated for the governance of the fisheries reserve.
3	Elect an Inland Fisheries Committee. An Inland Fisheries Committee is elected as per the constitution stipulating the composition, the roles, responsibilities and process to elect and remove members of the committee.
4	Conduct baseline surveys. Conduct baseline surveys to evaluate the impact protected areas have on the ecology in that area as well as communities residing there.
5	Identify the boundaries of the fisheries reserve. Community meetings are held to reach a consensus on the boundaries of the potential reserve. The entire community and Traditional Authorities (TA) must agree on these boundaries to avoid future disputes.
6	Develop a Fisheries Management Plan. A Fisheries Management Plan (FMP) is developed that outlines the rules and regulations of the fisheries reserves and stipulate the boundaries of the fisheries reserve.
7	Appoint fish guards and fish monitors. Appoint and empower fish guards to do rule enforcement and fish monitors as citizen researchers.
8	Official gazetting. Official recognition by Government through a gazette notice.

Community surveys are conducted at each potential reserve area to collect information for factoring into the respective Fisheries Management Plan (FMP). These surveys focus on the role fish play in the area, the current management practices in place, the species targeted, the livelihood strategies of the people and their willingness to establish and maintain FRs in their area. Workshops are held to discuss and outline the rules and regulations of each reserve, the benefit-sharing and the overall management strategy. These workshops provide the communities with the platform to ensure their (a) concerns are raised, (b) documented and (c) incorporated into the rules on how the FR should be managed. The FMP and the constitution are approved at the conservancy annual general meeting (AGM). The management plan contains information on the protocol to safeguard data, ensure data validation, data analysis and report writing. The Minister of MFMR prepares a formal gazette notice formalising each reserve once the appropriate documentation is received. All minutes of all meetings are documented; attendance registers kept; and letters of approval from the local villages, conservancy committee, the traditional authority and the Governor are included in the documentation provided to the Minister. The process is transparent and inclusive of all community members to ensure ownership of the process contributing to the sustainability of the reserve objectives.

Monitoring and evaluation processes are established after each FR is approved to ensure that the overall objectives are achieved as set out in the FMP. The elected fisheries committee members and the appointed *fish guards* and *monitors* are trained to ensure that they execute their functions effectively. Follow-up training workshops are held to ensure the continuity of governance, high-quality data recording and maintenance of effective patrols.

Areas in the Sikunga, Impalila, Lusese, Nakabolelwa, Balyerwa and Mayuni Conservancies were officially proclaimed as FRs through GRN gazette notices in 2020 and in Joseph Mbambangandu Conservancy in the Kavango Region in 2021 with other reserves planned inside and outside of established conservancies. The Sikunga Fisheries Reserve was the first to be proclaimed in 2015. Currently, 20 areas totalling 2,500 Ha have been gazetted as fisheries reserves.

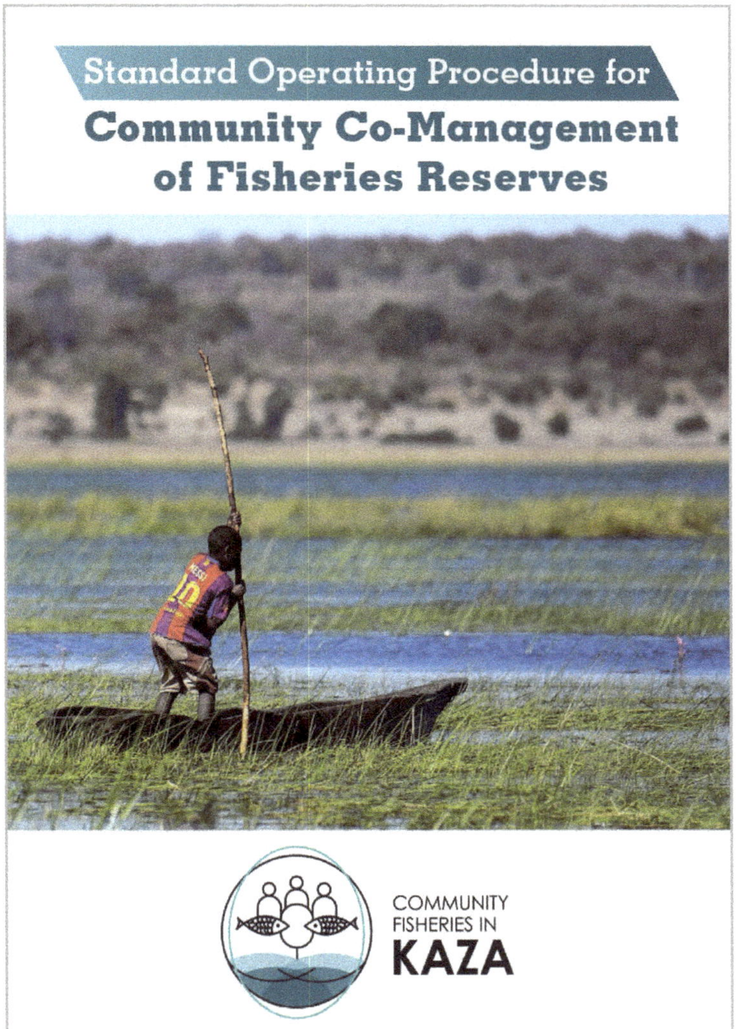

Figure 1. Guidelines to establish fisheries reserves along Namibian rivers (Namibia Nature Foundation).

2. Proclaimed Fisheries Reserves

Several areas have been declared as FRs in the Zambezi Region with others in the Zambezi and Kavango Regions at different stages of being gazetted (Fig. 2). Fisheries reserves are not necessarily 'no fishing zones'. They can all have different rules and regulation, but all must be in accordance to the Inland Fisheries Act and Regulations. The emphasis is on empowering communities to manage their resources while protecting the fishery resources for the benefit of present and future generations.

a. Fisheries reserves in the Kwando River: Three FRs were proclaimed along the Kwando River through MFMR gazette notices: the Luhingi Fisheries Reserve in the Mayuni Conservancy and the Qhuqhumupa and Kalume Fisheries Reserves in the Balyerwa Conservancy (Fig. 2). Only recreational fishing is allowed in these FRs. Recreational anglers practice catch-and-release and pay a levy to the conservancies. However, recreational angling is not a high priority along the Kwando River with tourists mainly interested in wildlife and birding. Large areas of the Kwando are protected through the Bwabwata, Mudumu and Nkasa Rupara National Parks. Fish guards appointed from the local communities patrol the FRs to enforce the rules and regulations.

Figure 2. A section of the Luhingi Fisheries Reserve in the Mayuni Conservancy (left). This area forms an isolated oxbow lake when the river level is low. Qhuqhumupa Fisheries Reserve in the Balyerwa Conservancy (right). Water lily species are visible in the backwater area of the reserve with marginal reeds and grasses (Photos: C. Hay).

b. Fisheries reserves along the Upper Zambezi-Chobe Rivers: The Impalila Fisheries Reserve is in the Impalila Conservancy along the Kasaya Channel connecting the Upper Zambezi and Chobe Rivers (Fig. 3). The Sikunga Fisheries Reserve is located in the Sikunga Conservancy where local fishers have reported higher catches both in and adjacent to the reserve. Nakobolewa Conservancy on the Chobe has three areas declared as fisheries reserves (Munga, Nsala and Makumbi FRs) that form isolated pools during low water periods.

Lusese Conservancy is situated on the Zambezi floodplains and has two lakes (Lwezuba and Kabweza) declared as FRs. The strategies of Nakabolelwa and Lusese Conservancies FRs differ in their management approach with the Sikunga and Impalila Conservancies in that (a) communities may harvest fish during predetermined periods and (b) catch and release angling is allowed. The latter FRs do not allow harvesting to take place except for catch and release angling.

Figure 3. Fish guard from the Impalila Conservancy (left). A fish monitor weighs a Western Bottlenose (centre). Fish monitors recording catch on the floodplains near Impalila Island. Members from the community are appointed to record catch (right). Data are used for management purposes (Photos: C. Hay).

c. Fisheries reserves along the Kavango River: Ten areas, including the entire river front, were proclaimed as fisheries reserves in the Joseph Mbambangandu Conservancy in 2021. Fishing gear types allowed are traps, hook and line, rod and reel and multifilament gillnets as are passive gears with mesh sizes three and four inch. People residing in the conservancy need permits to fish, valid for one year. Non-residents to the area must obtain permission from the Inland Fisheries Committee and pay a fee. Other non-fishing activities in the reserves are also controlled.

3. Basis of a Fisheries Management Plan (FMP)

Guidelines to assist communities and stakeholders in developing fisheries management plans (FMPs) for FRs, are defined as "*a tool to guide managers, institutions, local communities, etc. on how to manage a particular area to ensure that specific aims and objectives are achieved in the present as well as in future.*" A FMP is not a once-off event, but a process requiring regular review and update. The FMP should not be complex to avoid ineffective implementation. The objectives of the protected area, the threats (both natural and anthropogenic), the number of interested parties, the involvement of stakeholders and whether the protected area crosses international borders are valuable inputs. Underlying principles of Namibian FMPs for fisheries reserves are to:

- Minimize the duplication of procedures and functions;
- Promote consistency in policy, procedures and data collection;

- Respond to events not foreseen;
- Provide feedback to ensure continual improvement in the environmental performance; and
- Prevent, protect and conserve fish species, the different habitats and ecological processes within and adjacent to the FRs.

4. Efficacy of Fishery Reserves

Marine protected areas are well known globally with good track records concerning conservation of marine resources including fish stocks. Freshwater reserves, in contrast, are not well recognised as conservation management tools with few examples to learn from. The inland reserves of Thailand are notable exceptions, established within the past 25 years and arising without formal coordination.[1] Reserves created by 23 separate communities in Thailand's Salween basin exhibited markedly increased fish richness, density and biomass; the older and the larger the reserve, the more significant the results. Thus, creating protected areas for Namibia's inland fishes following the community-based management (CBM) model is expected to have the same resource benefits as observed for Namibian wildlife and Thailand fisheries, as well as for riparian communities reliant on fish for their livelihood and food security.

Defining habitat types and size of protected areas is problematic considering 80+ fish species inhabit these systems. Verifying fish movement is difficult, especially in crocodile and hippo infested waters. Fish movements must be studied across seasons to validate the impact fisheries reserves have in protecting fish species. Radiotelemetry can be applied to study the movement of fishes[2,3] and in the process determine home range and habitat preferences of fishes, critical in identifying potential areas for fishery reserves and their size.

Radiotelemetry was employed (Fig. 4) to evaluate the behaviour of (a) Threespot Tilapia, Greenhead Tilapia, Nembwe and Tigerfish in the Upper Zambezi River[4,5,6,7] and (b) Tigerfish in the Kavango River.[8] Data showed that Nembwe and Greenhead Tilapia are vulnerable to local overfishing due to their restricted movement patterns; conversely, these species do not require large reserves (Table 2). Threespot Tilapia demonstrated much higher mobility, thus seemingly would not be as vulnerable to local overfishing. However, intense fishing pressure is likely to impact Threespot populations and larger reserves will be required to manage this commercially important species. Hence, if reserves are established for Threespot Tilapia, all three cichlids will be afforded protection.

Tigerfish in the Kavango River exhibited two behaviour movements, one with high site fidelity with relatively short movements and those that had much longer movements up to 397 km upstream and 116 km downstream. Data suggested that a protected area of 10 km in the Kavango will protect at least 50% of the Tigerfish for 75% of the time. Seasonal shifts in water temperature and water discharge seemingly had no effect on Tigerfish movement.[8]

Tilapia and Tigerfish data combined demonstrated the value of radiotelemetry and annual gillnet surveys[9] to evaluate criteria for establishing FRs and managing fish stocks for the benefit of the local communities. Future studies are planned to evaluate site fidelity and home ranges of species relative to mature reserves.

Communities reported better catches and abundance also increased adjacent to protected areas. Fishers adhered to the official regulations adjacent to the FRs in contrast to non-compliance of those fishing some distance from protected areas. A sense of ownership was created, something that had eroded through the years.

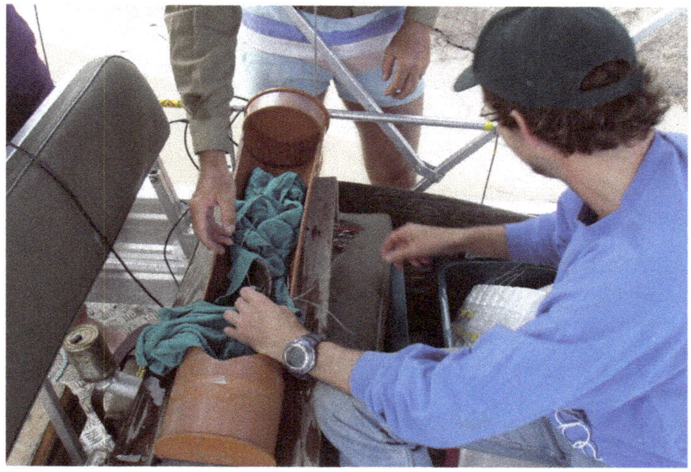

Figure 4. Tagging Greenhead Tilapia in the Upper Zambezi River (Photo: T. Naesje).

Table 2. Movement behaviour of three cichlid species and Tigerfish in the Upper Zambezi River.[4,5,6,7]

Fish species	Distance covered	Habitat visited
Threespot Tilapia	13.5 km (in 5.5 months) Range 2.6–44.6 km	Mainstream 39% Backwaters 26% Side channels 12% Permanent swamps 16% Floodplains 7%
Nembwe	3.2 km (in six months) Range 0.2–11.9 km	Mainstream 69% Side channels 17% Permanent swamps 12% Backwaters 1%
Greenhead Tilapia	4.1 km (in eight months) Range 0.3–19.7 km	Mainstream 46% Side channels 24% Backwaters 14% Floodplains 11% Permanent swamps 4% Mouth of backwaters 1%.
Tigerfish	26.5 km (in six months) Range 0.5–106 km	Mainstream 81% Side channels 7% Backwaters 7% Floodplains 2% Permanent swamps 2%

IV. TRANSBOUNDARY CO-MANAGEMENT

All of Namibia's perennial rivers as well as the Cuvelai have shared international borders, complicating the management of their fisheries. International borders demarcated by rivers are not clearly defined and can lead to casual movement of fishers between countries leading to potential conflict. Fishing grounds are also not always evenly divided between countries and this can lead to conflict when fishers seek better fishing sites. Namibia has worked with its neighbours to facilitate collaboration between governments through joint patrols and the setting up of research protocols (Table 3). This is also addressed at community level through exchange visits to engage in discussions and to promote joint management approaches of fish reserves.

Discussions between communities in the Zambezi Region continue through transboundary fora to expand Fisheries Reserves across rivers and to harmonise legislation to facilitate the management of the reserves. This has been followed up by meetings between scientists and management staff from the different fisheries departments. One outcome of such discussions was when Namibia introduced a closed fishing season between the closed fishing season from 1 December through the end of February annually in the Zambezi Region to complement legislation with that of Zambia. In the past, fishing activities during this period resulted in conflict when Zambian fishers moved across into Namibia to continue fishing during the fishing closure in Zambia.

A Transboundary Fisheries Management Plan for the Okavango River System was developed in 2013 between Angola, Namibia and Botswana to facilitate the co-management of the fish stocks to ensure their sustainable use for the benefit of the local communities. The management plan will contribute towards capacity building that will enable governments to better conserve and manage fisheries resources. Furthermore, the plan will expedite:

- Greater participation of communities dependent on the river for food security and income generation;
- Management of the resources; and
- Creation of a sense of ownership, critical for the sustainability of the process.

Aspects mentioned in the plan are the importance of collaboration and effective communication on a technical level, the standardisation of survey method protocols, capacity building both on community and government level, development of a protocol on data sharing, the socio-economic importance of the fish resource and the role it plays in communities' livelihoods. The harmonisation of policies and legislation is of critical importance to ensure joint management and some aspects relating to this have already been addressed. The potential role of tourism in the region is also referred to in the plan.

The Kavango-Zambezi Transfrontier Conservation Area (KAZA-TFCA) covers the area of the Okavango and Zambezi Rivers where Angola, Botswana, Namibia, Zambia and Zimbabwe meet. KAZA-TFCA is the world's largest terrestrial transfrontier conservation area covering 36 national parks, game reserves, forest reserves and wildlife management areas, a size larger than Germany and Austria combined. The main goal of this transfrontier area is:

> "*to sustainably manage its ecosystems and cultural components to ensure long term viability and diversity of its natural and cultural resources, while generating socio-economic benefits to improve and secure the livelihoods of communities within the ecoregion, by harmonising policies, strategies and practices of the five partner countries, building capacity and creating an enabling economic environment to promote investment.*"

A Freshwater and Fisheries Working Group was established to promote and coordinate transboundary fisheries activities and to integrate all government fisheries departments and fisheries project activities into a common framework that will contribute towards the sustainable usage of the fish resources within the KAZA TFCA region.

Some objectives of this Working Group are to 1) ensure the sustainable utilisation of the fish resources for the benefit of local communities, 2) harmonise legislation for shared fish resources, 3) standardise research and management protocols and 4) share information and best practices between neighbouring countries.

Initial joint surveys between government fisheries departments were conducted to move towards a standardised research protocol to collate databases from different parts of the river systems enabling a basin-wide approach to research. This is an ongoing process that will go far to manage the fish resources sustainably. Joint patrols were also conducted on some of the shared river systems.

Table 3. Water basin commissions in place to facilitate the sustainable utilisation of transboundary river systems mainly focused on shared water courses. These commissions provide platforms for discussions held between countries to address common fisheries related challenges.

Water basin commission	Acronym	Countries participating	Cross border cooperation
Cuvelai Watercourse Commission	CUVECOM	Angola and Namibia	
Permanent Okavango River Basin Water Commission	OKACOM	Angola, Botswana and Namibia	Some collaborative work on fisheries
Zambezi Watercourse Commission	ZAMCOM	Angola, Botswana, Malawi, Mozambique, Namibia, Tanzania, Zambia and Zimbabwe	Some initial work related to water and fisheries
Orange-Senqu River Commission	ORASECOM	Botswana, Lesotho, Namibia and South Africa	Some initial work
Permanent Joint Technical Commission – Kunene River	PJTC	Angola and Namibia	Hydroelectricity & water supply focus

V. REFERENCES

[1]Koning, A.A., K.M. Perales, E. Fluet-Chouinard and P.B. McIntyre. 2020. A network of grassroots reserves protects tropical river fish diversity. Nature 588: 631–635.

[2]Hocutt, C.H., S.E. Seibold and R.V. Jesien. 1994. Biotelemetry: A review of its potential for use in African waters, pp. 77–95. In: PARADI: Proc. International Symposium on Biological Diversity in African Fresh & Brackish Water Fishes, Dakar, 15–20 November 1993. Revue Hydrobiol. Trop. 27(2).

[3]Burnett, M.J., G.C. O'Brien, F.J. Jacobs, G. Jewitt and C.T. Downs. 2021. Fish telemetry in African inland waters and its use in management: A review. Rev. Fish. Biol. Fisheries: https://doi.org/10.1007/s11160-021-09650-2.

[4]Thorstad, E.B., C.J. Hay, T.F. Naesje, B. Chanda and F. Okland. 2002. Movements and habitat utilisation of Tigerfish (*Hydrocynus vittatus*) in the Upper Zambezi River. Implications for fisheries management. Norwegian Institute for Nature Research, NINA Proj. Rept. 19, Trondheim: 28 pp.

[5]Thorstad, E.B., C.J. Hay, T.F. Naesje, B. Chanda and F. Okland. 2003. Movements and habitat utilisation of Threespot Tilapia in the Upper Zambezi River. Norwegian Institute for Nature Research, NINA Proj. Rept. 23, Trondheim: 22 pp.

[6]Thorstad, E.B., C.J. Hay, T.F. Naesje, B. Chanda and F. Okland. 2005. Movements and habitat utilization of Nembwe, *Serranochromis robustus* (Günther, 1864), in the Upper Zambezi River. Afr. Zool. 40(2):253–259.

[7]Thorstad, E.B., C.J. Hay, F. Okland, N. Nickanor and T.F. Naesje. 2007. Spatial behaviour and management of Greenhead Tilapia (*Oreochromis macrochir*) in the Zambezi River, Namibia. Norwegian Institute for Nature Research, Trondheim, NINA Project Report 287: 26 pp.

[8]Jacobs, F.J., T.F. Naesje, E.M. Ulvan, O.L.F. Weyl, D. Tiyeho, C.J. Hay, G. O'Brien and C.T. Downs. 2019. Implications of the movement behaviour of African Tigerfish *Hydrocynus vittatus* for the design of freshwater protected areas. J. Fish Biol: 1–9.

[9]Jacobs, F.J., T.F. Naesje, E.M. Ulvan, C.J. Hay, F.H. Khaebeb and O.L.F. Weyl. 2022. Evidence for overfishing of tigerfish *Hydrocynus vittatus* (Castelnau 1861) in the Kavango River, Namibia. Afr. J. Aq. Sci. DOI: 10.2989/16085914.2022.2041387.

CHAPTER 9
TRADITIONAL AND MODERN FISHING METHODS

by
Ben C.W. van der Waal

I. INTRODUCTION

Fishing must be one of the oldest food collection activities of mankind. Friezes in tombs in Saqqara, Egypt dated 4,500 years ago show fishers using barbed hooks and a sophisticated dipnet to catch fish in the Nile. M. Bruton[1] nicely described the great variety of fishing techniques developed in Africa.

Fishing was employed in Namibia by the people migrating from the tropical north into the region who brought with them a variety of fishing techniques that were then adapted to suit the local hydrologic environments. The northern Silozi speaking people living in the Zambezi Region; the Thimbukushu, Rugciriku, Rusambyo and Rumanyo speaking people in the Kavango West and Kavango East regions; and Oshiwambo speaking people in the Cuvelai *iishana* catchment (the combined Namibian regions of Omusati, Oshana, Ohangwena and Oshikoto, across the border into Angola, Chapters 1 and 7) had a close association with water and fishing. Alternately, the Otjiherero speaking Himba and Herero people of northwest Namibia and Nama from the south were rather pastoral and lacked a fishing tradition.

Fishing can be separated into active and passive methods. Active fishing is similar to hunting, i.e., fishers seek their prey and use gear to capture targeted fish. In passive fishing, gears or plants (used as fish poison) are employed to hook, trap or stun fish indiscriminately. The various fishing methods and gear employed in regions of Namibia are discussed in this chapter. Additional information is found in Chapters 6 and 7.

Traditional gear today is replaced by modern gear, mainly the result of the availability of efficient nylon netting material that enables the construction of a variety of fishing gear and nets and of course gill nets. The modern transparent monofilament gillnet is 2–3 times as efficient as the previously available multifilament nylon nets. The widespread use of these modern efficient gears in the northern regions of Namibia has caused a drastic decline in fish population size and also the use of traditional fishing gear. If this trend continues, the knowledge of construction and use of age-old traditional gear may be lost.

Aside from fishing gear, a boat to operate from and enable travel over larger waterbodies of rivers and floodplains is essential for efficient fishing. Dugout canoes (*mikolo* (plural), *mukolo*, *mukoro* (singular in Silozi), *owato* (Cuvelai) or *wato* (Kavango) are required for fishing in the Zambezi and Kavango regions, but are very rare in the *iishana* region. Most *mikolo* are hewn from a single bole of trees. This has negatively impacted the availability of larger trees in these regions.

In order to make this book useful for local communities, reported local names are given for various fishing gear and methods.[2] Additionally, the language or the region is given in brackets for individual gear. Gears are categorised as (a) active, (b) passive or (c) adapted and modern.

A. Active fishing gear

Active fishing is undertaken by fishers looking for specific fish at a specific place and time. Fishing is practised by either targeting individual fish or straining water and vegetation with baskets, sieves and nets. A variety of instruments are used in different regions, including fishing spears, fishing bow and arrows, fish hook and line, thrust and pull baskets, and bunds of vegetation or more recently, seine nets. There is also evidence of constructed nets woven from local bark twine that were used for fishing before modern net material became available.

1. Fishing spear - *Ekupa* (Cuvelai); *Musha, Egonga* (Kavango); *Muwayo* (Silozi), *Muso* (Sisubia)

Multibarbed spears are commonly used by men in the Upper Zambezi floodplains and along the Kavango River for personal protection and to fish when opportunities arise (Fig. 1A). Smaller multi-barbed spears are also used in the Kavango and Zambezi regions when floods subside to impale migrating fish in shallow water. Spawning runs of catfish are opportunities where large spears are effectively used to collect large number of fishes (Fig. 1B). These multi-barbed spears are still made by skilled blacksmiths in certain villages.

Figure 1A. Business end of large spears used for both personal protection and fishing, Zambezi Region.
Figure 1B. Fisherman with spear and catch of large Sharptooth Catfish at inlet of small stream, Lake Liambezi,1976. (Photos: B. van der Waal).

2. Fishing harpoon - *Onkupa, Lukupa, Ekupa lyokukwata oohi* (Cuvelai)

Harpoons were known in the 1880s from the Cuvelai region, Namibia; the one illustrated here forms part of *Martti Rautanen's 1983 Ambo Collection, Helsinki: Julkaisija Museovirasto* (Fig. 2). According to an explanation given, it was swiped through the water to hook fish on the barb or impale larger migrating fish during *efundja*. It was also thrust into the water like a spear, impaling fish with the barb preventing loss. If the fish broke away, the iron tip came loose from the wooden shaft and was retrieved by the cord on the end. This gear is not seen today in the *iishana* region.

Figure 2. Front end of a fishing harpoon (**Onkupa/ekupa lyokukwata oohi**) from the Cuvelai region, Namibia (Photo: with kind permission from Ilan Jarvinen, National Museum of Finland).

3. Fishing bow and arrow - *Nguba (Ngumba), Uta Makanza* (Kavango); *Buta* (Silozi)

The extremely long arrows have a thin steel shaft of up to 30-cm long, making them sufficiently heavy to penetrate into the water (Fig. 3). The shafts have a single prominent barb up to 20-mm length that will prevent dislodgement of the arrow. The shaft is about 1-m long from a perfectly straight reed stem, offering enough flotation to bring a prey fish to the surface. These are reported mainly from the Kavango Region and are used from a dugout canoe to shoot tilapia and other large cichlids on their nests as well as catfish rising to the surface to breathe. In the *iishana* region, beautiful bows and arrows are commonly used for hunting birds and small mammals, but the *iishana* water is constantly murky, thus fish are not collected this way.

Figure 3. Fishing bow and arrows used in the Kavango West Region in 1981 from a dugout canoe (Photo: B. van der Waal).

4. Fishing hook and line - *Odjolo* (Cuvelai); *Egondo, Erowo* (Kavango); *Kashuto* (Silozi)

Most fishing lines and hooks widely used in Namibia today are factory derived; however, homemade barbed hooks, waxed fishing line made from local fibres and handmade lead weights are still in use (Fig. 4). Once baited, these rigs may be swirled around by hand to gain momentum before casting them. Or, they may be tied to a 'rod' made from cane or tree branch to dangle the bait in pools and eddies. Hook and line were seen effectively used in Angola to catch a variety of catfish, tilapia, largemouth bream and yellowfishes.

Figure 4. Homemade barbed fishing hook and waxed fishing line made from local fibres used in the upper Cuito River, Okavango River System, southern Angola in 2016. (Photo: B. van der Waal).

5. Pulling baskets - *Tambi*, *Rumkinda* (Kavango); *Lishino* (Silozi)

Pulling baskets are widely used in all northern regions by women, but with different design. They have a strong structure and can be used many years (Fig. 5A). They have diagonal handles across an oval shaped opening with which they can be pulled through aquatic vegetation to collect hiding fish. The baskets are made from thin sticks and tough local vine. This trap is very efficient in shallow overgrown areas (Fig. 5B, see also Chapter 6, Fig. 12). Fish scientists use a modern version, the D-frame net, to successfully collect most small fish specimens!

Figure 5A. Well-used pulling basket, Zambezi River, 2020 (Photo: B. van der Waal) and **Figure 5B.** Communal harvesting with pulling baskets, Barotse Floodplain, Zambia, 2003 (Photo: D. Tweddle).

A

B

6. Thrust (plunge, push) basket - *Oshongo* (Cuvelai); *Sididi* (Kavango): *Matuta, Singunde/Singundo* (Silozi); *Nchumba* (Sifwe)

This conical basket of straight small-diameter sticks has a round opening at the bottom about 50 cm in diameter and a long handle of 100–120 cm with an (often padded) opening on one side for the removal of fish from the top (Fig. 6A). They are used to plunge vertically over large catfish and tilapia swimming in muddy shallow pools in the *iishana* or *milapo* (Fig. 6B). A successful thrust is easily recognised from vibrations by the panicking fish. This method is used mainly by women and children, often as a social activity, together with small fish spears or small seine nets and a lot of fun.

Figure 6A. Thrust baskets used in *iishana* in the Cuvelai region (Photo: M. Ekandjo) and **Figure 6B**. Girls fishing in shallow *oshana*, May 2020 (Photo: B. van der Waal).

7. Open funnel basket - *Oshalala, Omudiva* (Cuvelai); *Kanguwa* (Kavango)

The funnel basket is conical, made of stiff grass stems or twigs with narrow slits a few millimetres wide. It is strengthened with hoops at the mouth and higher up, but without a valve (Fig. 7). It is used only for intercepting small barbs and other fish migrating upstream. To be effective, it has to be regularly lifted first by the mouth to collect fish. It has been seen only in the *iishana* region used by women.

A unique fishing technique is practised by women in some parts of the *iishana* region around Outapi. This involves digging small channels between *iishana* that have different water levels and collecting the steady stream of migrating fish. Baskets are regularly lifted, intercepting shoals of barbs and smaller catfish (Fig. 8).

A larger and more elaborate funnel basket is commonly used in the Kavango Region by women on the floodplain. This conical basket is much larger with a wide opening that can be more than 120-cm across and 150-cm long, made of stiff grass stems interspaced with groups of thin twigs as support, leaving slits of 2-4 mm that act as an efficient sieve (Fig. 7; see also Figure 7, Chapter 6). It can be hauled through weed beds like the pulling basket, but is mostly used in a group laid in a semicircle on the vegetated floodplain into which fish are herded. The funnels are then lifted in unison, catching mostly barbs and small cichlids, but with a great variety of species.

Figure 7. Funnel baskets (A) ready to be used for the next *efundja* in Angolan section of the Cuvelai in 2020, and (B) commonly used by women and girls in the Kavango Region, 2003 (Photos: B. van der Waal).

Figure 8. Women using funnel baskets to intercept fish migrations of barbs and small catfish in hand-dug furrows linking two *iishana* (Photo: B. van der Waal).

8. Fibre draw net - *Onete yokutela* (Cuvelai); *Ekwe, Makwe* (Kavango); *Tunyandi* (Lozi)

Draw nets were used in the Kavango Region by the Hambukushu people. There are also reports of nets made from plant fibre material in the Okavango Delta in the early 1900s.[2] Nets were knotted from the inner bark fibres of a common indigenous *Hibiscus* shrub growing in wetlands or thick leaves of *Sansevieria* spp. and treated with a red dye from Sweet Thorn (*Vachellia karroo*) bark to prevent rotting.[3] Historically, gill nets were rare due to the difficulty obtaining suitable yarn to braid them. A few self-made gill nets were seen in the 1970s in the Zambezi Region constructed from the inner threads of truck tyres. Later, when nylon twine became available, gillnetting became firmly established in the Zambezi River and Kavango regions (see section C.4 below).

9. Grass roll - *Uuluwa, Omudiva* (Cuvelai)

Women in the *iishana*, Kavango and Zambezi Regions make a roll of grass, plants and twine up to 10-m long to net shallow water areas during the rising and drawdown phases of the flood. Fishers push or pull the roll through shallow pools or along the margins. Small cichlids, barbs and other small fishes are caught by this method (no photo available).

10. Dugout canoe - *Mukolo, Mukoro, (Mikolo* plural) (Zambezi); *Wato* (Kavango); *Owato* (Cuvelai)

Figure 9. Finishing the outside of a dugout canoe with adzes, Zambezi Region, 2003 (Photo: B. van der Waal).

Dugout canoes are essential in the Zambezi and Kavango Regions; almost every household possesses a *mukolo* used extensively for the transport over water of people and goods and also for all forms of fishing. In the Zambezi and Kavango Regions with large deep waterbodies, many fishing techniques are only possible through the availability of *mikolo*. These methods include gill netting, seine netting, fishing with bow and arrow, and fishing in backwaters with hook and line. *Mikolo* are hewn with axes and adzes from the boles of large trees (Fig. 9). Such large trees have become so rare that a trade exists for *mikolo* in the Zambezi-Chobe floodplain that originates from the forest region in Zambia. Trees used for making *mikolo* in the 1970s in the Zambezi Region were valuable hardwood trees such as Kiaat (*Pterocarpus angolensis*), Zambezi Teak (*Baeikia plurijuga*) and Rosewood (*Guibortia coleosperma*). Less-preferred tree species, including the lower quality soft wood Mongongo nut tree or Manketti (*Schinziophyton rautanenii*) and the Sausage Tree (*Kigelia africana*) are often used presently. *Mikolo* measured during the 1970s in the Zambezi Region were from 5-7.3 m long and 35-79 cm wide with inner depths of 23-35 cm. *Mikolo* do not float and most paddlers do not possess flotation devices, resulting in a real risk of drowning if capsizing.

B. Passive fishing gear

1. Bund - *Uuluwa, Uuluga (*Cuvelai); *Mundo, Maalelo (*Silozi); *Bwaleleo*, *Bwelo* (Sisubia); *Ndoba* (Sifwe)

Figure 10. Bund constructed of earth and sticks with a number of adapted traps placed in an *oshana* to intercept migrating fish, Cuvelai Region, 2020 (Photo: B. van der Waal).

Low walls or fences are built in shallow inundated places with sods of earth and grass to act as a barrier and guide fish into corral entrances or valved traps (Fig. 10). Traps are placed in the deepest places in these bunds or fences (up to 60-cm deep) facing either upstream or downstream to catch migrating fish. Bunds or fences are used in all northern regions.

In the Namibian Zambezi Region and further upstream in the Barotse Floodplain of Zambia, a special type of bund, *maalelo* is built to keep receding floodwaters impounded for an extended period on the floodplain. The bunds form small ponds, or *maalelo*, that allow fishes in them to grow for a longer time. They are opened after the water has receded from the floodplain and traps placed to collect the residing fish (Fig. 11). Typical floodplain species such as Banded Tilapia (*Tilapia sparrmanii)*, Snake Catfish (*Clarias theodorae)* and Manyspined Climbing Perch (*Ctenopoma multispine)* as well as barbs, other tilapia and mormyrids are collected this way. The *maalelo* fishery is a type of traditional fish farming seen in the Kwando, Linyanti and Zambezi floodplains during the 1970s that now occurs rarely.

Figure 11. Earthen bund (*maalelo*) with the site constructed for the release of water where traps were placed to catch the fish left behind the bund, Barotse Floodplain, Zambezi, 2003. The long wall of another *maalelo* visible in background (Photo: D. Tweddle).

2. Fishing fence - *Uuluwa, Uuluga* (Cuvelai); *Sintunga, Marera, Masasa* (Kavango); *Mundo, Siyandi* (Lozi)

Fishing fences are used in the Kavango and Zambezi Regions in conjunction with valved traps, open funnel constriction traps or to guide fish to corrals (Figs. 12A and B). They are usually made of locally available sticks or reeds that are neatly spaced along two or more horizontal members held in place by sturdy stakes driven into the sediment.

Figure 12A. Fence with corrals in a back channel of the Kwando River, May 2008. **Figure 12B.** Two corrals fitted with reed covers to prevent avian predation, set to collect fish migrating upstream, Kavango River floodplain, March 2009 (Photos: B. van der Waal).

3. Foldable corral trap - *Marera, Masande* (Kavango); *Katamba* (Lozi)

Corral traps are widely used in the Kavango Region; they are neatly constructed mats made of slender straight sticks with pointed ends on one side (Fig. 13A). Women set them in a heart shape in shallow water to catch fish while doing other chores on the river edge such as washing themselves, children, pots and clothes (Fig. 13B). After use, the traps are easily removed and carried home. They are effective catching small tilapia, barbs and if left overnight, especially squeakers (*Synodontis* spp.), provided the trap is suitably baited with porridge or similar attractant.

Figure 13A. Corral traps rolled up for transport. **Figure 13B.** Corral traps set out on floodplain, Kavango Floodplain, 2009 (Photos: B. van der Waal).

4. Valved trap - *Omudiva, Otalate* (Cuvelai); *Muduwa* (Kavango); *Muono, Lukuko, Muhoho* (Silozi)

Figure 14. Valved fish trap from the Cuvelai region, 2020, made from local grass stems and bent twigs tied together with string made from woven bag strands, bicycle tubing keeping the valve tied to the body. The opening of the inward facing valve is at the bottom. It has an elegant flap on top for removing caught fish (Photo: B. van der Waal).

The valved trap (Fig. 14) is often used with a bund, low wall of soil, low fence of brush or grass (*uuluwa, uuluga*), or it can be used in submerged vegetation in shallower areas. In some regions north of Namibia, these traps are constructed with two valves behind each other and are often baited with balls of porridge or other food.

The trap's body is a conical structure of locally available stiff grass stems or small-diameter straight twigs from locally available shrubs, ~60-cm long with a diameter of 30-40 cm and slits 2-4 mm wide between the sticks or grass stems, allowing only the smallest fish to escape. It is kept open with two or more hoops made from bent sticks at the mouth to which a clever non-return valve of grass stems or twigs is also fitted, facing inwards to prevent fish from escaping. In the *iishana* region, fish are removed from the trap either through an opening above the funnel that is closed by a loose flap or elsewhere by untying a fastening knot at the other end of the trap (Fig. 15).

These traps are set separately amongst vegetation or placed in a low earth bund or brush fence, some facing downstream and others upstream, depending on the expected direction of fish movement. To improve catches, they can be baited with porridge or snail meat to attract tilapia and catfish. Catches consist of smaller fish like barbs, small tilapia, migrating mormyrids and catfish.

Figure 15. Valved traps set in a bund made of rocks in the Katima Mulilo Rapids in the Zambezi River in 2009, targeting mormyrids and mochokids. These traps all face downstream and have to be unbound at the narrow end to remove the catch (Photos: B. van der Waal).

5. Barrel valved trap - *Ngumba, Matumba, Njamba* (Silozi)

During the rising phase of the Zambezi flood, upstream and lateral migrations of large mature cichlids along river banks are intercepted by barrel-shaped traps with valves at opposing sides. These large sturdy traps are made from branches tied with bark rope and can withstand predation attempts by monitor lizards, crocodiles and otters (Figs. 16A and B). They have large slits, allowing smaller fish (<15 cm) to escape. The traps are placed inside inundated reed clumps, the preferred safe route of the target cichlids.

A new variety of this trap was seen on the Zambezi for the first time in 2009, similar to the double valved trap, but with an almost spherical shape and made from split bamboo and tied with truck tyre inner cords (Fig. 16C). This trap is used by Bemba fishermen from northern Zambia who had migrated to the Zambezi Region in order to fish.

Figure 16A and B. Giant barrel-shaped valved traps used to catch migrating cichlids in the main stream of the Zambezi River, near Kalambeza, 2008 and **16C** a spherical double-valved trap made from split bamboo and tied with truck tyre inner cords, Zambezi River, 2009 (Photos: B. van der Waal).

6. Constriction funnel - *Lifula* (Lozi)

Constriction funnels are only effective in strong flows to intercept downstream fish migrations. Strong currents are required to prevent fish from escaping once they are swept into the narrow end of the trap. The open funnels can be up to 4-m long and 1-m wide at the mouth (Fig. 17A). They are made from stout reed stems tied together into a long narrow funnel with spacing between the reeds becoming smaller, down to 3 mm. These traps are well known at major rapids of the Zambezi from Impalila Island and Katambora Rapids upriver into Zambia and also in major rapids of the Kavango.

These funnels are hung in fast current to be effective. This is accomplished at Impalila by stretching steel cables or logs across rapids on which the funnels are fastened (Fig. 17B). Sites to hang funnels are owned by families and inherited. Catch consists mostly of mormyrids during the dark new moon phase in June and July, but many other fish species are also captured depending upon downstream migrations (Fig. 17C).

Figure 17A. Constriction funnel, a long open funnel of reed stems. **Figure 17B.** Placing funnels in position by hanging them on a cable stretched across a side stream of the Chobe Rapids, Impalila Island, June 2008. **Figure 17C.** Catch of a constriction funnel (mostly *Petrocephalus longicapitis*) on 25 June 2008. Fish are jammed in the end of the funnel, filling a 20-l bucket (Photos: B. van der Waal).

C. Adapted and modern fishing gear

Materials such as steel, nylon, factory-made fish hooks, nylon fishing line, and especially multifilament and monofilament netting material, brought new possibilities to adapt and develop new or more efficient gear. Modern basic fishing gear such as gill nets and fish hooks were not common before these items became commercially available. In the 1970s, many self-made gill nets were seen in the Zambezi Region made from the inner cords of truck tyres and in 2019 home-made fishing hooks were seen in use in the upper parts of both the Cuito and Cubango Rivers in southern Angola (A4, above).

A surge of modernisation of traditional gear came with the availability of nylon netting to Namibia's northern fishing communities during the 1970s. There were two sources of netting. Discarded small mesh (8-15 mm stretched knotless) trawler nets from Walvis Bay found their way to the *iishana* region where various nets were constructed, including seine nets, funnel nets and even traps. Often the same piece of netting would be transformed from funnel nets into part of a large seine. The other source was nylon twine and later factory-made nylon multifilament and monofilament gillnets obtained from Malawi and Zambia, and more recently from China.

1. Modern hook and line - *Okalongolola, Okandjolo, Odjolo* (Cuvelai); *Egondo, Erowo* (Kavango); *Kashuto* (Silozi)

Fishing using a rod of 60-200 cm long with a nylon line, usually twice as long, small float, small sinker and hook, is very common in all the northern river systems where the water is deep and large fish occur. Fishing is usually by groups of boys, men or women, using earthworms, insects, snails, crabs or porridge as bait (Fig. 18). Small catfish, small cichlids and larger barbs are targeted. Fishing is undertaken from the banks or from anchored *mikolo*. However, Chapter 10 (Fig. 3) demonstrates that many local residents have now adapted to modern rod and reel to catch large fish, including Tigerfish.

Figure 18. Boys fishing with hook and line for cichlids, especially *Tilapia sparrmanii* from the river bank north of Nkurenkuru, Kavango in 2003. The boy on the right is attending an open funnel to collect small bait fish (Photo: B. van der Waal).

2. Long line - *Omlavalava* (Cuvelai); *Makoka* (Kavango); *Erowo, Kashuto* (Silozi)

In the *iishana,* long lines are set between two stakes in shallow water with a fishing line supplied with up to 15 small hooks (size 1), each dangling on a 50-cm long trace from the main line and baited with aquatic snail meat. These are used specifically to catch Sharptooth and Blunttooth Catfish (Fig. 19).

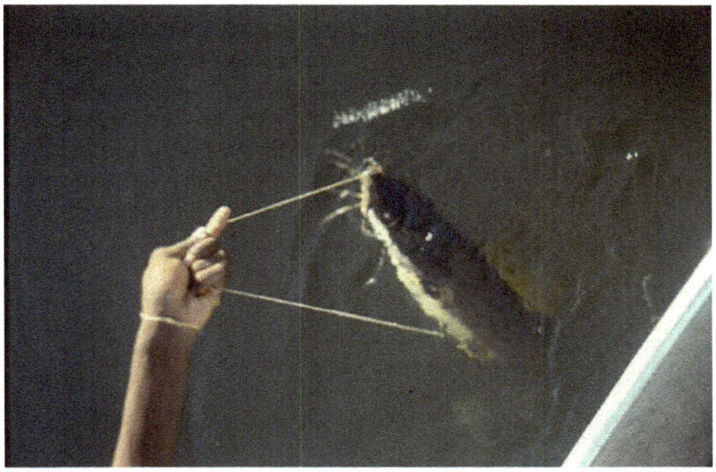

Figure 19. Longline with catfish, Kavango River (Photo: G. Merron).

3. Foul hooking

This uncommon method is practised in parts of the *iishana* region where up to seven hooks are tied on a nylon line with a small rock as weight at the end. This rig is jerked up in the flowing stream to jig fishes passing by.

4. Nylon gillnets - *Onete yokutela* (Cuvelai); *Ekwe, Makwe* (Kavango); *Tunyandi* (Silozi)

Figure 20. Modern small mesh gillnet set across a running *oshana* near the Omadhiya Wetland, 2008 (Photo: B. van der Waal).

Gillnetting is almost exclusively a male enterprise, whether for (principally) commercial purposes or own consumption (Fig. 20). Gillnets were rare up to 1970 as result of the difficulty obtaining suitable yarn to braid nets. When nylon twine became available, fishermen in the Zambezi River and Kavango, to a lesser extent, braided their own nets. Later, gillnetting material became available via Zambia in the 1980s from factories in Malawi and other countries, then directly from China. Gillnet material is usually available as 50 or 100-m long panels, 2.5-m deep. Since the end of the previous millennium, the gillnet fishery exploded with accompanying effects on fish stocks (Chapter 6). Fish harvest was further boosted by the more efficient monofilament gillnets becoming generally available by 2010.

Gillnets were originally mounted on a bottom rope made from the inner bark from local trees, including Mopane *(Colophospermum mopane*) and Silver Cluster-leaf (*Terminalia sericea).* Floats were made from folded *Sesbania*, sedge or papyrus leaf stalks or pieces of light wood, attached at intervals directly to the meshes on top (Chapter 6, Fig. 11). Weights to sink and stabilise the nets were originally made of baked clay with a hole for attachment. When twine and polystyrene became available, nets were accordingly modified with top and bottom ropes. Many fishermen still make strong thin double-twisted rope from the strands of common woven-plastic grain bags, rolled on the fisherman's thigh.

There was a drastic change in the use of multifilament to monofilament gillnets by the commercial fishers between 2010 and 2018. Monofilament gillnets are three times more effective than multifilament gillnets, intensifying the fishing pressure. MFMR promulgated legislation in 2016 prohibiting the use of monofilament gillnets, principally to protect large tilapia species. Although illegal, monofilament nets are still used as they are cheap and efficient. They have the disadvantage that they tear easily and are not easily mended, resulting in many discarded nets posing a serious threat to fish and other wildlife.

Gillnets form the backbone of the commercial fisheries that developed in the Zambezi-Chobe floodplain and especially Lake Liambezi during the 1970s to 1980s and 2010-2016 when large mesh gillnets with stretched meshes of 80-120 mm were employed. These nets selectively targeted the larger *Oreochromis* and *Coptodon*, fish that obtain a prime price on the markets and whose keeping properties are superior. Gillnets were not common in the other northern regions, although this is now changing, especially in the Kavango River. Gillnetting has occurred in Olushandja Dam for cichlids and catfishes since the 1990s.

In recent years with declining catches, gillnetting methods and techniques to increase catches were copied from further north or developed locally. The normal custom is to set nets late in the afternoon along or perpendicular to the bank, depending on flow and depth of the water, and collect the fish from the net very early the next morning. In some instances, gillnets are joined to form a very long net of up to 1.0 km. In very deep water, nets can be mounted in panels below each other, so the net stretches from top to bottom.

Unique modifications and handling practices of gillnets have been made by fishers, including:

a. Net washing and dying

It is generally known that gillnets kept stationary in one place for a few days catch almost no fish after the first two nights. Nets are then removed, washed and placed elsewhere. It was common practice at Lake Liambezi in the 1970s to wash nets thoroughly in an old mortar and pestle. Ash from burnt trees or some unspecialised herbal 'medicine' was added to the net before drying it thoroughly and setting it. Some nets were dyed with bark extracts, similar to those used for the traditional tanning of game skins, before the net was dried and set. Many of these practises have disappeared with the general commercialization seen in the regions.

b. Bashing or *Kutumpula*

To increase catches, nets are set in the shape of a bow alongside reeds or the river banks. Fish are chased into the net by beating paddles or more often using a special made wooden plunger on the end of a stick, causing a characteristic thumping sound and shock waves to chase panicking cichlids to flee into the set net (Fig. 21). This illegal method derives from Zambia and is now regularly practised in the Zambezi River during high water conditions at night.

Figure 21. Beer bottle fixed on the end of a stick acts as effective plunger for bashing or ***kutumpula***, Lake Liambezi, April 2008 (Photo: B. van der Waal).

c. Drift netting

Another technique developed in the Zambezi is to construct a deep long net from joined gillnet panels, up to 5-m deep and 400-m long usually attached to poles at either end. The constructed net has a float line and large floats are attached to each pole, which are also weighted at the bottom end to keep them vertical (Fig. 22). This large mesh gillnet is then set across the Zambezi River at night, guided for kilometres downstream with the current and eventually closed on a sand bank to encircle the catch, consisting mostly of Tigerfish and large tilapia. The method is deemed illegal by the MFMR, but continues to be practised.

Figure 22. Large mesh drift net packed out for drying on a fishing beach on the Zambezi River, 2008. Floats made from modern materials whilst rocks are used as weights. Manned dugout canoes will be tied to the ends (noted by the heavy rock weights) to guide the net down the river for kilometers (Photo: B. van der Waal).

5. Seine nets/drag nets - *Onete yokutela* (Cuvelai); *Ekwe* (Kavango); *Ituwa, Lituwa* (Silozi)

Figure 23. Pulling a large seine net (Photo: S. Tribelhorn).

The use of large seine nets, or drag nets, has occurred in the Omadhiya Wetland pans north of Etosha since the 1970s. Nets up to 300-m length with a proper bag in the middle, made from knotless pilchard netting commonly used in the marine fishing industry in Walvis Bay and set at least 200 m from the edge of the large pans, were pulled by teams of men and boys (Fig. 23). The catch consisted of catfish, barbs and some cichlids. When *iishana* start to dry out, smaller drag nets are used to collect most fish from the *iishana* before the water becomes muddy and fish succumb.

Figure 24A. A large dugout canoe with a dragnet, packed and ready for use in lagoons in the Chobe River, November 2008. The old-fashioned floats of *Sesbania* stems, round fired clay weights, formidable fish spear and neatly coiled self-made draw rope for hauling in the net are all in place. **Figure 24B.** Two fishermen removing their catch from a large seine net after a successful haul, at night time in a side channel of the Zambezi downstream of Kalambeza in 2009 (Photos: B. van der Waal).

The Kavango and Zambezi Rivers and their floodplains do not offer many suitable sites to use large seine nets, but this changed when gillnets with large meshes were modified for use in side pools of these rivers. Another development was the netting at night of sand banks, where tilapia and other cichlids congregate for nesting early in spring, by encircling nets around the lek and thus overharvesting the spawning populations, as is now reflected in declining catches (Fig. 24A and B).

6. Mosquito draw nets - *Esire, Masiri Makwe* (Kavango); *Moskito* (Lozi)

Figure 25A. Four mosquito nets, some impregnated with insecticide are used to collect small fish migrating onto the floodplain from the Zambezi River near Katima Mulilo, 7 April 2008. **Figure 25B.** Catch of two hauls of a mosquito net used at a culvert near Lisikile, Zambezi in April 2008 (Photos: B. van der Waal).

Commercially available with many provided free of charge by international donor organisations for the control of malaria, mosquito nets are widely used in the subsistence fisheries of all northern rivers to catch migratory fish. Although illegal, they are highly effective on the floodplains and backwaters once lateral migrations begin, capturing a wide variety of species and life stages. In that way, they are very destructive to the fish community. They are used in all regions at constrictions of water flow to virtually block the channel. They are also used on the shallow banks of the Zambezi River during the intense migration of small fish with dropping water levels (Fig. 25A and B). Most of these migratory fish catches consist of small mature barbs and other floodplain fish, forced off the floodplain with receding water levels. In a way, these nets simulate the use of open funnel fish baskets (*kanguwa*) and pulling baskets (*lishino*). However, when mosquito nets are used as seine nets or, especially, when joined into large drag nets, all fish and especially smaller cichlids are harvested, causing serious damage to fish communities and their recruitment.

A novel development became a serious issue in 2008 in the Zambezi Region when insecticide-impregnated nets donated by international organisations to combat malaria were used as fishing nets (Fig. 26). This problem is not restricted to the Zambezi Region and has been reported from many tropical countries[4] and coastal areas.[5]

Figure 26. Collecting migrating barbs with insecticide-impregnated nets held stationery below a culvert on the Zambezi floodplain near Lisikili, March 2009 (Photo: B. van der Waal).

7. Funnel net - *Onete* (Cuvelai)

Figure 27A. Nine funnel nets are tied next to each other below a culvert in an *oshana* crossed by a road, with at least four others further down, all intercepting the downstream fish migration during the 2008 efundja. **Figure 27B.** Fish passage completely blocked by funnels having large hoops to keep the nets open at a culvert crossing. The funnel net in the background has wings to guide fish (Photos: B. van der Waal).

Funnel nets are a recent innovation, only observed in the region since 1976. They are made from either small mesh (10-20 mm stretched) anchovy nets discarded at Walvis Bay or shade cloth, thus they capture all small fish. They are placed in *iishana* at constrictions such as culverts and bridges where velocities are high (>1 m/sec) (Fig. 27A). These small mesh nets are shaped like a funnel up to 20-m long with a mouth 1-3 m wide. They taper from the mouth to a closed purse end that can be untied to collect the catch. Some nets are held open with a large hoop of sticks while others have wings attached to guide fish (Fig. 27B). Nets are set facing upstream and tied to stakes or structures at culverts and bridges or placed directly next to each other to intercept downstream migrating fish. Nets are emptied every half hour or as required by wading in and untying the bottom end after chasing the fish down the net. Daily catches can be up to 100 kg. Later in the season when *iishana* start to dry, funnel nets are converted into seine nets! Funnel nets are fished by men and their operation can be dangerous in deep *iishana*.

8. Valved fyke net - *Omudiva* (Cuvelai)

A modification of the traditional valved trap is now regularly constructed in the *iishana* region from fishing net or shade net material, complete with wooden hoops and inner non-return valves (Fig. 28). They are used in the same way as the traditional *omudiva*.

Figure 28. Valved fyke net (*omudiva*) with hoops and covered with shade cloth and set in a bund made of sticks and clay, south of Oshakati, 2008 (Photo: B. van der Waal).

9. Plastic and glass bottle traps - *Otalate* (Cuvelai)

In all northern regions, glass bottles with conical bottoms (commonly used for brandy) and large plastic bottles, more recently, are modified to form fish traps. Glass bottles with conical hollow bottoms are turned into small valved fish traps by hitting a hole in the bottom. The top conical section of 2- and 5-l plastic bottles are removed and pushed upside down into the lower portion, forming elegant traps. Another recent modification (Fig. 29) is used in shallow floodplains where baited bottles are placed with the wide opening facing towards the bank. Fish are collected by rushing towards the trap and lifting it out of the water. Small holes are pierced into the bottles to allow drainage (Chapter 6, Fig. 8). These traps are efficient to catch small cichlids and barbs, which in turn may be used as a meal or bait to catch predatory fish.

Figures 29A and B. Open baited bottle traps used in Kavango East Region to catch mainly small cichlids. Drain holes are burned into the sides and bottom of the bottles to allow drainage when lifted up, imitating traditional open funnel baskets (see Figs. 7A, B above; Chapter 6, Fig. 8) (Photos: F. Jacobs).

10. Dipnet - *Okandjarala*, *Onete yombugu* (Cuvelai)

Handmade dipnets are round- or oval-hooped up to 1-m in diameter and 1-m deep. Typically, they are constructed from bent branches, lined with mosquito net material or shade cloth and used by individuals to collect barbs and other small fishes migrating at constrictions like bridges and culverts. They are mostly used by women at culverts in the *iishana* region to catch migrating barbs and catfish, especially in manually-dug channels connecting *iishana* (Fig. 30 and Fig. 8).

Figure 30. People digging parallel furrows between an oshana and overflow from canal to herd migrating fish into narrow channels where they are collected with dipnets and conical traps (see Fig. 8) (Photo: B. van der Waal).

11. Throw net

Commercial modern throw nets were first seen used effectively at bridges and culverts to catch migrating fish in 2008 (Fig. 31). This illustrates the effective adaptation of previously foreign fishing gear, some of which poses new management challenges.

Figure 31. Throw net with its catch on Mpundja Bridge, south of Oshakati, consisting of Three-spot Tilapia, barbs and Sharptooth Catfish, 15 April 2008 (Photo: B. van der Waal).

12. Lift net - a novel development

Lift nets were seen operated from dugout canoes in the Zambezi River below the large Katima Mulilo Rapids in 2007. The frame of sturdy wire prevents collapse in the current. The net, about 100-cm long, is baited with pieces of fish or porridge, hung over the side of a dugout for a few minutes and then drawn up (Fig. 32). It is effective catching darting fish in the current consisting of Silver Robbers and Barred Minnows.

Figure 32. Lift nets made from a steel wire frame and metal mosquito netting to be lowered in the current to catch small fish in the flotsam, Katima Mulilo Rapids, August 2007. Compare structure with pulling basket, Figure 5A (Photo: B. van der Waal).

REFERENCES

[1]Bruton, M. 2016. Traditional Fishing Methods of Africa. Cambridge Univ. Press, Creda Communications, South Africa. 96pp.

[2]Okeyo, D.O., G. Mubita, T.K. Harris, D.E. Sahombu, J. Namundjanga, S. Mulonga and S. Kapirika. 2004. Indigenous names of fish and fishing gear in the Cuvelai, Kavango and Caprivi regions of Namibia. Afr. J. Aq. Sci. 29(2): 249-258.

[3]Cowley, C. 1969. Volk van fabels. 'n Soektog na 'n verlore Boesmanstam in die Okavangomoeras. Kaap en Transvaal Drukkers, Kaapstad: 156 pp.

[4]Larsen, D.A., R. Welsh, A. Mulenga and R. Reid. 2018. Widespread mosquito net fishing in the Barotse floodplain: Evidence from qualitative interviews. PLoS ONE 13(5): e0195808. https://doi.org/10.1371/journal.pone.0195808.

[5]Jones, B.L. and R.K.F. Unsworth. 2019. The perverse fisheries consequences of mosquito net malaria prophylaxis in East Africa. Ambio 49: 1257-1267.

CHAPTER 10
FRESHWATER RECREATIONAL FISHING

by

Francois J. Jacobs

I. Overview of Freshwater Recreational (Sport) Fishing in Namibia

Fishing is probably as old as mankind with the earliest evidence of fishing dating to the Palaeolithic (i.e., Old Stone Age). Spears and harpoons were used to kill fish stranded from receding floodwaters or migrating in vast numbers, and hence easy to kill. In a true sense, this was more hunting and gathering than fishing, but developed to utilise a relatively easy food source compared to hunting for large dangerous game. Nets were invented long after harpoons and somewhere in between man used baited lines. There were no hooks in Palaeolithic times, instead gorges, which resembled a thick nail pointed at both ends, were used. They were embedded in bait to pass down the throat of a fish and when jerked, they jammed across the fish's gullet. The development of 'rods' expanded the area and depth to fish. 'Primitive' gears evolved over time to a variety of traps and push nets still used in Namibia's traditional fisheries (Chapter 9).

Although it is often blurred, there is a distinction between fishing for sport and fishing for food. The sport angler uses a method which gives the most pleasure even though it may not be the most effective way of catching fish, while persons fishing for food often join together with the goal of maximising yield. Today, sport fishing is one of the most popular outdoor activities in the world with 10% of the populace fishing for pleasure in developed nations, amounting to at least 220 million recreational fishers worldwide.[1]

Although freshwater fishing might not be synonymous with Namibia, mostly an arid country with vast dry landscapes, the waterbodies offer productive and important sport fisheries. Namibia's five major perennial river systems, including the Upper Zambezi-Chobe, Kwando-Linyanti, Kavango, Kunene and Orange Rivers, each offer a unique and world class sport fishery. The northeastern perennial rivers are arguably the premier Tigerfish angling destination in Africa. Local angling lodges on the Zambezi River received 4,000 anglers that caught an estimated 38,000 Tigerfish and 14,000 cichlids in 2010. This activity was responsible for up to 70% of revenue to these lodges that have major local economic importance in these rural areas of the country in the way of employment as fishing guides, chefs and cleaners.[2] Fishing remains one of the main tourism activities and has led to the establishment of more than 60 accommodation facilities (lodges and campsites) along Namibia's perennial rivers with almost all offering fishing trips as part of their portfolio.

In addition, the interior of Namibia is scattered with intermittent rivers that feed numerous impoundments, most of which support recreational fisheries. These are populated largely by introduced (e.g., Largemouth Bass and Common Carp) or translocated indigenous species (*Oreochromis* spp.) and are considered important sport fisheries for income generation and employment creation by attracting tourists to the respective regions.

The Namibian Bass Angling Association (NBAA) conducted a survey in 2016 to estimate the socio-economic value of recreational fishing in Namibia. They estimated that Namibians spent over N$ 77 million (US$ 1.00 = N$ 15) on boats, vehicles, marine electronics and fishing equipment and an additional N$ 8.5 million annually on recreational fishing activities.

From a conservation perspective, recreational anglers largely practice catch and release. Fishing competitions require that all catches be handled humanely and returned to the water after weigh-in or photographic recording. Thus, recreational angling in Namibia is largely a non-consumptive activity that promotes game fishes as renewable resources.

Inland recreational fisheries and angling are managed by the Ministry of Fisheries and Marine Resources (MFMR) under the Inland Fisheries Resources Act, 2003 (Appendix 1).

II. Freshwater Fishing Associations in Namibia

A. History

The South-West Africa (SWA) Anglers Union, formed in October 1962, was the first angling association in Namibia. Competitions were originally restricted to rock and surf angling, but activities were extended to embrace all facets of the angling sport as membership increased. A freshwater section was established in 1965 as a component of the union. The number of clubs affiliated to the union in subsequent years increased from two to nine and the sections extended to include rock and surf, freshwater, deep-sea, ski-boat, distance casting and light-tackle boat angling. Various sections had reached a stage by 27 January 1973 whereby they became independent self-governing associations under the auspices of an umbrella body known as the SWA Angling and Casting Board of Control.

B. Namibia Federation for Freshwater Anglers

The Namibia Federation for Freshwater Anglers (NFFA) was formed in 1991 apart from sea water angling and became one of the first sporting bodies in Namibia to be recognised by and affiliated to the Namibia Sports Council. With the advent of the Namibia Sports Act of 2003, NFFA was again one of the first sporting bodies to be officially affiliated to the newly established Namibia Sports Commission and recognised by the Ministry of Youth, Sports and Culture.

NFFA and its sub-affiliated member associations host about 50-55 national, international and developmental angling events on Namibian waters where approximately 3,000 tournament anglers participate annually. NFFA currently has approximately 450 individual anglers belonging to six associations formed under its umbrella. These associations were established at various times as more specialised angling techniques were established internationally. There is some overlap among the associations and the species that they target, but each (1) uses sanctioned techniques to angle, (2) handles all catch as humanely as possible and (3) releases catch after weigh-in. Competitions and events take place across Namibia including on the Kavango, Kwando and Upper Zambezi-Chobe Rivers and on Swakoppoort, Von Bach, Daan Viljoen, Oanob, Hardap and Naute Dams. The competitive angler in Namibia is driven by committed sportsmanship.

NFFA is also a member of the International Sport Fishing Confederation (CIPS) established in 1952, and specifically its sub-affiliated freshwater body, namely the Fédération Internationale de la Pêche Sportive en eau Douce (FIPSed). International competitions are attended on a regular basis and five member associations of the NFFA participated in the CIPS Sport Fishing World Championships in South Africa in 2019.

The six NFFA sub-affiliated member associations include:

1. Namibia Artificial Lure Angling Association (NALAA): NALAA is a powerboat based, individual participation sport in which competitors fish only with artificial lures for a variety of species. The aim is to catch as many different species (and the heaviest for each) during the allocated tournament time. The length of each species is measured and recorded by photo. Individual species are allocated factors, based on their level of difficulty to catch. Fish length is multiplied by the respective factor to determine the best angler of the day.

2. Namibia Bass Angling Association (NBAA): NBAA is a powerboat based, individual participation sport in which competitors fish only with artificial lures. The heaviest five bass in a bag is the winner. The NBAA hosts 5–8 national tournaments, national trials and other qualification events annually. In addition, the NBAA hosts a number of open events annually for the promotion of the sport with the general public, novice and prospective anglers participating in large numbers.

Namibian competitions are quite small, but are fierce requiring weeks of preparation. During these events, the top anglers qualify for the Namibian National Team that competes annually in the Region 5 Bass Federation Championships against South Africa, Zimbabwe, Zambia, Eswatini and Mozambique. The top anglers also have the opportunity to participate in the Black Bass World Championships (CIPS) and other championship events in the USA, hosted by Bass Anglers Sportsman Society (BASS) and with Major League Fishing (MLF) in the U.S.A.

3. Namibia Freshwater Bank Angling Association (NFBAA): NFBAA is a shore angling and team participation sport in which competitors fish mostly with organic baits. Teams consist of six anglers, paired anglers per peg/station in three zones. Points are accumulated individually and then consolidated to determine team results. These anglers fish for various species. The aim is for each angler to catch as many fish as possible during an 8-hour (h) event with a 20 fish/species bag limit. An angler is allocated catch points based on the number and weight of all fishes caught.

4. Namibia Light Tackle Boat Freshwater Angling Association (NLTBFAA): NLTBFAA is a powerboat based, individual participation sport in which competitors (up to three per boat) fish, mostly with organic baits, for a variety of species. Only light tackle monofilament fishing line (maximum 4-kg breaking strain) is allowed for the sport. The aim is to catch five fish of each species. NLTBFAA anglers photograph each fish being weighed with a calibrated scale and that day's weigh-tag. All photos are then transferred to the weigh-in official for verification and determining the daily scores.

5. Species Angling Namibia (SAN): SAN is a shore angling, peg participation sport with two competitors per peg fishing mostly with organic baits only for Common Carp. The aim is to catch the most carp (accumulated weight) over a 72-hr period. Total weight is accumulated per peg to determine the best peg team of the tournament. Catch-and-release is strictly adhered to and all fish are released live.

6. River Angling Namibia Association (RAN): RAN is a powerboat-based team participation sport in which a predetermined number of anglers fish the perennial rivers of Namibia with organic baits and artificial lures. RAN anglers fish for a predetermined number of targeted species that may change from time to time. RAN anglers take videos of fish length measurement and live release, and then present these videos at weigh-in. Individual species are allocated factors based on their level of difficulty to catch. Fish length is multiplied by the "catch-difficulty" factor to determine the best angler of the day.

C. Freshwater Fish Records

NFFA and its member associations currently have two types of freshwater fish records: (1) closed records, which are set in a tournament or an association-sanctioned event within the borders of Namibia and (2) open records, set outside of an association-sanctioned event within the borders of Namibia or set in a sanctioned event outside Namibia. Catches made inside Namibia's borders and registered as National records are reported in Table 1.

Table 1 also includes (a) the largest fishes caught in Namibia irrespective of open/closed/junior/men or women class that have been officially registered as records by club members and (b) the largest known freshwater fish caught in Namibian waters (i.e. during competitions or social events), but not registered under the NFFA. It is likely that larger fishes have been caught in Namibian waters, but have not been officially registered or reported (Fig. 1). The International Game Fish Association (IGFA) recognises four World Records in Namibia (Table 2).

Chobe Fishing Competition, 1985 (Photo: G. Merron).

Table 1. The largest fish caught in Namibia and officially registered as records under open/closed/junior/men or women class, and the largest known freshwater fish caught in Namibia waters (i.e., during competitions or social events), but not registered as records under the NFFA (shaded in grey).

SPECIES	CLASS	Wgt.	LOCATION	CATCH	ANGLER
African Pike	Men's Closed	0.300	Zambezi	27-Aug-07	I Grobbelaar
Banded Tilapia	Men's Closed	0.110	Von Bach	03-Apr-07	C Postma
Common Carp	Junior Open	12.50	Upper Fish River	09-May-10	J D Oberholzer
Greenhead Bream	Men's Closed	2.165	Zambezi	26-Aug-06	C Postma
Humpback Largemouth	Zambezi Classic	3.900	Zambezi	26-Aug-18	H. de Villiers
Largemouth Bass	Men's Closed	4.699	Lake Oanob	13-Oct-12	J Geiger
Longfin Largemouth	Men's Closed	0.008	Zambezi	26-Aug-06	A Brink
Mozambique Tilapia	Men's Closed	1.970	Oanob	01-Oct-05	T Bohme
Moggel	Men's Closed	0.550	Hardap	02-Feb-05	J Malherbe
Nembwe	Men's Closed	2.545	Zambezi	24-Aug-06	C du Plessis
Nembwe	Crockango Club Report	3.500	Kavango River	-	D. du Plessis
Orange River Mudfish	Open Inside Namibia	2.470	Hardap Dam	09-Feb-08	P Eastes
Purpleface Largemouth	Men's Closed	0.140	Zambezi	26-Aug-07	P Opperman
Pink Bream	Crockango Club Report	3.240	Kavango River		P de Wet (RSA)
Rainbow Bream	Juniors Closed	0.240	Zambezi	27-Aug-07	P Andreas
Redbreast Tilapia	Nwanyi Club Report	1.940	Zambezi	-	S. Coertzen
Redbreast Tilapia	Men's Closed	1.300	Swakoppoort Dam	02-Feb-13	A Garces
Sharptooth Catfish	Men's Closed	25.70	Swakoppoort Dam	20-Oct-07	B Oberg
Silver Catfish	Men's Closed	0.120	Zambezi	27-Aug-07	C du Plessis
Smallmouth Yellowfish	Men's Closed	2.400	Hardap Dam	05-May-02	P Fox
Threespot Tilapia	Juniors Closed	2.700	Zambezi	24-Aug-06	C P Bothma
Threespot Tilapia	Nwanyi Club Report	4.050	Zambezi	11-Dec-11	R. Bester
Tigerfish	Juniors Closed	6.000	Zambezi	07-Dec-93	T Figaji
Tigerfish	Lake Liambezi Competition	9.980	Lake Liambezi	1986	L. Lensing
Tigerfish	Crockango Club Report	8.270	Kavango River	16-Oct-12	E Atkinson

Table 2. Currently there are four entries in Namibia registered as World Records under the International Game Fish Association (IGFA).

SPECIES	LINE CLASS	WEIGHT	LOCATION	CATCH DATE	ANGLER
	Line Class World Records (Conventional Tackle)				
Nembwe	M-02 kg (4 lb)	2.81 kg (6 lb 3 oz)	Zambezi River	25-Sep-07	Bradley Kidd
	Junior & Smallfry World Records				
Nembwe	M-Junior	0.99 kg (2 lb 3 oz)	Zambezi River	26-Sep-07	Nicholas Kidd
	Tippet Class World Records (Fly Tackle)				
Tigerfish	Tippet M-06 kg (12 lb)	6.57 kg (14 lb 8 oz)	Chobe River	26-Oct-06	Alan K. Craig
	All-Tackle World Records				
Redbreast Tilapia	All-Tackle	1.62 kg (3 lb 9 oz)	Zambezi River	14-Nov-94	Bill Staveley

Figure 1. The official Namibian Largemouth Bass record of 4.699 kg caught in Oanob Dam by Jurgen Geiger on 13 October 2012 (left). Ernest Atkinson with the current largest known Kavango River Tigerfish of 8.27 kg taken in 2012 (right) (Photos: F. Jacobs).

III. Popular Freshwater Fishing Competitions

A. Zambezi Classic

The Zambezi Classic is one of the largest sport fishing competitions in Africa and is hosted by Nwanyi Angling Club under the NFFA. The Nwanyi Angling Club was established in the 1970s.

The beginning of the prestigious Zambezi Classic had its roots in 2000 when parents of the Zambezi Private School together with the Nwanyi Angling Club organised a fishing competition as a fundraising event. Prior to the Classic, the Kalizo Lodge had an annual fishing competition on the Zambezi River with participation by club members in the early 1990s.

The first Zambezi Classic grew from 9 teams in 2000, mostly from Katima Mulilo, to 18 teams in 2001 and 75 teams in 2012. It became logistically difficult to accommodate all the teams. More anglers resulted in complex organisational arrangements and more prizes requiring sponsorships to support the competition. There was also a concern about the negative impact on the riverine habitats and fish. A decision was made to cap the entries at ~50 teams. Growth in membership and Classic participation motivated the club in 2013 to secure land 30 km downstream of Katima Mulilo on the Zambezi River to use as the official competition grounds.

Zambezi Classic teams are predominantly from Namibia, but South Africa, Botswana, Zambia, Zimbabwe and even Europe and USA have participated throughout the years. The Classic takes place every year in late August. It is one of the biggest angling competitions in the country and participants can fish a 148-km long section of the Zambezi River to win first prize, which is usually a fully equipped boat and trailer (Fig. 2).

Figure 2. Boats lined up in the Sikunga Channel to weigh their catches (left). A Redbreast Tilapia measured by the marshal and fisheries biologist Denis Tweddle (right) (Photos: B. van der Waal).

The Zambezi Classic, as with all organised fishing events in Namibia, has evolved with conservation efforts to preserve the fisheries. For example, the heaviest bag in earlier years consisted of 10 fish/species that were entered at a central weigh-in station, but was changed to only 3 fish/species. Today, however, participants are not allowed to weigh-in fish that are smaller than the previous day or they will be penalised with points. Digital and/or video cameras that record continuous footage are compulsory for every boat, showing (a) the fish in a landing net, (b) live fish on a measuring mat and (c) the live release (swimming away on its own in its natural state) or release into a live-well if the fish needs to be weighed. The format of the competition may change from year to year, but in general fisherman aim to catch the three heaviest Tigerfish, catfish, squeaker, largemouth and smallmouth bream (see Rainbow Bream, P. 148).

Historically, recreational angling was a major income generator for the region. Other fishing competitions were regularly held on the Zambezi River in addition to the Zambezi Classic. However, between 2010 and 2020, the perception developed that populations of angling species diminished to the point that recreational fishing declined, tournaments were cancelled and lodges suffered, all affecting the local economy. Much of the decline in game fish numbers over the past several years was attributed to an increase in commercial gillnetting. Also, the poor economic performance of Namibia since 2016 led to a significant decline in team entries, numbering less than 40 and decreasing to only 23 teams in 2020 due to Covid-19 that resulted in social distancing, restricted intra-country travel and closed international borders.

The Nwanyi Angling Club was concerned about this decline in recreational species and began collaborating with researchers, local businesses, community members, lodges, angling clubs and various other stakeholders. The Club in concert with the MFMR/NNF played a significant role in the establishment of the Sikunga Fisheries Reserve in 2015. Together with sponsorships from various companies and private individuals, this was a major success for fisheries conservation in Namibia (Chapter 8). The Zambezi Classic continues to support this conservation initiative and a portion of each team's entry fees is earmarked for the Sikunga Fisheries Reserve. The Reserve has achieved considerable conservation status with fish numbers steadily increasing.

B. Crockango Angling Bonanza

The Crockango Angling Bonanza hosted by the Crockango Angling Club, a member of the NFFA and similar to the Zambezi Classic, is another prestigious angling competition in Africa. The Crockango Angling Club was founded in 1998 by local anglers interested in artificial lure fishing in the Kavango River. The Club promotes sport fishing by information dissemination, protecting the environment and supporting conservation measures.

The Bonanza attracts anglers from throughout southern Africa and abroad and offers prizes worth more than N$ 500,000. The Bonanza also had a grand prize winner for indigenous fishers who use traditional canoes to encourage the benefits derived from community-based management (CBM) ecotourism. The event is usually held the last weekend of April. The Club has a dedicated site next to the Kavango River that centralises sport fishing competitions into a more efficient and effective social gathering and data generating atmosphere (Fig. 3).

The Bonanza grew from about 81 participants and 27 teams in 2005 to a maximum of 214 participants and 76 teams in 2016. The Bonanza is now capped to accommodate approximately 60 teams. Similar to most of the organised fishing events in Namibia, this competition has evolved with conservation efforts to preserve the Kavango River fisheries. For example, fish caught in earlier years were weighed at a central station, but there were concerns about (a) the negative effects of translocating fish from their habitats and releasing them in other areas, (b) potentially diseased fish being moved around the system, and (c) oxygen starvation of catch kept in live-wells for extended periods. Therefore, the rules and regulations regarding the competition were adjusted and are similar to those of the Zambezi Classic to promote the well-being of caught fish.

In general, anglers aim to catch the three heaviest Tigerfish, clariid catfish, Nembwe and squeaker. The format of the competition changed in 2019 because of the low catches of Nembwe over the past few years and the inability of fishers to fill their allowable bags. Bream species were divided in 2019 into two categories: (1) big bream, including Nembwe, Threespot Tilapia and Largemouth and (2) small bream, including Redbreast Tilapia, Greenhead Tilapia and Pink Bream.

Figure 3. The Crockango Angling Bonanza annually attracts anglers from around the world and the local community. Gabriel Hikerwa won a fully rigged aluminium boat with an engine as first prize for catching the largest Tigerfish during the competition (Photos: B. van der Waal).

The Bonanza has provided an important data base for fisheries managers since 2005. There were 791 teams and 2,178 anglers that participated from 2005–2018 (Table 4, Fig. 4). In total, 1,568 Nembwe and 911 Tigerfish were caught with a total weight of 2,291 and 2,392 kg, respectively (Table 4). However, it was difficult to calculate an actual catch per unit effort (CPUE) from tournament data since many undersized fish were captured and immediately released.

It is estimated that the number of fishing hours per competition ranged between 1,215 h in 2005 to a maximum of 4,280 h in 2016. The 2005 results indicated a success rate of 53 h to catch a Nembwe large enough for weigh-in. Success rate or CPUE improved significantly through 2015 with effort decreasing to 15–35 h, presumably due to anglers becoming more experienced in targeting this species. However, CPUE decreased dramatically after that. The number of hours to catch a single Nembwe large enough for weigh-in increased to 100 h in 2016 and 123 h in 2018 (Table 4, Fig. 4). Contrary to common belief, there was no statistically significant relationship between water discharge and the number of hours to catch a single Nembwe large enough to weigh-in.

The number of hours to catch a Tigerfish large enough for weigh-in fluctuated during the same time period, but there was no clear difference of catch rate among the years. This suggests that populations of Tigerfish in the Kavango River are more stable than Nembwe (Table 4), a territorial species susceptible to overfishing.

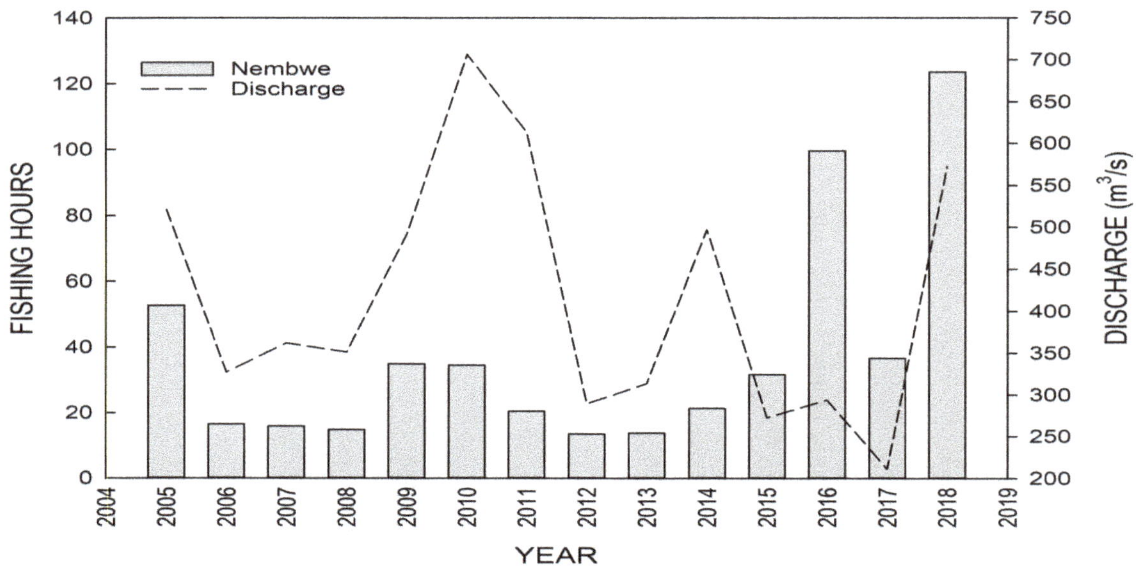

Figure 4. Catch rate (number of hours required to catch an individual large enough for weigh-in) of Nembwe during 2005-2018 at the Crockango Angling Bonanza, Kavango River, Namibia (Discharge data provided by the Okavango Research Institute).

Table 4. Summary of the catch statistics for the Crockango Annual Bonanza from 2005–2018

					------------- NEMBWE-------------				--------------------TIGERFISH------			
Year	A	B	C	D	E	F	G	H	I	J	K	L
2005	27	81	3	1215	23	52.83	1.75	45.41	89	13.65	2.04	49.55
2006	45	118	3	1770	107	16.54	1.68	45.58	105	16.86	1.78	49.13
2007	43	118	3	1770	111	15.95	1.40	42.59	44	40.23	2.85	56.73
2008	55	139	3	2085	141	14.79	1.54	43.84	40	52.13	2.98	57.27
2009	65	182	3	2730	78	35.00	1.62	44.76	48	56.88	2.99	57.17
2010	65	182	3	2730	79	34.56	1.62	44.73	45	60.67	3.06	57.67
2011	58	157	3	2355	115	20.48	1.63	44.88	46	51.20	3.22	58.24
2012	60	163	4	3260	242	13.47	1.66	44.51	64	50.94	3.03	57.76
2013	65	173	4	3460	251	13.78	1.54	43.98	83	41.69	3.41	60.60
2014	63	158	4	3160	148	21.35	1.51	-NA	84	37.62	3.43	-NA
2015	60	184	4	3680	116	31.72	1.77	45.49	61	60.33	3.28	60.61
2016	76	214	4	4280	43	99.53	1.78	45.65	77	55.58	3.33	61.19
2017	56	167	4	3340	91	36.70	1.56	44.49	48	69.58	3.19	60.00
2018	53	142	4	2840	23	123.48	2.08	46.09	77	36.88	3.44	58.55

(A. Teams; B. Fishermen; C. Fishing days; D. Fishing hours; E. Nembwe caught; F. Fishing hours/Nembwe; G. Average weight (kg); H. Average length (cm); I. Tigerfish caught; J. Fishing hours/Tigerfish; K. Average weight (kg); L. Average length (cm)).

IV. Popular Freshwater Recreational Fish Species

The most targeted sport fish species in Namibia are Tigerfish, bream/tilapia, yellowfish, Largemouth Bass, Common Carp and Sharptooth Catfish.

A. Tigerfish

There are currently five known Tigerfish species, all endemic to Africa, of which *Hydrocynus goliath* in the Congo River basin is the largest, reaching up to 50 kg. *Hydrocynus forskahlii* and *H. brevis* are West African species; *H. tanzaniae* occupies the Ruvu and Rufiji-Ruaha drainage basins in Tanzania; and *H. vittatus* is considered mostly a southern African species, although they do occur in other parts of the continent in large river systems. These species are top piscivores in African freshwater ecosystems and play an important role as predators in the transfer of energy between floodplains and the main river, crucial in maintaining ecosystem functioning.

Tigerfish (*H. vittatus)* occur in Namibia in the Upper Zambezi-Chobe, Kwando-Linyanti and Kavango Rivers. The MFMR has conducted both conventional tagging programmes and radiotelemetry studies of Tigerfish since 2002, documenting their preferred habitats. They are generally an open-water species not attracted to structure unless prey is present. Tigerfish demonstrate strong territoriality with home ranges of ~1 km, although individuals have migrated 397 km upstream in the Kavango River. This southern species does not grow as large as those in the Congo, Ruvu and Rufiji-Ruaha drainage basins; the current world record of 16.1 kg was caught in Lake Kariba (IGFA)

Namibia's Tigerfish is one of the premier sport fish species in the world, known for acrobatic leaps when hooked (Fig. 5). Any fish >5 kg (5–20 years of age) can be considered a *trophy.* Tigerfish are opportunistic predators feeding on a variety of fishes, small crocodiles, snakes, birds and insects. Tigerfish prefer prey <40 % of their body length. For example, a 55-cm fish can eat prey up to ~22 cm although food size varies greatly. Tigerfish mostly hunt in packs of similar sized individuals and feeding frenzies can often be observed in open water habitats. The teeth of Tigerfish are set in a bony jaw and are made for biting and holding (Fig. 5).

F. Maartens

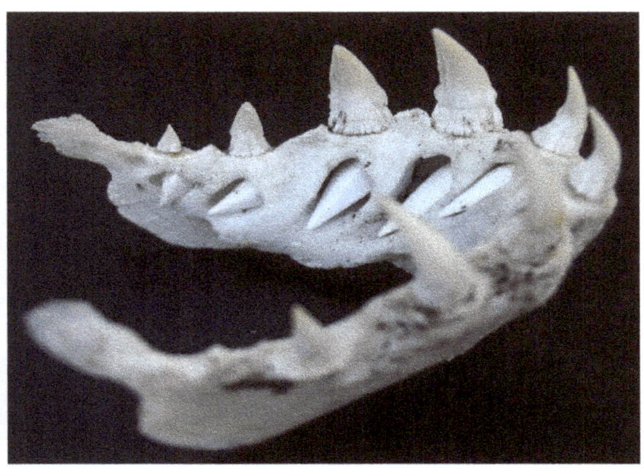

F. Jacobs

Figure 5. A trophy Tigerfish explodes from the water after being hooked (left). The bony interlocking jaws of Tigerfish carrying a series of 10 large, protruding, sharply pointed teeth with developing replacement teeth embedded in the jaws (right).

B. Bream/Tilapia (Largemouth and Smallmouth)

The most popular native bream species targeted by recreational fisherman in Namibia are Nembwe, Threespot Tilapia, Greenhead Tilapia, Pink Bream, Redbreast Tilapia, Humpback Largemouth and Thinface Largemouth. The Nembwe, Humpback and Thinface Largemouths occupy trophic levels similar to the Largemouth Bass and have comparable feeding behaviour by using structure to launch an ambush on prey. Larger specimens prefer clay banks, sunken trees, deep main channels, rocky areas and perennial lagoons, whereas smaller fishes occur mainly in lagoons and secondary channels. Similar to Largemouth Bass, these species will sometimes take to the air when hooked.

The Kavango and Upper Zambezi-Chobe Rivers are the top destinations in Namibia for Nembwe, Humpback and Thinface Largemouths, Threespot Tilapia (Fig. 6) and Pink Bream, while the Kwando River produces plentiful Redbreast and Greenhead Tilapia with the occasional African Pike in between. Non-native Mozambique Tilapia are targeted in most state dams throughout Namibia.

C. Yellowfishes (Smallmouth and Largemouth)

Seven yellowfish species belonging to the genus *Labeobarbus* are currently recognised in southern Africa with three of these occurring in Namibia (Chapter 4). These fish are generally omnivorous feeders and prefer clear flowing water with rocky or sandy substrates, but some can be found in large numbers in dams and lakes and even farm dams.

The Vaal-Orange Smallmouth, particularly, and Vaal-Orange Largemouth Yellowfish are targeted in the Orange River drainage in Namibia for sport fishing (Fig. 6). Generally, they feed on insects and aquatic organisms under rocks and in the shallow river beds. The Namibian record for Vaal-Orange Smallmouth Yellowfish is 2.4 kg as compared to the World Record of 7.837 kg from South Africa.

The Vaal-Orange Largemouth Yellowfish is the largest of the two species and an apex predator eating a variety of prey from aquatic insects, crabs, dragonflies, small baitfish/minnows, frogs, mice, lizards and birds. The current South African angling record is 22.2 kg; no sanctioned record is reported for Namibia.

D. Catfishes

Catfish species are common to all Namibian perennial rivers and state waters, including seasonal impoundments and ephemeral rivers. They are readily caught by recreational anglers in all waterbodies where they occur and are considered very important in the subsistence fisheries. They feed mainly on fish, frogs, reptiles, small birds and mammals, and occasional fruits fallen from surrounding riparian trees. They capture their prey by gulping them in with a rapid opening of the mouth that creates a vacuum and prey are swallowed whole.

The Sharptooth Catfish is considered Namibia's largest freshwater fish and can grow to 1.4 m SL and weigh 60 kg. The Lower Orange River is a popular destination for targeting large specimens (>25 kg), but this species is known from all Namibian perennial rivers and many seasonal waterbodies.

E. Largemouth Bass

Largemouth Bass are well established in Von Bach, Oanob and Friedenau Dams. They occur rarely in Swakoppoort and Hardap Dams, and a few private dams in central Namibia that are mostly inaccessible to the public. The NBAA has made significant investments in the bass fishing heritage by initiating restocking programmes and construction and deployment of artificial fish habitats.

Figure 6. The Threespot Tilapia and Nembwe (top row, left and right, respectively) are popular angling species in Namibia, targeted mostly in the Upper Zambezi-Chobe and Kavango Rivers. Vaal-Orange Largemouth Yellowfish and Smallmouth Yellowfish (left and right, 2nd row). Largemouth Bass and Common Carp, mirror phase (bottom, left and right, respectively).

Bass spawn in Namibia when water temperatures reach about 18 °C, which is usually in late August to September in S. von Bach Dam, and around October in Oanob Dam. Spawning is triggered by temperature and the continued warming of water associated with the lengthening of days (i.e., increased sunlight hours) at the beginning of spring. Bass are at their most vulnerable during the breeding season and effort has been made by the NBAA to protect spawning areas during the breeding season to ensure sustainable populations for future generations

F. Common Carp

Carp are found in all of Namibia's state impoundments, Orange and Fish Rivers, and small private impoundments throughout the country. They are a prime species targeted by sport and subsistence fisherman in southern Namibia, and sanctioned tournaments are highly competitive. Carp are general omnivores with plant material often found in their stomachs.

V. FUTURE OF FRESHWATER RECREATIONAL FISHING

The future of recreational and competitive sport fishing in Namibia is bound to the limited resources present, i.e., the paucity of dams and rivers and their respective water levels in relation to natural rainfall conditions and seasonally high temperatures. Namibia's rains have been linked to La Niña with drought in years that El Nino dominates. There is concern over the effects of global warming that favours El Nino and decreased regional rainfall might prove to be catastrophic (Chapter 11). The other major threats to recreational and competitive angling are over- and illegal-utilisation of fishery resources. The former implies overharvest of the resources to the point of their collapse, whereas the latter infers illegal fishing for capital gain and not adhering to size, season, location and gear restrictions. The illegal use of monofilament gill nets, especially fine-meshed nets and mosquito nets/seines (as opposed to traditional gear), threatens Namibia's fish heritage. Unfortunately, this issue is not a simple cause:effect relationship given the expanding population in the rural North and need for food security (Chapter 8), nor is the solution bound to a single Government Ministry.

Monofilament gillnets rapidly replaced multifilament from the late 2000s and were used nearly 100% in the northern rivers of Namibia by 2012, being purportedly three times more effective. With the eventual decline in commercial catch following the introduction of monofilament gillnets, fishermen changed their traditional method of setting static gillnets to using the nets for drag/seine netting, drift netting and *bashing*. As a result of these destructive illegal fishing methods, it was estimated that only 10% of the large cichlid stocks remained in the Upper Zambezi River by 2012.[3] A similar decline of cichlids occurred in the Kavango River where the number of hours required to catch a Nembwe large enough for weigh-in increased from 13.47 h in 2012 to 123.48 h in 2018. The marginal increase in the average weight of Nembwe caught during the Crockango Annual Bonanza also suggested that the recruitment and population structure of this species had been affected with only the largest, most fecund fishes being caught. Commercial exploitation targets these same fish, thus greatly diminishing their successful recruitment and population recovery.

To reach the goals of responsible management, utilisation and conservation of the Namibian freshwater fish resources, the Namibian Government, through appropriate legislation, must implement a more effective and efficient management regime involving traditional, regional and central authorities. This may mean a cooperative cross-ministeral approach with a common agenda. Benefits will include increased catches; ecotourism and sport angling; and commercial fishing and trade with the objective of self-sustaining management to ensure food security for rural communities and the protection of Namibia's biodiversity heritage.

The continued monitoring of fish competitions is akin to the use of community-based fish monitors to augment the capabilities of MFMR, especially targeting the larger fish species that comprise the upper levels of the food web. As referenced above, the value of tournament monitoring is displayed in Table 4 that demonstrated the increased effort required to catch a single trophy Nembwe between 2012 and 2018, a sure sign of stress on the fishery if not the ecosystem. These data strongly infer the need for an expanded, standardised protocol for data collection, e.g., including counts/unit time of undersized discards of targeted species and bycatch. This will ensure that all fish caught by anglers during competitions are recorded and in turn this will provide a more complete database for the long-term assessment of ecosystem health.

Fishing competitions in affiliation with MFMR offer an opportunity for large-scale tagging of fish. Such studies, termed 'mark and recapture', allow fishery scientists to evaluate long-distance movement; age and growth; and population estimation if conducted over consecutive years. This can generate valuable data for population estimates, angling management and development of specialist angling tourism. Complementing conventional tagging projects, radio-tagging programmes have proven valuable for evaluating species' behaviour daily and seasonally; evaluating home ranges and critical habitat; and assessing criteria for catch and release.[4,5,6,7,8,9,10]

Admirably, all recreational fishing associations are motivated to participate in expanded monitoring. It is emphasised that conservation measures are a part of virtually every fishing competition in Namibia. Efforts are made to preserve breeding stocks of targeted angling species through humane handling and catch and release. Too, recreational angling is largely a non-consumptive sport that promotes fish as a renewal resource.

Measures for maintaining both recreational and subsistence fisheries of Namibian waters are linked rather than being mutually exclusive. An unhealthy ecosystem benefits no one. The most effective measure that can be expedited in a relatively short time remains community-based natural resource management (CBNRM) schemes whereby there is buy-in from local communities and the sportfishing membership. Local communities recognise and desire the need for stronger management practices to sustain fishery resources.[11]

VI. REFERENCES

[1]Arlinghaus R., R. Tillner and M. Bork. 2015. Explaining participation rates in recreational fishing across industrialised countries. Fish Manag. Ecol. 22: 45–55.

[2]Cooke, S.J., V.M. Nguyen, J.M. Dettmers, R. Arlinghaus, M.C. Quist, D. Tweddle, O. Weyl, R. Raghavan, M. Portocarrero-aya and E.A. Cordoba.2016. Sustainable inland fisheries–perspectives from the recreational, commercial and subsistence sectors from around the globe, pp. 467–505. In: G.P. Closs, M. Krkosek and J.D. Olden (Eds.). Conservation of Freshwater Fishes. Cambridge: Cambridge University Press.

[3]Tweddle, D., I.G. Cowx, R.A. Peel and O.L.F. Weyl. 2015. Challenges in fisheries management in the Zambezi, one of the great rivers of Africa. Fish. Manag. Ecol. 22(1): 99–111.

[4]Hocutt, C.H., S.E. Seibold and R.V. Jesien. 1994. Biotelemetry: A review of its potential for use in African waters, pp. 77–95. In: PARADI: Proc. International Symposium on Biological Diversity in African Fresh & Brackish Water Fishes, Dakar, 15–20 November 1993. Revue Hydrobiol. Trop. 27(2).

[5]Thorstad, E.B., C.J. Hay, T.F. Naesje, B. Chanda and F. Okland. 2002. Movements and habitat utilisation of Tigerfish (*Hydrocynus vittatus*) in the Upper Zambezi River. Implications for fisheries management. Norwegian Institute for Nature Research, NINA Proj. Rept. 19, Trondheim: 28 pp.

[6]Thorstad, E.B., C.J. Hay, T.F. Naesje, B. Chanda and F. Okland. 2003. Movements and habitat utilisation of Threespot Tilapia in the Upper Zambezi River. Norwegian Institute for Nature Research, NINA Proj. Rept. 23, Trondheim: 22 pp.

[7]Thorstad, E.B., C.J. Hay, T.F. Naesje, B. Chanda and F. Okland. 2005. Movements and habitat utilization of Nembwe, *Serranochromis robustus* (Günther, 1864), in the Upper Zambezi River. Afr. Zool. 40(2): 253-259.

[8]Thorstad, E.B., C.J. Hay, F. Okland, N. Nickanor and T.F. Naesje. 2007. Spatial behaviour and management of Greenhead Tilapia (*Oreochromis macrochir*) in the Zambezi River, Namibia. Norwegian Institute for Nature Research, Trondheim, NINA Project Report 287: 26 pp.

[9]Jacobs, F.J., T.F. Naesje, E.M. Ulvan, O.L.F. Weyl, D. Tiyeho, C.J. Hay, G. O'Brien and C.T. Downs. 2019. Implications of the movement behaviour of African Tigerfish *Hydrocynus vittatus* for the design of freshwater protected areas. J. Fish Biol: 1–9.

[10]Burnett, M.J., G.C. O'Brien, F.J. Jacobs, G. Jewitt and C.T. Downs. 2021. Fish telemetry in African inland waters and its use in management: A review. Rev. Fish. Biol. Fisheries: https://doi.org/10.1007/s11160-021-09650-2.

[11]Tiyeho, D.C. 2020. Assessment of fishers' catches on the Kavango River floodplains, Namibia. M.Sc. Thesis, Univ. Namibia, Windhoek: 82 pp.

Appendix 1. Namibia Freshwater Fishing Regulations (2021)

The Minister of Fisheries and Marine Resources has under Section 29 of the Inland Fisheries Resources Act, 2003 (Act No. 1 of 2003) made the regulations applicable to the recreational angler (Annexures refer to those included in the Act).

Application for recreational license

24. The Minister shall for the purposes of an application for recreational licence provide forms which substantially correspond with Annexure C.

Form of recreational licence

25. The recreational license issued by the regional council must substantially correspond with Annexure D.

Duration of recreational license

26. A recreational license shall be valid for the period of twelve months from the date on which it was issued.

Duty to possess and show recreational license

27. A recreational license holder shall at all times of engagement to recreational activities keep in possession the recreational license for the purposes of presentation to an inland water fishery inspector.

Fees for recreational licenses

28. The fees payable for a recreational license are as follows:

(a) Namibians:	N$ 20.00	monthly license
	N$ 240.00	yearly license; and
(b) Non-Namibians:	N$ 240.00	monthly license
	N$ 2,500.00	yearly license.

Fishing gear restrictions

29. A recreational license holder is not allowed to use more than 2 rods and 2 lines with 2 hooks attached.

Bag limit

30. (1) A recreational license holder is not allowed to fish more than 10 fish in the aggregate of any species in one day.

(2) A person is not allowed to catch more than 2 Tigerfish in one day for recreational purposes.

Size limit

31. Any angler shall not possess-

(a) a Tigerfish or *Hydrocynus vittatus* less than 30 cm in length;
(b) a Sharptooth Catfish or *Clarias gariepinus* less than 41 cm in length;
(c) a Greenhead Tilapia or *Oreochromis macrochir* less than 22 cm in length;
(d) a Threespot Spot or *Oreochromis andersonii* less than 30 cm in length;
(e) a Nembwe or *Serranochromis robustus* (now *S. jallae*) less than 34 cm in length;
(f) a Silver Catfish or *Schilbe intermedius* less than 22 cm in length;
(g) a Banded Tilapia or *Tilapia sparrrmanii* less than 13 cm in length;
(h) a Redbreast Tilapia or *Coptodon rendalli* less than 23 cm in length;
(i) an African Pike or *Hepsetus cuvieri* less than 28 cm in length;
(j) a Purpleface Largemouth or *Serranochromis macrocephalus* not less than 20 cm in length;
(k) a Humpback Largemouth or *Serranochromis altus* not less than 30 cm in length;

(l) a Thinface Largemouth or *Serranochromis angusticeps* not less than 33 cm in length;
(m) a Blunttooth Catfish or *Clarias ngamensis* not less than 37 cm in length;
(n) a Green Happy or *Sargochromis codringtonii* not less than 24 cm in length;
(o) a Pink Happy or *Sargochromis giardi* not less than 36 cm in length;
(p) a Longfin Largemouth or *Serranochromis longimanus* not less than 16 cm in length;
(q) a Brownspot Largemouth or *Serranochromis thumbergi* not less than 14 cm in length;
(r) a Rainbow Happy or *Sargochromis carlottae* not less than 20 cm in length;
(s) an Orange River Mudfish or *Labeo capensis* not less than 28 cm in length;
(t) a Moggel or *Labeo umbratus* not less than 28 cm in length;
(u) a Smallmouth Yellowfish or *Labeobarbus aeneus* not less than 21 cm in length; or
(v) a Largemouth Yellowfish or *Labeobarbus kimberleyensis* not less than 30 cm in length.

Sale of fish
32. A recreational license holder is not allowed to sell any fish which was caught under a recreational license.

Form of license for recreational fishing charter boat
33. A license for a purpose of using the recreational fishing charter boat must correspond with Annexure E.

Daily catch report for recreational fishing charter boat
34. The owner or operator of a recreational fishing charter boat shall submit daily catch report to the Ministry on or before the 15th day of the following month.

Fishing from a recreational charter boat
35. A person who is not in possession of a valid recreational fishing license is not allowed to fish from a recreational charter boat.

CHAPTER 11
MONITORING OF FRESHWATER SYSTEMS AND CLIMATE CHANGE

by
Ben J. van Zyl

I. Introduction

Freshwater systems in a functioning context supply provisioning, regulatory and cultural services[1,63] to human communities dependent upon these waterways (Fig. 1). In the first instance, provisioning services are considered consumables (fisheries, food, water, firewood). These benefits can be locally highly significant, particularly where fishes are often the only source of animal protein to which communities have access.

Many regulating services (e.g., flow regulation, ground water recharge, waste assimilation and sediment transport) are associated with specific elements of the flow regime and can be negatively impacted by modifications to that regime.

Freshwater systems also provide strong cultural services. These services generally come in two forms: spiritual and commercial. Various spiritual activities of communities are linked to water courses with baptism being an obvious example. Another regional example is the Nyami Nyami, considered to be a giant water snake and one of the most important gods of the Tonga people of the Middle Zambezi River of Zimbabwe and Zambia, who believe it controls life in and on the river. By contrast, cultural activities from a commercial perspective include recreation and tourism.

It is prudent to expand on the perceptions above[1] of freshwaters acting as 'services' to also think of waterways acting as 'political services', especially in Namibia where every perennial river of Namibia serves as an international border.[2] Unfortunately, borders divide ancestral lineages and are disputed for water and fishing rights, land ownership and international boundaries. Even within a drainage, communities partition themselves based upon common ancestry and heritage, e.g., five traditional Kingdoms are recognised along Namibia's Kavango River.

Improving the wellbeing of fishing-dependent people has been proposed as a potential key objective of fisheries governance. There are three key components of wellbeing: the material (economical), relational (social, political and cultural) and subjective (individual satisfaction). Thus, a 'wellbeing scale' ranges from the individual to household, community, fishery to human-ecological systems as a whole.[63] The wellbeing of fishers and communities is intertwined with the above services provided by waterways (Fig. 1).

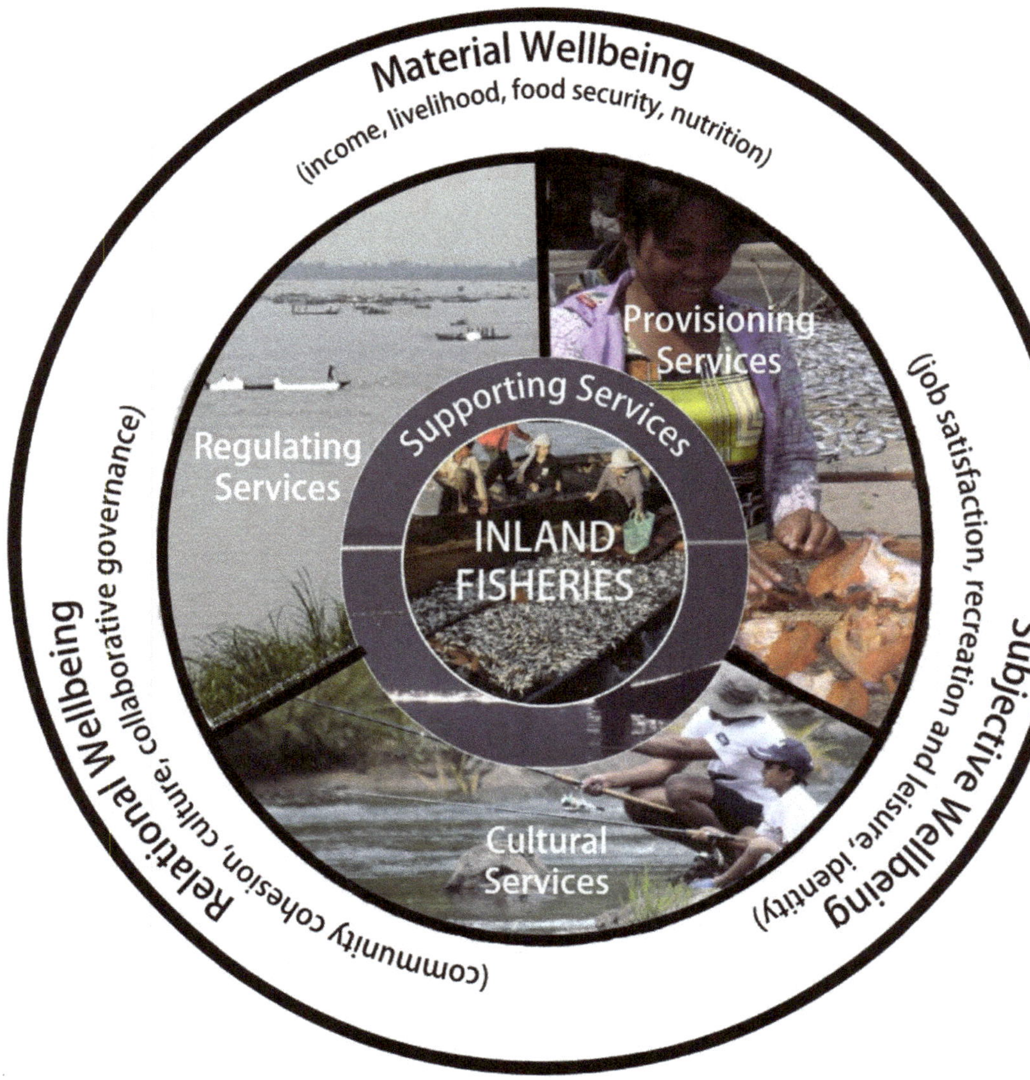

Figure 1. The relationship between inland fisheries, ecosystem services[1,63] and the three dimensions (subjective, material, and relational) of the human wellbeing framework.[64] Ecosystem services provided by inland fisheries support material wellbeing such as practical welfare and standards of living (e.g., food, nutritional and economic security; poverty alleviation); relational wellbeing including personal and social relations (e.g., community building, social security and cohesion, social capital, mutual respect); and subjective wellbeing of individual concerns, values, perceptions and experience (e.g., sense of identity, traditions, livelihood, culture and hope, individual satisfaction and happiness).[63]

Service provisions are especially critical in light of the many subsistence families in the rural communities in Namibia. The United Nations indicated that most rural people rely on activities within the freshwater ecosystems for their livelihoods, most prominently primary and secondary production. Hence, the important role of freshwater ecosystems is increasingly being recognised. The role healthy freshwater systems play, in terms of both ecosystem functioning and as a resource upon which a range of freshwater services are based, is unfortunately often identified only when these systems have been degraded or lost.[1]

Many of the benefits derived from the protection of freshwater ecosystems are indirect and cannot be valued easily in economic terms. This means that a triple bottom-line approach is required where the benefits are measured in social, environmental and economic terms.[1] The importance is that environmental outcomes are not separate from other benefits, but should be seen as having a legitimate call on water resources when trade-off decisions are made.

One of the most significant provisioning services provided by freshwater systems is their fisheries. Fisheries in Africa are probably one of the most underappreciated natural resources on the continent, as the fisheries sector as a whole (i.e. marine artisanal, marine industrial, inland fisheries and aquaculture) contributed US$ 24 billion to the GDP of all African countries with the inland fisheries sector contributing an estimated 26% (i.e., US$ 6.3 billion) in 2014. Nearly 12.3 million people are employed Africa-wide in the sector with half of those being fishers and 27% being women.[3] For example, the fishing industry in sub-Saharan Africa provided >70% of animal protein consumed in Malawi in 2018,[4] whereas Lake Victoria has historically supported the world's largest freshwater fishery, yielding 1 million tonnes of fish a year, supporting about 4 million persons in the value chain.[5] The inundation of floodplain wetlands as part of the regulating services provided by the river systems might not receive the recognition that they merit and is related to the maintenance of elements of the flow regime. Namibia's Upper Zambezi-Chobe floodplain fisheries, with Lake Liambezi included, currently yields 5,000 tonnes of catch valued at around US$ 6.25+ or N$ 100 million/annum (N$ 16.00 = US$ 1.00).

II. THREATS TO THE FRESHWATER ECOSYSTEMS

Freshwater ecosystems are particularly vulnerable to climate change because they are relatively isolated, physically fragmented within a largely terrestrial landscape and already heavily exploited by humans.[6] Furthermore, freshwater biodiversity is disproportionately at risk on a global scale; while freshwaters cover only 0.8% of the Earth's surface, they are home to an estimated 6% of all species.[7] Freshwater ecosystems in Namibia are under threat from a number of anthropogenic influences as well as climate change and variability.

A. Anthropogenic Stressors

Both the quantity and quality of water resources are influenced by land-use change, pollution and the direct consumption of biodiversity.[8] Water extraction is determined by changes in population, food consumption, economy, lifestyle and societal views regarding the value of freshwater ecosystems. Freshwater ecosystems in Namibia have already been detrimentally altered from the perspective of riparian communities dependent upon them for fisheries, water supply, sanitation, agriculture and tourism.

B. Fisheries Overexploitation

The overexploitation of fish not only reduces productivity and resilience of those resources, it also affects other species directly dependent on those resources and the integrity and recovery ability of ecosystems. Such loss of integrity has been observed in the Kavango, Kwando and Upper Zambezi catchments. For instance, the selective gillnet fishery in the Zambezi-Chobe Rivers targeting the larger cichlids caused a decline in the catch rates between 2010 and 2018, while the fisheries of the Kavango had been optimally exploited by the mid-1980's.[9] No decline was observed in smaller species that were not targeted by the gillnet fishery. An unbalanced fishery where selected fish species are targeted will impact the natural structure and functioning of the system.

Backwater habitats that are favourite areas for fishers are cleared of aquatic plants with dragnets to facilitate fish catches, but harm fish productivity in the process. Floodplains are critical for fish nursery areas to enhance juvenile survival and facilitate the natural food web. The combined effects of overexploitation and habitat alteration will in the long-term deplete the very resources upon which subsistence living and food security are dependent.

In addition to regional population growth, the Upper Zambezi River system also experiences encroachment from migrant fishers due to a high demand for fish in Zambian urban areas and the Democratic Republic of the Congo (DRC).[10,11,12] These migrant fishers lack interest in long-term sustainability and they compete with local fishers, who depend on fish for food security as a vital component to their livelihoods.[10,12] The commercialised industry indirectly forces communities to make use of environmentally destructive and unsustainable fishing methods (e.g. drifting gillnets and beach seine netting), both contributing to declining catches for subsistence families.[10] The increased use of these methods has diminished the larger bodied fishes such as Tigerfish and cichlids valued as highly important subsistence and recreational species.[13,14,15] The depletion of these charismatic species may also have caused changes in food web structures, which could influence the productivity of the river in the long term.[16] This will have not only an impact on local communities who depend on fish as a source of food security, but also on the angling-tourism industry and in turn the economy of the country.[12]

Namibia's Upper Zambezi and Kavango Rivers are arguably the premier Tigerfishing destinations in Africa.[11,17,18] Local angling lodges on the Zambezi River for example received 4,000 anglers that caught an estimated 38,000 tigerfish in 2010. The recreational angling industry provides up to 70% of revenue to local lodges that have major local economic importance in these rural areas, e.g., employment as fishing guides, chefs and cleaners (see Chapter 10).[12]

The consequences of a declining fishery to subsistence fisherman, riparian communities and the economy surrounding the tourism industry, fast became a concern as conflicts increased between the stakeholders.[11] Recently, the first community-based fish protected areas were established in the Sikunga (Sikunga channel) and Impalila (Kasaya channel) conservancies on the Zambezi River. Their effect has yet to be fully evaluated, but shows great promise (Chapter 8).

C. Annual Flood Cycle

The annual flood cycle is the main driver of productivity in Namibia's floodplain rivers (Chapter 1) and any modification in the hydrology will cause a change in ecosystem services and production, including fish stocks. The annual flood cycle is the mechanism transferring nutrients from the terrestrial ecosystem to the aquatic ecosystem and is likely to be altered with global warming influencing a more variable and erratic flood cycle with annual runoff expected to decline over the long-term. Climate change models predict >20% flow reduction by 2050 accompanied by a 17.3% increase in evapotranspiration.[4] Large commercially-important species are more likely to be impacted by climate change variability than smaller *r*-selected species that have shorter life cycles and are better adapted to environmental variability.

D. Habitat Loss

Anthropogenic stresses on the freshwater ecosystems from other sectors like water supply for agriculture and urban usage, damming, urban development along the rivers and pollution are further challenges that need to be addressed. Urban developments, including agriculture projects and deforestation, can damage or destroy key riverine and riparian habitats with negative impacts on biodiversity unless responsibly planned and implemented. The riparian zones along the northern rivers have been deforested by 70% owing to human and animal population pressure.[19] This has led to destabilisation of the river banks, soil erosion, reduced water quality, threats to biodiversity and a noticeable reduction in available resources.

E. Climate Change and Variability

Climate change and variability are additional challenges that may potentially push freshwater ecosystems beyond their natural recovery limits. Recent studies in Africa[8] highlight the vulnerability of local communities that depend primarily on natural resources for their livelihoods, indicating that their resource base, already severely stressed and degraded by overuse, is expected to be further impacted by climate change.[20] Furthermore, the vulnerability of freshwater ecosystems to climate change also depends on national and international water management institutions and their capacity to respond.

Climatic variations occur naturally over seasonal to millennial time scales, yet there have been unprecedented rates of warming observed in recent decades.[21] The climate system has a number of dominant patterns of natural variability that directly influence the hydrological cycle. Regional climates may vary out of phase owing to the action of 'teleconnections' often associated with droughts, flood and other changes which have significant impacts on global ecosystems.[8] For instance, there is evidence that La Nina patterns (cooling) in the Pacific Ocean are strongly correlated months later with rainfall in Namibia, while El Nino (warming) is related to drought, with the dominance of either influenced by solar activity.

Global warming impacts on terrestrial ecosystems in Africa are expected to be high with a greater risk of droughts[22,23] in mid-continental areas where there is already an inherent tendency for drying during summer. However, precipitation forecasts in sub-Saharan Africa vary widely with some models projecting increases[8] even though regional variability is expected.[23] Average winter rainfall in sub-Saharan Africa is projected to decrease by ~6 mm by mid-century and ~7-9 mm by the end of the century, whilst mean summer rainfall may decrease ~17-23 mm by mid-century and ~19-40 mm by the end of the century. Rainfall is almost entirely experienced in the summer season for most regions of Namibia excluding the southwest where winter rains dominate.

The Cubango/Kavango River is one of southern Africa's major rivers and natural resources for human water supply. Too, it maintains a rich and important ecological function throughout the basin, notably within the Okavango Delta wetland region[24] where climate change impacts will be a function of the upstream hydrology and, thus, river inflows.[25] The best estimate is that there is a relatively high probability of large changes to the extent and duration of inundation within the Delta during the 21st Century, but the degree is unpredictable to any degree of confidence.[24]

In synergy with climate change, vegetation cover across the continent is expected to decrease, largely driven by anthropogenic land-use change including the expansion of agriculture, soil degradation, livestock grazing and fuel wood usage. These complex interactions make it difficult to determine the role of climate change in isolation from the other drivers. Thus, assumed impacts should be viewed in the context of this substantial uncertainty.[26] The challenge is to mitigate the influence of anthropogenic stresses on freshwaters against the background of natural variability.

1. Building resilience to climate change: The climate change agenda has been embraced to varying degrees on the ground and is firmly placed in the National policy of Namibia. The Republic ratified the United Nations Framework Convention on Climate Change (UNFCCC) in May 1995 and submitted their Initial, Second and Third National Communications.[27,28] The Climate Change Unit (CCU) is located institutionally within the Directorate of Environmental Affairs (DEA) and was set up in 2001 under the Ministry of Environment, Forestry and Tourism (MEFT). This unit oversees the coordination of climate change issues in Namibia, and thus the development and submission of National Communications to the UNFCCC. The multi-sectoral National Climate Change Committee (NCCC) supports the CCU and reports to the Cabinet of Namibia. Namibia's National Climate Change Policy (NCCP) of 2011 encourages the implementation of adaptive measures and recognises the cross-cutting nature of climate change and the importance of mainstreaming policy across sectors.

It is expected that changes in water availability, water quality, precipitation, and the likelihood and magnitude of flooding events will interact with climate change to impact human settlements and infrastructure.[29] As well, the integrity, resiliency and productivity of freshwater ecosystems will be negatively affected, threatening the food security of those small-scale communities dependent on fisheries.

2. Sensitivity to climate change: People's vulnerability in Namibia to climate is fundamentally driven by broad social, political, economic and structural factors that shape their ability to maintain secure livelihoods.[30] These factors, including a history of inappropriate economic policies, gender disparities and colonisation, have led to chronic poverty and inequality in the country, as well as marginalisation.[31] This has resulted in a strong reliance on natural resources and dependence on rain-fed crops and fisheries in the freshwater ecosystems.[32]

3. Adaptive capacity: Much of the adaptive capacity that exists in Namibia, especially at the regional and National level, is generic, being inadequate to deal with both current and future climate risk and the inequality and poverty that exist. Ironically, remote chronically-disadvantaged communities at the local level may be more resilient than urban residents or informal settlements to cope with climate variability since they already battle to secure resources on a daily basis.[33] However, marginalised communities are concerned that they will not be able to deal with an increasing frequency of extreme events.[34] Limited capacity to respond to the effects of change leads to increased vulnerability of communities and sectors.

4. Key barriers to increasing resilience and reducing vulnerability: Key barriers to increasing resilience and reducing vulnerability to climate impacts include:

Barrier 1: Limited knowledge and understanding of the impacts of climate change on freshwater ecosystems and the vulnerability of riverine populations to climate change.

Inadequate knowledge and understanding of the impacts and implications of change on (a) Namibia's freshwater ecosystems and (b) those whose livelihoods centre on the ecosystems are fundamental barriers to adaptation. Limited information is available in terms of seasonal forecasts, early warning systems, climate scenarios and adaptation options.[27,35] Without that information it is impossible to make meaningful and widespread decisions by regional and local authorities with regards to adaptation to climate threats and risks.

The importance of building on traditional fisher's knowledge linked to stressors experienced by riverine communities has been highlighted as one of the most important adaptation requirements, indicating the need for its incorporation into climate change policies to ensure the development of effective adaptation strategies that are cost-effective, participatory and sustainable.[36]

Barrier 2: Widespread lack of awareness of the risks of climate change

Namibia at the policy level has identified the threats and possible opportunities of climate change through the Second National Biodiversity Strategy and Action Plan 2013–2022 and the National Policy on Climate Change.[27] Still, a lack of information, awareness and networks of communication provide substantial gaps in responding to climate change in Namibia. In short, there is poor understanding of the problems, adaptive options and potential solutions.[37] Knowledge and awareness amongst most stakeholders, including those that have already been impacted and are likely to be further impacted, remain low or absent.

Barrier 3: Prevailing sectoral-based approaches to development and climate change

Environmental change generally has a low priority in Namibia relative to other issues. This means that insufficient resources are allocated to address the complex problem in an effective way.[27] In common with many other countries and regions in the world, fragmented, sectoral-based management and planning is the norm typically leading to isolated, uncoordinated and therefore sub-optimal decisions and actions. Freshwater ecosystems, and the lives and vulnerabilities of humans who make use of their services, are affected by many different sectors and activities and any robust long-lasting solutions to reducing vulnerability need to take all of those into account. This requires improved coordination and cooperation between all relevant sectors, government departments and stakeholders.

Barrier 4: Capacity constraints

Limited and frequently insufficient capacity to address climate change impacts and implement adaptation plans within the framework of an ecosystem approach is a further constraint in Namibia. Unless addressed and improved, this forms a serious barrier to progress. More specific technical constraints include limited capacity to access, interpret, translate and communicate climate change information to (a) policy makers and the public for decision making and (b) technical staff to provide advice and implement measures.[27] The lack of experience with and capacity to implement effective co-management is a particularly important constraint.

On a regional level the paradigm of 'integrated water resources management' is vital. With reference to Namibia's perennial rivers, it is essential that the vulnerability of freshwater systems to climate change be addressed at the international level with neighbouring states in addition to National inter-Ministerial water management efforts.

5. Adaptation options: Environmental changes due to population growth, urbanisation and/or climate shifts pose a conceptual challenge to water managers by introducing uncertainty in future hydrological conditions. It may also be very difficult to detect an underlying trend, meaning that adaptation decisions may have to be made before it is clear how hydrological regimes are actually changing. Integrated water resources management should be an instrument to explore adaptation measures to climate change, but so far it is in its infancy in Namibia.[8] Successful integrated water management strategies include, among others: capturing society's views, reshaping planning processes, coordinating land and water resources management, recognising water quantity and quality linkages, conjunctive use of surface water and groundwater, protecting and restoring natural systems, and consideration of climate change.

The existing adaptation measures affected in Namibia are not strategic long-term interventions and are unlikely to be able to address the threat posed by the long-term nature of the climate change challenge.[22] With a substantial proportion of the Namibian rural population living in uncertain climatic conditions and at great distances from basic municipal services, there is a history of using traditional approaches for coping with climate variability. Relief measures that have been used in the past were mostly geared toward dryland agriculture and have the potential to increase the resilience of communities, but are not always adequately employed or sufficient. These responses would be considered autonomous as they have been carried out independently and have not necessarily been applied with climate change in mind.[22]

Ecosystem-based adaptation (EbA) can be a vital part of the solutions to tackle adverse impacts of climate change in inland fisheries.[38] EbA interventions can help to secure ecosystem services and reduce people's vulnerability in ecosystem contexts ranging from largely natural (e.g. inland fisheries ecosystems) to modified landscapes. Subsistence fisheries are a major food production system that relies entirely on wild resources and their natural production cycles. The integrity and health of the ecosystems that sustain these resources is essential to the services they provide. To the extent of conserving or restoring ecosystems and enhancing conservation for climate resilience while promoting sustainable aquatic food production, EbA is well aligned with the FAO Code of Conduct for Responsible Fisheries (CCRF) and the principles of the ecosystem approach to fisheries (EAF).

The application of the EAF in a particular system foresees the thorough identification of key issues that affect, or are likely to affect in the future, the (a) ecological well-being of the system under consideration, (b) well-being of the people it supports and (c) the ability to achieve these well-beings, whether this ability is related to governance or to external factors such as climate change.[38] The first step is to identify issues under the ecosystem well-being component of EAF that relate directly to EbA. These issues would require the adoption of measures aimed at:

- Preserving or restoring ecosystems (including their genetic and species diversity) and ecosystem services of relevance to inland fisheries (e.g. rehabilitation of aquatic habitats, landscape management and river rehabilitation); and
- Creating buffers against the adverse impacts of climate change.

The second step is to provide an operational framework for the application of EbA in the fisheries sector.[39,40] The approach allows for a participatory process, whereby different stakeholders engaged in inland capture fisheries carry out vulnerability assessments, adaptation prioritisation, implementation, and monitoring and evaluation.[38]

Examples of EbA measures in connection with inland capture fisheries include:

- Restoration and conservation of critical aquatic ecosystems that sustain fish production;
- Improved fisheries management to minimise environmental impacts; and
- Integrated approaches to deal with multiple ecosystem degradation processes.

III. Monitoring and Surveillance

The British chemist James Lovelock advanced the 'Gaia hypothesis' that the Earth as a whole is alive, its functioning being a sum of all the living biota at one time integrated with all inorganic matter to form a single self-regulating environment.[41] The premise was further promoted by viewing Earth's atmosphere as "a component part of the biosphere rather than as a mere environment for life".[42] The hypothesis remains controversial;[43] however, a number of scientists have called for global climate/ecosystem monitoring with the primary objectives of (1) maintaining balanced biological communities, (2) protecting the structural and functional integrity of ecosystems, and (3) protecting biodiversity.[44] The *World Conservation Strategy* by the International Union for the Conservation of Nature and Natural Resources (IUCN), United Nations Environment Programme (UNEP) and World Wildlife Fund in 1980 stressed the critical need to balance the protection of the natural environment with the equally vital need to raise the standard of living for the world's poor.[45] These references date to the beginning of the era of 'environmental assessments' and 'environmental impact statements' and laid a foundation for a message that has equal relevance today, i.e., conservation of resources and development are not incompatible, but rather they are mutually dependent.

Any alteration of the natural state of water bodies will shift the interaction of fish and other biota within the aquatic regime. In order to measure these shifts over time, an adequate database is required to understand the historical situation (closest to unaltered state) as opposed to future shifts.

A. Databases

An active fisheries monitoring programme of state dams and rivers started in the 1980s and was continued by the MFMR though the mid-1990s. Basic parameters such as 'catch per unit effort' (CPUE), species composition, gonad stages and parasite infestation were logged annually into a computerised data base. Although there was a break in data sampling for the state dams, river monitoring has continued annually since 1996. Subsequently, all perennial rivers are sampled at selected stations facilitating MFMR to identify changes in the fish community. Approximately 500,000 individual fish have been entered into the database through these monitoring surveys, noting date collected, location, species, size, maturity, gear used, habitats and personnel involved.

Furthermore, catches from local fishers are recorded along the Zambezi and Chobe Rivers providing valuable information on the fishers' catches and their fishing patterns, providing very important information for management purposes. These surveys have been expanded to the Kwando and Kavango Rivers as well.

Fish movement, whereby individual fishes from different species were tagged with radio-transmitters, has been selectively studied. Migratory behaviour and species interactions are especially important when considering fisheries reserves. Fisheries reserves are more effective if the movement patterns of different fish species are known. A valuable database on these movements is available and was also considered during the demarcation of fisheries reserves in the Zambezi-Chobe, Kwando and Kavango Rivers (Chapter 8).

Frame surveys documenting the importance of fish in the daily livelihoods of the communities have been conducted by MFMR. Another important database is from the fish market in Katima Mulilo (initiated in 2002) where data on species, sizes, price, origin of fish and fish processing are recorded (Chapter 6). All these databases are housed by the MFMR, contributing towards the sustainable management of fishery resources.

Complementing the MFMR activities were surveys conducted by (now) the Ministry of Agriculture, Water, and Land Reform (MAWLR).[46,47,48] Much of MAWLR's efforts focused on environmental assessments of the Eastern National Water Carrier (ENWC), a 700+ km open canal to supply water from the Kavango River to Windhoek, the Republic's capital. Principle concerns were: (1) the unintentional transfer of fish from the Kavango to interior drainages; (2) possible transmission of schistosomiasis (bilharzia); (3) algal growth and deterioration of water quality within the canal; (4) effects of the open canal on wildlife migration routes; (5) effects of groundwater withdrawal on regional aquifers; and (6) highly significant evaporation from the canal and storage reservoirs. MFMR has recorded different water quality parameters from perennial rivers and the state reservoirs although not consistently. Water levels from the rivers and the state dams are available from MAWLR.

B. Biological Monitoring

Biological monitoring protocols should be considered as an extension of biodiversity surveys, which are normally aimed toward natural heritage programmes.[49] Fish communities are a viable means for assessing the effects of change on aquatic ecosystem *'health'*, be it from natural shifts or those caused by man, e.g., pollution, changes in hydrology, climate change, overfishing, etc. A diverse fish community and their various life stages represent several different trophic levels within an aquatic food chain, being primary consumers, detritivores, insectivores or tertiary piscivores. Hence, fish communities both directly and indirectly can provide evidence of water quality in that they incorporate all the local perturbations into the stability of the communities themselves.

This is not to say that fish community assemblages or other biotic data can be used as indicators of ecosystem health for all classes of pollutants. Some ecosystems might have structurally diverse and functionally complex interactions only to have heavy metals, pesticides, petrochemical or radiological contaminants present at unacceptable levels with potential of long-term chronic effects.[50] Such inputs can be measured through biochemical, radiological and tissue testing, but these procedures are expensive and time consuming. In Namibia, the poorest water quality conditions are experienced mostly in the Orange-Fish River system, highly modified for water supply, agriculture and mining. Main inputs into the Kunene and Kwando-Linyanti Rivers are mostly organic in nature; however, increased inorganic loading is expected within the larger, more developed Kavango and Upper Zambezi River basins, especially the latter with its extensive watershed that drains agricultural lands in western Zambia.

Realising the value of fish communities as a biological monitoring tool, the Index of Biotic Integrity (IBI) was proposed as a means to rapidly measure aquatic ecosystem health.[51,52] The IBI has now been adapted or modified to local conditions globally using fish communities[e.g., 53,54,55,56] and benthic organisms.[57] The IBI approach was first employed in Namibia in 1994 for the Kavango

River[49], expanded on in later studies.[58,59] These studies serve as baselines to adapt to other riverine systems. To emphasise that such information is invaluable for monitoring and predictive baselines, as this book goes to press environmentalists are opposing licenses for a Canadian-led consortium to explore oil and gas reserves in a 35,000 km^2 section along the Namibia-Botswana border in the Kavango-Okavango Delta basin. The exploration sites are reported to be within the Kavango-Zambezi (KAZA) Trans-frontier Park, the largest conservation area in the world incorporating protected and communal land in Namibia and neighbouring Angola, Botswana, Zambia and Zimbabwe.[60] This area is bounded by three National Parks, the Kavango River and the Okavango Delta.

C. Repositories of Reference Collections

E.J. Crossman[61] highlighted the value of reference collections and reminded that Carl Linnaeus's taxonomy foundation in the late 1700s was virtually the first environmental impact assessment. That is, Linnaeus's work for the King of Sweden was to include an interpretative analysis from his collections on what type of economic development could be undertaken in different parts of the country. Unquestionably, reference collections of fishes are fundamental to the role of bio-monitoring.

Repositories come in different forms.[61] As addressed in Chapter 2, one typically thinks of museums such as the National Museum of Namibia, which indeed are the usual recipients of type specimens for taxonomic purposes. In today's world of environmental assessments, however, a repository may be a university, consulting firm, private collection, government agency or non-government organisation that holds and curates reference collections. Such collections if properly maintained include at a minimum the date of collection, location (facilitated by GPS) and collector(s), preferably with a description of the capture gear, habitat, environmental parameters (e.g., temperature, dissolved oxygen, pH, etc.) and notes on colour of live specimens.

D. Local and International Collaboration

The need for (1) inter-ministerial cooperation within Namibia and (2) transboundary management across borders with neighbouring countries cannot be overemphasised. At the national level, various ministries have direct mandates to manage either environmental, water, wetland or fishery resources, while others are responsible for the rural populace and its well being, including heath and food security. Others have a direct responsibility to oversee and mitigate the effects of development, be it in the form of roads, infrastructure or mining. Often, the missions of these ministries are in direct conflict. What is lacking is a cohesive plan of action and direct leadership to ensure the sustainability of the Nation's aquatic resources, especially in the north where over 50% of the populace lives at an extreme poverty level (Chapter 1). Environmental degradation whether in the form of natural or man-related causes will only exacerbate a deteriorating situation.

From a transboundary perspective, all the perennial rivers traversing Namibia and the vastly important, but ephemeral, Cuvelai River are shared resources with Angola, Zambia and South Africa, while the common border with Botswana harbours some of the most environmentally sensitive waterways on the continent. A number of transboundary water commissions are in place (Chapter 8), but their overall impact on co-management of aquatic resources, including fisheries, has been minimal. Although a Transboundary Fisheries Management Plan exists within the Okavango River Basin Water Commission, joint strategies have been lacking. Yet, local and distant movement of Tigerfish up to

379 km (as measured by radiotelemetry[62]), both along and across shared borders with Angola and Botswana, effectively demonstrates the necessity of active transboundary management.

IV. REFERENCES

[1]Le Quesne, T., J.H. Matthews, C. Von der Heyden, A.J. Wickel, R. Wilby, J. Hartmann, G. Pegram, E. Kistin, B. Geoffrey, G.K. de Freitas, E. Levine, C. Guthrie, C. McSweeney and N. Sindorf. 2010. Flowing forward: Freshwater ecosystem adaptation to climate change in water resources management and biodiversity conservation. In: The World Bank (Ed.). Water Working Notes: Note No. 28, August 2010, Washington, D.C.

[2]Hocutt, C.H. 2021. *Pers. comm.*

[3]de Graaf, G. and L. Garibaldi. 2014. The value of African fisheries. FAO Fish. Aquacult. Circ. No. 1093, FAO, Rome: 76 pp.

[4]Government of Malawi. 2019. Annual economic report 2019. Ministry of Finance, Lilongwe: 191 pp.

[5]Ainsworth, R., I.G. Cowx and S. Funge-Smith. 2021. A review of major river basins and large lakes relevant to inland fisheries. FAO Fish. Aquacult. Circ. No. 1170. Rome: 314 pp.

[6]Woodward, G, D.M. Perkins and L. E. Brown. 2010. Climate change and freshwater ecosystems: Impacts across multiple levels of organization. Phil. Trans. R. Soc. B 365: 2093–2106.

[7]Dudgeon, D., A.H. Arthington, M.O. Gesser, Z, Kawabata, D.J. Knowler, C. Leveque, R.J. Naiman, A. Prieur-Richard, D. Soto, M.L.J.Stiassny and C.A. Sullivan. 2006. Freshwater biodiversity: Importance, threats, status and conservation challenges. Biol. Rev. 81: 163–182.

[8]Bates, B.C., Z.W. Kundzewicz, S. Wu and J.P. Palutikof (Eds.). 2000. Climate change and water. Technical Paper of the Intergovernmental Panel on Climate Change, IPCC Secretariat, Geneva: 210 pp.

[9]van Zyl, B.J. 1988. 'n Vergelykende studie van die groeivermoëns van vier tilapia spesies in natuurlike and intensiewe produksie-eenhede. M.Sc., University of the Orange Free State. South Africa.

[10]Abbot, J. G., C.J. Hay, T.F. Naesje, D. Tweddle and B.C.W. van der Waal. 2015. Rain and copper: The evolution of a fish marketing channel in a rapidly changing region of southern Africa. J. South. Afr. Stud. 41: 29–45.

[11]Tweddle, D., I. Cowx, R. Peel and O. Weyl. 2015. Challenges in fisheries management in the Zambezi, one of the great rivers of Africa. Fish. Mngt. Ecol. 22: 99–111.

[12]Cooke S.J., V.M. Nguyen, J.M. Dettmers, R. Arlinghaus, M.C. Quist, D. Tweddle, O. Weyl, R. Raghavan, M. Portocarrero-aya and E.A.Cordoba. 2016. Sustainable inland fisheries– perspectives from the recreational, commercial and subsistence sectors from around the globe, pp. 467–505. In: G.P. Closs, M. Krkosek and J.D. Olden (Eds.). Conservation of Freshwater Fishes. Cambridge University Press, Cambridge.

[13]Marshall, B. 1987. On the relationship between tigerfish and sardines in Lake Kariba. J. Limnol. Soc. South. Af. 13: 78–79.

[14]Thorstad, E.B., C.J. Hay, T.F. Naesje, B. Chanda and F. Okland. 2004. Effects of catch-and-release angling on large cichlids in the subtropical Zambezi River. Fish. Res. 69: 141–144.

[15]Okland, F., E. Thorstad, C. Hay, T. Naeje and B. Chanda. 2005. Patterns of movement and habitat use by Tigerfish (*Hydrocynus vittatus*) in the Upper Zambezi River (Namibia). Ecol. Freshwat. Fish 14: 79–86.

[16]Taylor, G.C., O.L.F. Weyl, J.M. Hill, R.A. Peel and C.J. Hay. 2017. Comparing the fish assemblages and food-web structures of large floodplain rivers. Freshwat. Biol. 62: 1891–1907.

[17]Hay, C.J., T. Naesje, J. Breistein, K. Harsaker, J. Kolding, O.T. Sandlund and B.J. van Zyl. 2000. Fish population, gill net selectivity and artisanal fisheries in the Okavango River, Namibia. Recommendations for sustainable fishery. NINA-NIKU Project Report 10: 1–105.

[18]Tweddle, D. 2010. Overview of the Zambezi River system: Its history, fish fauna, fisheries, and conservation. Aquat. Ecosyst. Health Manag. 13(3): 224–240.

[19]Namibia Vision 2030. 2004. Policy Framework for long-term national development (main document). Office of the President, Windhoek.

[20]Leary, N., J. Adejuwon, W. Bailey, V. Barros, M. Caffera, S. Chinvanno, C. Conde, A. De Comarmond, A. De Sherbinin, T. Downing, H. Eakin, A. Nyong, M. Opondo, B. Osman, R. Payet, F. Pulhin, J. Pulhin, J. Ratnasiri, E. Sanjak, G. von Maltitz, M. Wehbe, Y. Yin and G. Ziervogel. 2006. For whom the bell tolls:

Vulnerabilities in a changing climate. AIACC Working Paper No. 30, International START Secretariat, Washington, DC: 31 pp.

[21]Lundqvist, J., C. de Fraiture and D. Molden. 2008. Saving water: From field to fork – curbing losses and wastage in the food chain. SIWI Policy Brief. Stockholm International Water Institute.

[22]Spear, D., M.A.H. Zaroug, D.D. Daron, G. Ziervogel, M.N. Angula, E.N. Haimbili, S.S. Hegga, M. Baudoin, M. New, I. Kunamwene, C. Togarepi and J.E. Davies. 2018. Vulnerability and responses to climate change in drylands: The case of Namibia. CARIAA-ASSAR Working Paper. University of Cape Town, Cape Town, South Africa. http://www.assar.uct.ac.za

[23]IPCC. 2014. Climate Change 2014: Synthesis Report. Contribution of Working Groups I, II and III to the Fifth Assessment Report of the Intergovernmental Panel on Climate Change [Core Writing Team, R.K. Pachauri and L.A. Meyer (Eds.)]. IPCC, Geneva, Switzerland: 151 pp.

[24]Hughes, D.A., D.G. Kingston and M.C. Todd. 2011. Uncertainty in water resources availability in the Okavango River basin as a result of climate change. Hydrol. Earth Syst. Sci. 15: 931–941.

[25]Wolski, P., H.H.G. Savenije, M. Murray-Hudson and T. Gumbricht. 2006. Modelling of the flooding in the Okavango Delta, Botswana, using a hybrid reservoir-GIS model. J. Hydrol. 331: 58–72.

[26]Thiaw, I. 2015. Is the changing climate changing African ecosystems? Ecosyst. Health Sustain. 2, Volume 1(2): 1–3.

[27]Ministry of Environment and Tourism (MET). 2011. Second national communication to the United Nations Framework Convention on Climate Change. Ministry of Environment and Tourism, Windhoek: 125 pp.

[28]Ministry of Environment and Tourism (MET). 2015. Third national communication to the United Nations Framework Convention on Climate Change. Ministry of Environment and Tourism, Windhoek: 173 pp.

[29]Sherbinin, A., A. Schiller and A. Pulsipher. 2007. The vulnerability of global cities to climate hazards. Environ. Urbaniz. 12(2): 93–102.

[30]Ribot, J. 2014. Cause and response: Vulnerability and climate in the Anthropocene. J. Peasant Stud. 41: 667–705.

[31]Jauch, H., L. Edwards and B. Cupido. 2011. Inequality in Namibia, pp. 181–255. In: H. Jauch and D. Muchena (Eds.). Tearing us apart: Inequalities in southern Africa. Open Society for Southern Africa, Johannesburg.

[32]Ministry of Environment and Tourism (MET). 2014. Namibia rural food and livelihood security vulnerability assessment and analysis report 2014/2015. Office of the Prime Minister - Directorate Emergency Management, Windhoek: 22 pp.

[33]Maru, Y.T., M.S. Smith, A. Sparrow, P.F. Pinho and O.P. Dube. 2014. A linked vulnerability and resilience framework for adaptation pathways in remote disadvantaged communities. Global Environ. Change. 28: 337–350.

[34]Newsham, A.J. and D. Thomas. 2009. Agricultural adaptation, local knowledge and livelihood diversification in north-central Namibia. Research Technical Report, Tyndall Climate Change Research Centre, Norwich: 48 pp.

[35]Giorgis D. 2011. Design of an improved climate risk management early warning system (EWS) and EWS information centres. Inter South Development Consultants. Ministry of Environment and Tourism and United Nations Development Programme as part of the Africa Adaptation Project: Namibia, Windhoek: 55 pp.

[36]Osman-Elasha, B., N. Goutbi, E. Spanger-Siegfried, B. Dougherty, A. Hanafi, S. Zakieldeen, A. Sanjak, H.A. Atti and H.M. Elhassan. 2006. Adaptation strategies to increase human resilience against climate variability and change: Lessons from the arid regions of Sudan. AIACC Working Paper No. 42, Assessment of impacts and adaptation to climate change in multiple regions and sectors programme.

[37]Dirk E., C. Hager, M. Tadross, S. Bethune and B. Curtis. 2008. Climate change vulnerability and adaptation assessment: Namibia. Desert Research Foundation of Namibia and Climate Systems Analysis Group for the Ministry of Environment and Tourism, Windhoek: 149 pp.

[38]FAO. (in press). Ecosystem-based adaptation in fisheries and aquaculture. Rome.

[39]FAO. 2003. Fisheries Management. 2. The ecosystem approach to fisheries. FAO Technical Guidelines for Responsible Fisheries. No. 4, Suppl. 2, Rome, FAO: 112 pp.

[40]FAO. 2010. Aquaculture development. 4. Ecosystem approach to aquaculture. FAO Technical Guidelines for Responsible Fisheries. No. 5, Suppl. 4, Rome, FAO: 53 pp.

[41]Lovelock, J.E. 1972. Gaia as seen through the atmosphere. Atmosph. Environ. 6: 579–580.

[42]Lovelock, J.E. and L. Margulis. 1974. Atmospheric homeostasis by and for the biosphere: The Gaia hypothesis. Tellus 26: 1–10.
[43]Kirchner, J.W. 2003. The Gaia hypothesis: Conjectures and refutations. Cli. Change 58: 21–45.
[44]Hocutt, C.H., R. Bally and J.R. Stauffer, Jr. 1992. An environmental assessment primer for less developed countries, with an emphasis on Africa, pp. 39–61. In: J. Cairns, Jr., B.R. Niederlehner and D.R. Orvos (Eds.). Predicting Ecosystem Risk. Princeton, Scientific Publ. Co.
[45]IUCN. 1980. World Conservation Strategy: Living resource conservation for sustainable development. Internatl. Union Conserv. Nat. Res. (IUCN), United Nat. Environ. Prog. (UNEP), World Wild. Fund (WWF).
[46]Bethune, S. 1991. Kavango River wetlands. Madoqua 17: 77–112.
[47]Bethune, S. and E. Chivell. 1985. Environmental aspects of the Eastern National Water Carrier. S. West Afr. Ann., Windhoek: 23–27.
[48]Bethune, S. and K.S. Roberts. 1991. Checklist of the fishes of Namibia for each wetland region. Madoqua 17 (2): 193–199.
[49]Hocutt, C.H., P.N. Johnson, C.J. Hay and B.J. van Zyl. 1994. Biological basis of water quality assessment: The Kavango River, Namibia. Rev. Hydrobiol. Trop. 27(4): 361–384.
[50]Hocutt, C.H. 1981. Fish as indicators of biological integrity. Fisheries 6(6): 28–31.
[51]Karr, J.R 1981 Biological integrity: A long-neglected aspect of water resource management. Ecol. Appl. 1: 66–84.
[52]Karr, J.R. and D.R. Dudley. 1978. Biological integrity of a headwater stream: Evidence of degradation, prospects for recovery, pp. 3–25. In: J. Lake and J. Morrison (Eds.). Environmental impact of land use on water quality. USEPA Final Rept. EPA-905-9-77-007-D.
[53]Belpaire, C., R. Smolders, I. van den Auweele, D. Ercken, J. Breine, G. van Thuyne and F. Ollevier. 2000. An index of biotic integrity characterizing fish populations and the ecological quality of Flandrian water bodies. Hydrobiologia 434: 17–33.
[54]Joy, M.K. and R.G. Death. 2004. Application of the index of biotic integrity methodology to New Zealand freshwater fish communities. Environ. Mgmt. 34(3): 415–428.
[55]Kotze, P.J., G.J. Steyn, H.H. du Preez and C.J. Kleynhans. 2004. Development and application of a fish-based sensitivity-weighted index of biotic integrity (SIBI) for use in the assessment of biotic integrity in the Klip River, Gauteng, South Africa, Afr. J. Aq. Sci. 29(2): 129–143.
[56]Kadye, W.T. 2008. The application of a fish assemblage integrity index (FAII) in a southern African river system. Water S.A. 34(1): 25–32.
[57]Li, Z. and B. Zeng. 2020. Health assessment of important tributaries of Three Gorges Reservoir based on the benthic index of biotic integrity. Sci. Rept. 10 (18743): 1–11.
[58]Hay, C.J. 1995. The development of a data base for the assessment of the biotic integrity and sustainable utilisation of the Okavango River, Namibia. Ph.D. Diss., Rand Afrikaans Univ., Johannesburg: 236 pp.
[59]Hay, C.J., B.J. van Zyl and G.J. Steyn. 1996. A quantitative assessment of the biotic integrity of the Okavango River, Namibia, based on fish. Water S. Afr. 22(3): 263–284.
[60]Shikongo, A. 2020. Environmentalists rebuff Okavango oil exploration. The Namibian, 24 November 2020: p5.
[61]Crossman, E.J. 1980. The role of reference collections in the biological monitoring of fishes, pp. 357–378. In: C.H. Hocutt and J.R. Stauffer, Jr. (Eds.). Biological monitoring of fish. Lexington Books, D.C. Heath & Co., Lexington, MA.
[62]Jacobs, F.J., T.F. Naesje, E.M. Ulvan, O.L.F. Weyl, D. Tiyeho, C.J. Hay, G. O'Brien and C.T. Downs. 2019. Implications of the movement behaviour of African Tigerfish *Hydrocynus vittatus* for the design of freshwater protected areas. J. Fish Biol. 96: 1260–1268.
[63]Cooke, S.J., E.H. Allison, T.D. Beard, Jr., R. Arlinghaus, A.H. Arthington, D.M. Bartley, I.G. Cowx, C. Fuentevilla, N.J. Leonard, K. Lorenzen, A.J. Lynch, V.M. Nguyen, S-J. Youn, W.W. Taylor and R.L. Welcomme. 2016. On the sustainability of inland fisheries: Finding a future for the forgotten. Roy. Swed. Acad. Sci., Ambio: DOI 10.1007/s13280-016-0787-4.
[64]Weeratunge, N., C. Be´ne´, R. Siriwardane, A. Charles, D. Johnson, E.H. Allison, P.K. Nayak and M.C. Badjeck. 2014. Small-scale fisheries through the wellbeing lens. Fish and Fisher. 15: 255–279.

CHAPTER 12
AQUACULTURE

by

Francois J. Jacobs and Renier Burger

I. History of Freshwater Aquaculture in Namibia

Aquaculture in the Namibian context is the farming and ranching of aquatic organisms which include fish, crocodiles, molluscs, crustaceans and aquatic plants in freshwater bodies and in the ocean. This chapter focuses on freshwater fish aquaculture in Namibia.

Different forms of 'fish ranching' were carried out in Namibia under customary (traditional) law and date back centuries. Traditional leaders insisted on the protection of important water bodies that became inundated during the annual floods for the sole purpose of harvesting larger fishes. The riparian communities learned through observation that fishes from the mainstreams migrated into floodplains during the annual floods to feed and in turn grow to larger sizes. This led to very basic forms of aquaculture as communities constructed small bunds, dug out small areas, and identified natural depressions in floodplains which could keep water for longer periods than the surrounding floodplain (Fig. 1). Communities only started harvesting the "larger" fishes once the water level receded to a specified level. Many years ago these methods were especially evident along the Kavango, Kwando-Linyanti, Upper Zambezi-Chobe River systems and in Lake Liambezi (Chapter 1, Fig. 1, Regions 6, 7 & 8).[1] In the Cuvelai system of Namibia (Chapter 1, Fig. 1, Regions 2, 3 & 4), local inhabitants were familiar with the concept of stocking and managing fishes long before aquaculture was introduced. Small catfish (~20 cm TL) were collected and stocked into home-dug wells or pits. Fish were either left to grow to larger sizes to be eaten or subsequently released to reproduce in local pools once the new rainy season filled depressions. The management and stocking of water bodies to enhance the growth of fishes can be seen as the first type of extensive aquaculture in Namibia and is still practised on a small scale today.

The Common Carp (Fig. 2) was probably the first fish species that was intentionally introduced into Namibia for aquaculture purposes by the early German settlers sometime after 1884 when Namibia was a German colony, German South-West Africa. The district commander of Windhoek, Dr. Max von Eschstruth, obtained 50 Common Carp in September 1904 from an institution in Jonkershoek, South Africa (*Fischerei-Zeitung* 11 August 1906), which were transported by ship to Swakopmund, later by train to Windhoek and hand-carried the last 35 km. They experienced no mortality and were introduced into an artificial pond intended to become part of an agricultural experimental station, established at Neudamm five years later.[2] The idea of fish farming, however, must have begun prior to this, as experts had already planned wells and dams in view of their desire to turn the arid landscapes into self-sustaining agricultural settler environments.[3] The establishment of better water provision was a priority and there was a concern regarding the difficulty of transporting seed fish. The Mole harbour in Swakopmund was completed by 1903 and settlers and cargo could disembark easier (Fig. 3). By this time, new colonialists could also make use of train transport, a much simpler journey compared to relying on ox wagons, which shortened travelling time and promoted the idea of transporting fish for farming.[2]

Figure 1. Bunds made from rock, clay and dung to retain water after the annual flood season to grow out trapped fish are some of the earliest forms of aquaculture in Namibia (Photo: G. Merron; See also page 258, Chapter 9).

Following the successful introduction of Common Carp, a German colonial official (probably the same Dr. Max von Eschstruth) on his return from Germany in 1909 brought thousands of Brown Trout (*Salmo trutta fario,* Fig. 4*)* eggs from a fish-breeding institution in Saxony for release in the new colony.[2] Given that Namibia was 11,100 km away, eggs were packed on ice and covered with cloth, careful not to expose the precious eggs to temperatures higher than 10 °C. The incredible journey took three weeks at sea and finally landed in Swakopmund where the eggs were repackaged. A further three days inland by train and the eggs arrived at Grootfontein (named after the large natural springs in the area) where only a few hundred eggs had not survived (*Windhuker Nachrichten* 30 July 1910). The trout eggs were unpacked and introduced into a stream (probably one of the natural springs). However, many eggs stopped moving and by the next day all were dead; the shock of releasing them into 21 °C water had killed them (*Allgemeine Fischerei-Zeitung* 1 May 1910; *Windhuker Nachrichten* 30 July 1910). Another attempt was made in 1911 when approximately 1,000 Brown Trout eggs were still alive at arrival in Grootfontein. This time the trout actually hatched and even swam into the artificial pond. Yet once again the water temperature reached more than 20 °C killing the trout (*Südwestbote* 25 February 1911).

Figure 2. Common Carp was the first fish species that was intentionally introduced into Namibia for aquaculture purposes by the early German settlers sometime after 1884.

Figure 3. Cattle are brought ashore in the early 1900s at the Mole breakwater in Swakopmund, which is probably the same place where fish would have been offloaded in wooden boxes after their long voyage from Germany (left). The stamp (right) shows an early Simmentaler import (Source: National Archives).

Largemouth Bass were obtained from Jonkershoek Hatchery, South Africa, in 1932 and introduced into Narubis Municipal pools, followed by another introduction in 1934 into the Windhoek Municipal Dam.[4] The bass obtained from Jonkershoek were originally spawned in Arnhem, Holland in the summer of 1927. The fingerlings from Holland were shipped via England in the care of Mr. P.J. Neville, who handed over 45 survivors to Mr. F.G. Chaplin at the Jonkershoek Hatchery. These fish were spawned in the hatchery in October 1929 before being introduced into Namibia.

Mozambique Tilapia were also obtained from Jonkershoek and introduced into the country in 1947 and 1949.[5] Numerous introductions of Largemouth Bass and Mozambique Tilapia from Jonkershoek took place during 1944 to 1949. It is believed that by this time Common Carp were also readily available for stocking and populations were well established in impoundments throughout the country. In the autobiography "*The Sheltering Desert*", Henno Martin[6] reported carp from pools in the Kuiseb Canyon at the border of Namib Desert in the 1940s, which incidentally shows how this species infiltrated new ranges beyond their initial stocking areas.

Figure 4. A German official brought thousands of Brown Trout eggs from a fish-spawning institution in Saxony, central Germany, for intended release into the new colony in 1909. The trout eggs were released at Grootfontein, but none survived the unsuitable water temperatures.

Modern aquaculture in Namibia was probably conceptualised with the building of Hardap Dam 260 km south of Windhoek. Although Hardap Dam was already envisaged in 1897, construction only started in 1960 and the dam was completed in 1963. The Hardap Fisheries Institute was established in 1968 after which it was responsible for the commercial production and distribution of Common Carp, Largemouth Bass and Mozambique Tilapia to farmers between 1974 and 1983.[5] After 1983, however, no permits were issued for the importation of these species and supply to farmers was limited.[5] During the decades following their first introduction, these species were widely distributed throughout the country and became important recreational and subsistence species.

There was negligible interest in aquaculture up to the mid-1980s, when the government and the private sector started to promote aquaculture development and launched numerous investigations into the feasibility of inland aquaculture and small reservoir fisheries. Interest in aquaculture during this period grew and some of the earliest feasibility experiments included:

- Stocking and harvesting of *iishana* (see *oshana*) in northern Namibia in the mid-1980s was carried out by the Rural Development Centre (RDC), Ministry of Agriculture, Water and Rural Development (MAWRD), Ongwediva, which operated on a contract basis from 1988 by a private consulting company.[7] RDC undertook a wide range of rural development activities in then Owamboland, including training of farmers and extension through a limited number of rural extension officers. The RDC was responsible for aquaculture development and extension in Owamboland and it focussed on producing fingerlings and stocking *iishana* with Sharptooth Catfish from Ongwediva and Threespot Tilapia and Redbreast Tilapia from Ongwediva and Mahanene. In 1989, approximately 50 small scale fish farmers were supplied annually with fingerlings at a nominal price for stocking *iishana*. Stocking rates were low and no fertilisers or feeds were applied. Since most *iishana* dry up after the rainy season, all fish were harvested within 3–5 months after stocking.[7]

- The Rossing Foundation, a local NGO, stocked Mozambique Tilapia fingerlings in the saline waters of a reservoir supplied by an artesian well at the Agricultural Training Centre, Okashana, on the southern edge of the Cuvelai system in (before then) Owamboland. They experienced high mortality, but those that survived grew to over 500 g, although they reportedly had a sulphur taste. The Foundation considered the experiment a success, but the results were never documented nor applied elsewhere. Additionally, the Rossing Foundation experimented on an integrated aquaculture-vegetable production at the Horticulture Training Centre, Okahandja, based on a design by an outside consultant with a view that farmers could adopt the system. This experiment did not yield good results and the Foundation realised that the technology was inappropriate for rural farmers in Namibia.[7]

- A private businessman used floating cages in the late 1980s to grow Mozambique Tilapia in Hardap Dam.[8] The cage culture operation purchased about 10,000 fry from the Hardap Fisheries Institute annually for N$ 500 (US$ 1 = N$ 2.30 at that time), which grew from 2 to 150 g each, yielding about 1.5 tonnes/year. Three (3) tonnes of food were required based on a feed conversion ratio of 2:1 (total cost N$ 2,400). Other costs included overhead (N$ 150), repairs (N$ 300) and miscellaneous items (N$ 600), for a total production cost of N$ 3,950 per 1.5 tonnes, which equals N$ 2.63/kg, virtually the same price as frozen marine fish (N$ 3/kg). Processing costs for fillets increased the price to N$ 6.57/kg, and if frozen and packaged, production cost further increased an additional N$ 1.25. These fillets sold for N$ 10/kg; thus, the market was very selective, e.g., processed beef prices were about N$ 10/kg.

- In 1986, a private individual commissioned a feasibility study on the public health aspects of growing fish in the Windhoek municipal sewage works. Although the study's outlook was positive, the investor did not proceed.

- Oxfam Canada-Namibia Cooperation Programme (CANAMCO), Okavango initiated a community-based development programme in 1992 planned for five years. Activities were started during the first year in seven villages, including vegetable gardening in five riverside villages using water pumped from the Kavango River. There was an aquaculture component to this programme. Fish were stocked in 5-m diameter irrigation cisterns, but Oxfam and the local communities needed technical advice on how to manage their aquaculture system, including training for rural development promoters. The aquaculture aspect of this project was unsuccessful.

II. Institutional Development of Aquaculture

A. Legal Framework

Despite the early setbacks in the development of aquaculture, the Government of the Republic of Namibia (GRN) acknowledged the need to continue with aquaculture after Independence. During the 1990s and early 2000s, aquaculture was specifically addressed as a development priority from Namibia's Second National Development Plan (NDP-2) through to the current NDP-5 and in the Government's VISION 2030 document, wherein it is envisaged that by the year 2030 aquaculture will have grown into a thriving industry. The Namibian Government actively sought to create an "enabling environment" for aquaculture development which required specific policy, legislative and institutional interventions. With the help of aquaculture experts from around the world, Namibia was highly motivated to develop its aquaculture policy and legislation and laid a sound institutional foundation for sector development. This commanded the development of the Aquaculture Policy of 2000, Aquaculture Act of 2002 and Aquaculture Master Plan (NAMP) which can be viewed as a synthesis of international best practises. The mandate for aquaculture was deposited with the Ministry of Fisheries and Marine Resources (MFMR) who created the Directorate of Aquaculture in 2003.

B. Inland Aquaculture Centres for Intensive Aquaculture

There are currently two main types of freshwater aquaculture in Namibia: (1) 'Intensive' fish culture where fingerlings are introduced into ponds or cages and intensively fed with artificial food until harvest. This type of fish culture practice in Namibia has been incorporated into commercial inland aquaculture. Commercial farms are market driven, hence they aim to grow high-value fish species which could potentially be exported as well as sold locally. Intensive to semi-intensive systems use commercial feeds and high-tech systems which require higher levels of production management. Currently all intensive aquaculture farms in Namibia are government owned; and (2) 'Extensive' culture where fingerlings are introduced in natural or artificial ponds to grow on natural or limited supplemented foods for harvest. This is mainly community-based aquaculture using home-based systems or cooperative-run farms in rural areas, the aim being food security and to a lesser extent income generation. There is less reliance on commercial feeds as the fish are fed naturally or on household wastes. Low-cost small-scale systems and practices are applied.

The GRN anticipated that freshwater aquaculture could enhance food security, generate income and improve rural livelihoods with production primarily targeted to ensure food security in local communities as well as for local and international markets. This led to the establishment of six freshwater inland aquaculture centres (IACs) and four fish farms (FF) that now fall under MFMR's Directorate of Aquaculture (Table 1, Fig. 5).

Table 1. Inland aquaculture centres (IACs) and fish farms (FF) in Namibia.

1. Kamutjonga Inland Fisheries Institute, KIFI (Kavango Region)
2. Katima Inland Aquaculture Centre, KIAC (Zambezi Region)
3. Rundu Inland Aquaculture Centre, RIAC (under construction, Kavango Region)
4. Hardap Fisheries Institute changed to Hardap Inland Aquaculture Centre, HIAC (Hardap Region)
5. Onavivi Inland Aquaculture Centre, OIAC (Omusati Region)
6. Ongwediva Inland Aquaculture Centre, OIAC (Oshana Region)
7. Mpungu, MFF (Kavango Region)
8. Leonardville Fish Farm, LFF (Omaheke Region)
9. Fonteintjie Fish Farm, FFF (Karas Region)
10. Epalela Fish Farm, EFF (Omusati Region)

1. Kamutjonga Inland Fisheries Institute (KIFI): KIFI is presently the main centre for inland fisheries and aquaculture research in Namibia (Fig. 5). This institute acts as a research, training and capacity building facility for scholars, students and prospective freshwater aquaculture farmers (Fig. 6). KIFI is also responsible for the routine monitoring of fish stocks in the northern perennial rivers. KIFI currently produces Threespot Tilapia, Redbreast Tilapia and Sharptooth Catfish fingerlings for research purposes, but surplus fingerlings are sold to small-scale farmers. The possibility of this facility acting as a hatchery for the production of fish fry or fingerlings to be introduced to natural waters for enhancement of dwindling natural fish stocks is being explored.

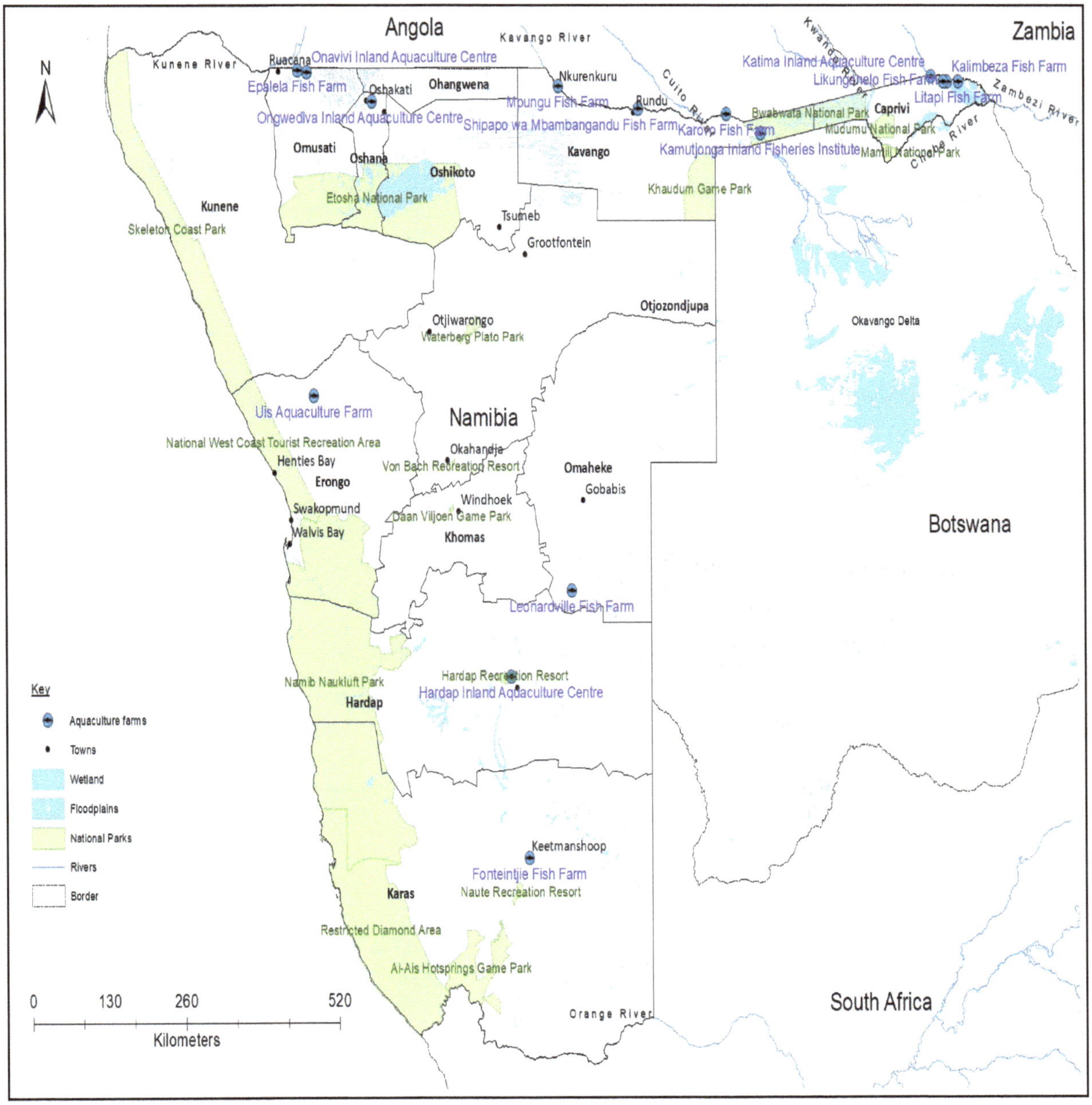

Figure 5. Map depicting locations of Government, cooperative and private aquaculture centres in Namibia (prepared by N. Jacobs).

2. Katima IAC (KIAC): KIAC is responsible for aquaculture development in the Zambezi Region and current farming includes Threespot Tilapia and Redbreast Tilapia. KIAC is also responsible for the routine monitoring of fish stocks in the Zambezi, Kwando and Chobe Rivers.

3. Rundu IAC (RIAC): RIAC is responsible for the routine monitoring of fish stocks in the Kavango River and aquaculture development in the Kavango East and Kavango West Regions, including Mpungu Fish Farm. RIAC receives fingerlings from KIFI for grow out, including Threespot Tilapia, Redbreast Tilapia and to a lesser extent Sharptooth Catfish.

4. Hardap IAC (HIAC): HIAC is responsible for aquaculture development in the Omaheke, Hardap and Karas Regions and partly in the Erongo and Khomas Regions (Fig. 5). The Leonardville and Fonteintjie Fish Farms are provided with fingerlings from HIAC; current species include Mozambique Tilapia, hybrid Nile Tilapia, Sharptooth Catfish and Common Carp (Fig. 6). HIAC is also responsible for the routine monitoring of Namibia's impounded water bodies and the Orange and Fish Rivers.

5. Other IACs and fish farms: Onavivi is currently the main aquaculture centre responsible for fish spawning, fingerling distribution, fish feed production (research) and extension services in northwestern Namibia (Fig. 5). Ongwediva IAC is mainly for extension services and fish spawning. Epalela Fish Farm is the only GRN fish farm in the country responsible for growing fish under semi-intensive open-pond conditions to market size and sale to the community (Fig. 6). Onavivi IAC, Ongwediva IAC and Epalela Fish Farm produce Threespot Tilapia, Redbreast Tilapia and Sharptooth Catfish fingerlings.

C. Cooperative Fish Farms for Extensive Aquaculture

In addition to MFMR-owned fish farms, a decision was made by the Ministry of Trade and Industry in 2001 to proceed with establishing six fish farms in the Zambezi and Kavango Regions forming the Freshwater Fish Farming Project (FWFFP) (Turnaround Strategy 2006). This resulted in three cooperative fish farms in the Kavango Region which included Karovo Fish Farm at Kangongo, Shipapo wa Mbambangandu Fish Farm at Kaisosi and Mpungu Fish Farm at Nkurenkuru (Fig. 5), the latter being the only farm currently operational and now under control of the MFMR with an annual production of around 5 tonnes of tilapia in open plastic-lined ponds. Kalimbeza Fish Farm at Isize and Litapi Fish Farm at Musanga in the Zambezi Region were flooded several times and had limited operation. The Likunganelo Fish Farm at Lisikili was renovated by the MFMR, but was also hampered by flooding and poor drainage with no viable fish production.

D. Private Dams

There are an unknown number of small dams, perhaps totaling over 10,000 of variable size on over 6,337 private commercial farms in the freehold title land area of Namibia, used mainly for cattle and wildlife watering. Many of these dams dry up after the rainy season. MFMR has historically stocked dams with Common Carp, Redbreast Tilapia, Nile Tilapia, Mozambique Tilapia, Sharptooth Catfish and even Largemouth Bass at subsidised prices. Although a stocking database is maintained, no other management or harvesting information is available. There is a modest record of commercialising catch, but the vast majority of stockings is for recreation.

Figure 6. KIFI is the main centre for inland fisheries and aquaculture research (top left) and has state of the art research facilities (top right). HIAC is responsible for aquaculture development in the Omaheke, Hardap and Karas Regions and partly in the Erongo and Khomas Regions (middle left). Leonardville (middle right) and Fonteintjie Fish Farms are provided with fingerlings from HIAC. Epalela Fish Farm is responsible for growing fish under semi-intensive open-pond conditions (bottom left) to market size and sale to the public (bottom right) (Photos: F. Jacobs).

III. Aquaculture Species

A. Candidate Species

The need to use only indigenous fishes in aquaculture operations in Namibia cannot be over emphasised. The physical, chemical and biological effects of aquaculture include a plethora of potential problems, including the introduction of nonindigenous species into the wild; hybridisation and the breakdown of isolating mechanisms of closely-related species; competition for preferred food resources and spawning sites; alteration of the structural and functional integrity of the ecosystem; and the introduction of new disease and parasite vectors.[8]

Currently, five fish species are commonly farmed within the borders of Namibia. In the northern regions close to the perennial rivers, indigenous Redbreast Tilapia, Threespot Tilapia and Sharptooth Catfish are allowed in aquaculture practices. In the central and southern regions, however, non-native Mozambique Tilapia and Common Carp are produced and distributed as fingerlings. Nile Tilapia was previously introduced as a culture species at HIAC, but the use of this species is not encouraged due to the threat of hybridisation.

1. Tilapia. The Threespot Tilapia is currently the most abundant aquaculture species in Namibia. This herbivore-omnivore species utilises insects and smaller crustacea as an adult, thus accepting a wide variety of (inexpensive) natural foods. It has been extensively used in aquaculture in southern Africa. Its spawning and management are well documented in the aquaculture industry and it is readily accepted as a source of protein by rural communities.

The Redbreast Tilapia and Greenhead Tilapia are also prime candidates for aquaculture in Namibia according to research conducted at the HIAC in 1988. These species require little or no supplemental feeding. Each is a herbivore utilising algae as a food source with the Redbreast Tilapia graduating to a macrophyte diet as it matures. Both species are indigenous to northern Namibia, reproduce readily with minimal care and can be harvested at juvenile stages (70-100 mm TL) within weeks. Each species is also an accepted food fish within the traditional fishery. Redbreast Tilapia were reported to spawn up to eight times a year under controlled conditions in Zimbabwe.[9] Research conducted by KIFI showed that each spawning event contains on average 3,000 individual eggs. Therefore, a single female when managed correctly could produce at least 16,000 juveniles per year. If fingerlings are allowed to grow to at least 100 g, the combined offspring from each female could be as high as 1.6 tonnes/annum.

Mozambique and Nile Tilapia are species often associated with aquaculture in Africa and elsewhere. However, these species are not native to the northern rivers of Namibia and there is concern that escapees from a culture facility might either hybridise with the Threespot Tilapia, a phylogenetically closely-related species, or otherwise alter the natural faunal composition.[5] The current use of Mozambique Tilapia is restricted to the central and southern areas of Namibia and no permits have been issued for movement to northern Namibia.

2. Sharptooth Catfish. Much attention has been focused in southern Africa on the use of various catfish species of the genus *Clarias* in aquaculture programmes.[10] The Sharptooth Catfish has great potential in Namibian aquaculture by virtue of its suitable reproductive strategy, its nutritional efficiency, ability to tolerate unfavourable water conditions and breathe atmospheric oxygen.[9] In contrast with South Africa where there is no demand for this unattractive looking fish, they are a preferred species in the northern regions and a ready market already exist.

3. Common Carp. The production of Common Carp in Namibia has seen a steady decline since 1983; however, private farms in the south and central Namibia still prefer this species as they can tolerate a wide range of water temperatures. Their growth, however, is slowed by the cold winters.

B. Production Trials

All of the MFMR centres succeeded in establishing climate-adapted aquaculture husbandry practices for the farmed freshwater species (tilapia and catfish) in Namibia. Climate-adapted fish farm management plans are in place that cover spawning (brood-stock management, hatchery management and nursing of fish fry), grow-out of fish in ponds, fish farm biosecurity and health management. Research will aim to improve production efficiency of freshwater aquaculture. The biggest hindrance to increased income/m^3 for culture of tilapia under semi-intensive pond conditions is the limited stocking potential of approximately 5 fish/m^3. The current research production trials on mixed-sex Threespot Tilapia at farms like Epalela, resulted in growth of around 250 g/fish and 1.125 kg/m^3 (10% mortality). Tilapia in 2020 sold at around N$ 60/kg (US$ 1 = N$ 16) on the Katima Mulilo Fish Market with a potential income for the fish farmer of around N$ 20–25/kg of tilapia with feed conversion ratios (FCR) of 2–2.5:1. Fish feed costs make up 70% of the production cost. Namibia currently imports fish feed from Zambia with prices varying between N$ 8–12/kg (32% protein pellets) over the last 5 years. Fish feed production in Namibia is expensive at about N$ 13/kg in 2019 due to higher cost of raw materials compared to Zambia.

Air-breathing Sharptooth Catfish were identified for increased stocking/m^3 and therefore increased income. The MFMR engaged in a stocking trial of 25 catfish/m^3 at KIFI during 2019 (Fig. 7). Catfish were stocked at an average weight of 45 g/fish and grown over a 353-day period to an average weight of 307 g/fish. A production cost of N$ 29/kg was achieved with a FCR of 1.81:1 and a feed price of N$ 8/kg. Catfish sold for around N$ 50–80/kg in central and north Namibia, thus resulting in a possible net profit of N$ 28/kg catfish produced. There is a six-fold increase in income of N$ 219/m^3 using catfish compared to the N$ 15–35/m^3 for tilapia, due to increased stocking densities of 25 fish/m^3 vs. 5 fish/m^3, respectively. Catfish farming can therefore greatly contribute to food security and household income if managed properly. There is a great consumer demand for fish in Namibia. In the Zambezi region alone, more than 2,300 tonnes of freshwater fish were traded during the 2016/17 season. More than 238,000 fingerlings were supplied annually 2015–2017 with small scale farmers increasing from 55 to 88 during that period.

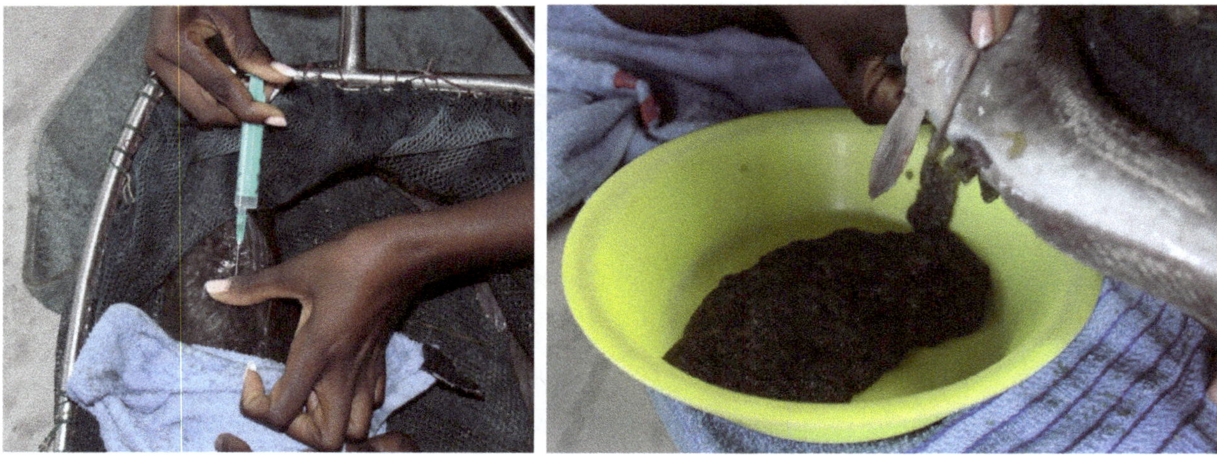

Figure 7. A highly valued aquaculture species is Sharptooth Catfish which are typically injected (left) with Ovaprim, an analog of salmon gonadotropin releasing hormone (sGnRHa), and a brain neurotransmitter (dopamine) inhibitor which aid artificial spawning by stimulating ovulation and spermiation in sexually mature fish (right) (Photos: F. Jacobs).

IV. Challenges of Aquaculture Development in Namibia

A. Limitations

ALCOM, a regional aquaculture and fisheries programme of the FAO based in Zimbabwe, was commissioned to carry out a comprehensive feasibility study on inland aquaculture in the Republic of Namibia.[6] This study identified the limits that the resource base and the socio-economic setting have on the development of the aquaculture sector; these remain unchanged today and include, but are not limited to:

- Low and erratic rainfall throughout all Namibia's regions;
- Widespread sandy and saline soils;
- Short growing season for the species present;
- Abundant and cheaper coastal marine fish resources and northern inland fish resources;
- Low intensity agriculture and livestock practices with very few by-products;
- Community ownership of land and water which results in uncontrolled activities;
- Specific freshwater fish preferences of people living in different regions; and
- Preference for beef and poultry throughout the country.

In addition, Namibia is situated in the sub-tropics with very cold winters from May to September with water temperature going down to below 10 °C in the south and around 14–16 °C in the north. The aquaculture species farmed within Namibia do not grow sufficiently during these cold months. Warm conditions suitable for fish grow out occur from October to April with water temperature above 25 °C. The climate of Namibia therefore only allows for one growing season of seven months per year. Fish farm management plans developed by the MFMR are climate adapted and make provision for spawning and grow out during suitable climatic conditions. The nature of the rivers and water bodies and in general the scarcity of water and the sandy nature of soils in Namibia limit large-scale commercial aquaculture development with high cost of pond construction and water extraction.

Within these limitations, however, opportunities using the annual inundated *iishana*, canals, small-scale rural irrigation schemes, impounded waters within state and private farm dams and maturation ponds of sewerage treatment systems in urban areas, were identified as possible areas that could be used for small-scale aquaculture projects.[7]

B. Food Security vs. Rural Development

Foreign aid generally targets the historically disadvantaged countries with the intention of enhancing livelihoods through several avenues including food security. In practice, foreign aid aquaculture projects in Africa have not lived up to expectations. There are several reasons for this, but a major factor has been the replacement of traditional farming techniques with new methodologies or concepts. Numerous aquaculture projects were launched in the early 2000s, but these projects had their objectives dressed in terms like "increased living standards", "FOREX development" and "rural development", rather than protein production for basic food security.[8] Several authors considered that a fundamental flaw in developmental aid for subsistence aquaculture was the tendency to measure success by western world standards, e.g., the desire to grow fish to marketable size.[8,11,12] They reasoned (1) there had been a move away from simply providing protein supplements to the daily diet of persons living at the subsistence level; (2) aquaculture should not be thought of as a formula for achieving instant affluence as such ventures were virtually doomed for failure unless there was a strong profit motivation; and (3) commercial aquaculture ventures were capital, technology and labour intensive. Building upon the above issues, MFMR research and management activities will focus on the cultivation of freshwater species to enhance local food security and market growth.

C. Fish Feed

As previously mentioned, fish feed represents 60%–70% of the production cost for intensive and semi-intensive aquaculture and is why MFMR invested in a fish feed pellet plant situated in the Omusati Region (Chapter 1, Fig. 1, Region 2; Fig. 8). The plant, however, lacks extrusion capabilities (heating/cooking of ingredients for improved digestion/utilization of the feed). The MFMR opted to use locally sourced ingredients to economically empower rural communities, a commendable objective. However, quality aquafeeds can only be made from quality feed ingredients and not from inferior, spoiled or otherwise damaged or contaminated ingredients. Protecting both human and animal health are also prime considerations in the production of quality cost-effective aquafeeds. It follows, therefore, that the production of feeds for aquatic species requires extrusion, quality ingredients and a high degree of precision to measure and mix as many as four dozen constituents in a feed formula.

D. Predators and Wildlife

Several species of birds, otters, wild cats, snakes, frogs and even crocodiles can become problematic for aquaculture farms in Namibia and cause severe economic losses when left unchecked. Fish-eating birds are probably the most common predator at Namibian aquaculture facilities and hatcheries. The open water found at aquaculture farms are attractive to migratory and wintering birds and are also considered important to many species whose habitat has been lost to development or other land uses. Unfortunately, fish-eating birds can cause problems for producers, although the economic impact these birds have can vary widely among species and from one region to another.

KIFI assessed fish losses to bird predation in 2018–2020 in an effort to develop, implement and recommend control measures for aquaculture facilities. Species-specific information was obtained on feeding rates, sizes of fish consumed and the value of these fish. The most common predatory bird species recorded at KIFI included the Great White Egret (*Ardea alba*), Hamerkop (*Scopus umbretta*), Striated Heron (*Butorides striata*), Black-headed Heron (*Ardea melanocephala*), Reed Cormorant (*Microcarbo africanus*) and Pied Kingfisher (*Ceryle rudis*) (Fig. 9). An estimated 184,000 fish fingerlings (<100 g) were preyed upon annually by these species combined, resulting in an estimated economic loss of at least N$ 1.5 million annually. The results revealed that predation occurred both day and night. The most preferred prey size were fish fingerlings (<100 g) which were consumed whole. Other wildlife predator species included otters (*Aonyx capensis, Lutra maculicollis*), frogs (*Pyxicephalus adspersus*), African Wildcat (*Felis lybica*), snakes (*Python natalensis*), Monitor Lizard (*Varanus niloticus*) and Nile Crocodile (*Crocodylus niloticus*), but estimates on their predation were not evaluated (Fig. 10).

Aquaculture farmers must constantly adapt their predator management programmes to effectively prevent and reduce damage. All fish-eating wildlife are protected under the Nature Conservation Ordinance administered through the Ministry of the Environment, Forestry & Tourism (MEFT). Farmers can protect their fish from predation using both nonlethal and lethal methods with the appropriate permits and depredation orders issued by MEFT. These permits are species specific.

Nonlethal management options include harassment with pyrotechnics and other scaring devices. Because some birds are nocturnal, bird harassment patrols should continue on a 24-h basis. Some birds such as herons are hazed when fish are fed floating feeds or when the fish are concentrated. If ponds are drawn down for seining to depths of 0.6 m or less, all wading birds should be hazed. Netting, fencing and gridwire are effective for small-scale farmers using ponds and raceways.

Figure 8. The MFMR invested in a fish feed pellet plant at Onavivi IAC situated in the Omusati Region of Namibia. The MFMR opted to use locally sourced ingredients to economically empower rural communities (Photos: E. Ndivayele).

Figure 9. Species of birds such as Reed Cormorants (top left), Striated Heron (top right), Great White Egret (bottom left) and Pied Kingfisher (bottom right) are some of the most common fish predatory bird species at KIFI (Photos: P. Ackerman).

Figure 10. Wildlife species identified as predators of fish at KIFI include otters (top left), crocodiles (top right); snakes (bottom left), and monitor lizards (bottom right). All captured predators were relocated back to the Kavango River unharmed (Photos: F. Jacobs).

E. Uis Aquaculture Farm: A Case Study

The Uis Aquaculture Farm, a family-run business and the only privately-owned commercial fish farm in recent years, was situated close to the town of Uis in the Erongo Region, Namibia (Chapter 1, Fig. 1, Region 9; Fig. 5). Similar to previous aquaculture projects on Hardap Dam, the Uis Aquaculture Farm used cage culture whereby fish were kept in a series of 50 cages or fish pens in an abandoned mine quarry approximately 11 Ha in size (Fig. 11). Construction of the Uis facility (i.e., cages and a fish hatchery) started in 2010 and the spawning of Mozambique Tilapia by 2011. In the beginning of 2012, the farm produced between 1–2 tonnes/month after which their production increased dramatically to an average of 4–6 tonnes/month at an average price of N$ 25/kg (US$ 8.50 to N$ 1.00 in 2012).

Figure 11. The Uis Aquaculture Farm used cage culture whereby fish were kept in a series of 50 cages or fish pens in an abandoned tin mine quarry (Photo: L.P. van der Westhuizen).

In February of 2014, a freak flash flood in the quarry caused significant structural damage to the aquaculture operation and the loss of its entire capacity of approximately 400,000 fish (Fig. 12). The water in the quarry rose at an incredible rate of 5 m/h during the flood destroying many cages, walkways and equipment. During the subsequent months, approximately 2 tonnes of escaped fish were captured using gillnets in the quarry. Aquaculture operations resumed by April 2015 with production averaging 4–6 tonnes of fish per month at prices between N$ 25–32/kg.

The Uis farm experienced another major natural catastrophe in 2016 losing their entire production of 400,000 fish. Fish were observed gulping air at the surface and exhibiting lethargic behavior, signs of low dissolved oxygen levels. This had been preceded the previous week by extreme air temperatures (>50 °C) with very little wind to facilitate the movement and mixing of water in the quarry. It was concluded that the lack of wind combined with extreme air temperatures caused a phenomenon of *thermal stratification* of quarry waters followed by *turnover*. The lack of wind responsible for the mixing of the water in the quarry and providing oxygen at all depths initially resulted in extremely high surface temperatures and low oxygen levels in the bottom water where fish, no doubt, were barely surviving. Once winds resumed and turnover occurred, fish were exposed to both hypoxic conditions and high ambient temperatures, either of which would result in mortality. Large algae/phytoplankton blooms followed during the subsequent days and further deteriorated the water quality.

The Uis farm made significant financial investment into the operation in 2018 by constructing a new and improved hatchery together with a fish feed plant (Fig. 13). After only a few months of operation, a disease outbreak in June of 2018 caused the loss of approximately 200,000 fish. Despite numerous management efforts to identify and control the disease outbreak, the loss was approximately 400,000 fish by August 2018. In addition, an unknown number of Mozambique Tilapia and Sharptooth Catfish living in the quarry succumbed to the disease. The disease could not be identified by various aquatic veterinarians and fish health specialists and remains prevalent at the Uis facility. As a result of all these challenges, the Uis Aquaculture Farm stopped all operations in 2019.

Figure 12. Uis Aquaculture Farm before (left) and during (right) a freak flash flood in February 2014 causing significant structural damage and the loss of approximately 400,000 fish (Photos: L.P. van der Westhuizen).

Figure 13. The Uis Aquaculture Farm erected its own fish feed plant (left) and added structures such as a new hatchery and a climate-adapted broodstock building to their facilities in 2018 (right) before ceasing all operations in 2019 (Photos: L.P. van der Westhuizen).

V. Short- to Mid-term Objectives of KIFI

The focus at KIFI has moved from merely producing fish to becoming a research demonstration facility and conducting trials to improve the economic viability of aquaculture. A primary objective is to achieve the lowest possible production cost with highest possible yield as achieved through stocking Sharptooth Catfish at higher densities, this being possible due to their air-breathing capability and tolerance to poor water quality.

A second complementary objective is to demonstrate the successful rearing of certain cichlid species to a consumable size with an emphasis on food security rather than commercial production. This means harvesting juveniles at approximately 70–100 mm TL, which is a size acceptable to the populace and commonly caught in traditional gear. Under maintained conditions, tilapia can yield 4 tonnes/Ha of fingerlings. Additionally, the practice of juvenile harvesting minimizes the impact of natural mortality, reduces the requirements for advanced feeding (diet) technology and significantly reduces the space requirements for growing fish to larger sizes. Since tilapia can spawn several times a year, a rather continuous supply of cultured fish should be available for local consumption.

VI. Ornamental Fish Trade

Ornamental fish trade is a multibillion-dollar industry globally and contributes positively to many developing countries, but the ornamental fish trade sector has received little attention in the country. Namibia however, makes provision for the sector in its legislation [Regulation 21, Section 43, Aquaculture Act, 2002]. Any introduction, export or transfer of organisms for the purpose of aquaculture must be approved by the MFMR or any other person designated by the Minister. Briefly, the Act states that companies/entities must register with the MFMR as an Importer and/or as an Exporter if live aquatic organisms or aquaculture products are to be imported into or exported from Namibia. It further states that "an importer must comply with additional importing requirements that are required by law." This includes legislation such as Veterinary Import/Export permit and Health Certificates; relevant legislation under Ministry of Environment and Tourism regarding CITES and IUCN listed species; and legislation of the Ministry of Trade and Industry regarding Imports and Exports. KIFI is currently designated by the MFMR for issuing permits for the import/export and/or transfer of aquatic organisms (Fig. 14). In regards to permit applications received:

> (1) *A person may not, without written permission granted by the Minister, introduce or cause to be introduced into Namibia or any Namibian waters any species of aquatic organism or any genetically modified aquatic organism or transfer any species of aquatic organisms from one aquaculture facility in Namibia to another or from any location in Namibia to another; and*
>
> (2) *The Minister must not issue any approval under this section unless the impact of any introduction or transfer of any aquatic species or genetically modified aquatic organism has been assessed, if so required, in accordance with the legislation or policy dealing with environmental assessments.*

The MFMR supports and promotes sustainable utilisation of freshwater fishes and therefore protects the well-being of indigenous fish species. Popular ornamental fish species such as Arapaima,

Arowana, Snakehead Catfishes, Piranha, Pacu and Peacock Bass are among species not approved for importation into Namibia. These species are considered highly invasive and a threat to indigenous fish species. This means that some species have the potential to alter aquatic habitats, compete with and/or contaminate the genetic integrity of indigenous stocks. In addition, the importation of new diseases and parasites associated with these species, which are often understudied, may pose a risk to indigenous fauna. Chapter 5 documents the presence of 12 fish species introduced into Namibian waters from ornamental or aquaculture sources and identifies the threat of other aquatic invasive species.

The estimated value of the imported ornamental fish trade in Namibia was approximately N$ 5 million annually from 2016 to 2018. Namibia, however, experienced a sharp decline in the ornamental fish trade sector. A fact-finding mission in 2019 revealed that only 7 of 18 registered ornamental fish traders were still operational. Of these registered ornamental fish traders, only one breeds Malawi cichlids for the local ornamental market. It was concluded that the ornamental fish traders were heavily impacted by the poor economic performance of the Republic and thus many were forced to close down. This situation was exacerbated by COVID-19 in 2020.

Figure 14. Ornamental fish traders being inspected for compliance with the importation regulations of Namibia (Photos: MFMR).

VII. REFERENCES

[1]Khaebeb, F.H, F.J. Jacobs and G. Hamutenya. 2018. Proceedings of the Kavango River fisheries stakeholder workshop. Report Number MFMR-KRFSW-01-27/28-03-2018. Sarasunga Lodge, Rundu, Kavango Region.

[2]Kalb, M. 2017. Naturalizing Trout? Fish farming in German Southwest Africa. Environ. Soc. Portal, Arcadia (Autumn 2017), no.33. Rachel Carson Center for Environment and Society. doi.org/10.5282/rcc/8158.

[3]Rode, K. 1904. Südwestafrika: Flora, Fauna, Ackerbau, Viehzucht, aber im praktischen standpunkt betrachtet. Zeitschrift für Kolonialpolitik, Kolonialrecht und Kolonialwirtschaft 11, no. 10: 749–759.

[4]Harrison, A.C. 1936. Black bass in the Cape Province: Second report on the progress of American largemouth black bass. Department of Commerce and Industries, Fisheries and Marine Biological Survey Division, Union of South Africa. Investigational Report No 7, 1–119.

[5]Schrader, H.J. 1985. Invasive alien fishes in South West Africa/Namibia, pp.35–40. In: C.J.Brown, I.A.W. MacDonald and S.E.Brown (Eds.). Invasive Alien Organisms in South West Africa/Namibia. S.Afr. Natl. Sci. Prog. Rept. No. 119: 35–40.

[6]Martin, H. 1957. The sheltering desert. William Kimber, London. 324 pp.

[7]Haight B. and E. Ersdal. 1994. Mission Report: Feasibility of inland aquaculture and small reservoir fisheries, Namibia. ALCOM Field Document No. 23. Harare, Zimbabwe.

[8]Hocutt, C.H. and P.N. Johnson. 1993. Fisheries resource assessment of the Kavango and Caprivi Provinces, Namibia. USAID Compl. Rept., Grant No. DAN-5053-G-00-1048-00: 271 pp.

[9]Bell-Cross, G. and J.L. Minshull. 1988. The fishes of Zimbabwe. National Museums and Monuments of Zimbabwe, Harare, Zimbabwe: 294 p.

[10]Haylor, G.S. 1989. The case for the African catfish, *Clarias gariepinus* Burchell, 1822, Clariidae: A comparison of the relative merits of Tilapiine fishes, especially *Oreochromis niloticus* (L.) and *C. gariepinus* for African aquaculture. Aquacul. Fish Manage. 20: 279–285.

[11]Little, D.C., et al. 2012. Alleviating poverty through aquaculture: Progress, opportunities and improvements, pp. 719–783. In: R.P. Subasinghe, J.R. Arthur, D.M. Bartley, S.S. De Silva, M. Halwart, N. Hishamunda, C.V. Mohan and P. Sorgeloos (Eds.). Farming the Waters for People and Food. Proc. Global Conf. Aquacult. 2010, Phuket, Thailand. FAO, Rome and NACA, Bangkok.

[12]Kaminski, A.M., D.C. Little, L. Middleton, M. Syapwaya, M. Lundeba, J. Johnson, C. Huchzermeyer and S.H. Thilsted. 2022. The role of aquaculture and capture fisheries in meeting food and nutrition security: Testing a nutrition sensitive pond polyculture intervention in rural Zambia. Foods 2022 (11). https://doi.org/10.3390/foods11091334.

CHAPTER 13
DISEASES AND PARASITES OF NAMIBIA'S INLAND FISHES

by

Francois J. Jacobs and Kyle J. McHugh

I. Introduction

Freshwater fish, like other animals, can suffer from a wide range of *diseases*. Unlike livestock where the clinical signs of disease are more evident, the symptoms of disease are harder to observe in fish living underwater. Fish disease diagnosis is therefore a complicated process and requires a high level of expertise. Fish typically occupy several trophic levels within the aquatic food chain; thus, a diverse community is generally regarded as representative of overall ecosystem *health*. Fish are often exposed to various natural environmental fluctuations such as in temperature, turbidity and dissolved oxygen concentrations. The ability to tolerate natural fluctuations and anthropogenic impacts on the aquatic environment will reflect the health of an individual fish, community or ecosystem. As with all living biota, increased stress can lead to an increase in diseases.

But what is a disease? A disease is an abnormal condition affecting the normal functioning of an organism. Broadly, there are two major groups of fish diseases:

1. Pathogenic fish diseases: These are caused by viruses, bacteria, algae, fungi and animal parasites. These diseases are usually contagious; and

2. Non-pathogenic fish diseases: These diseases may be caused by degrading conditions of the environment (e.g., low dissolved oxygen; high water temperatures); alimentary (e.g., lack of food; eating an indigestible or toxic item); heredity (genetics); or mechanical-traumatic in origin. Such diseases are not contagious although several individuals may exhibit the same symptoms, such as multiple fish gulping at the surface under conditions of low dissolved oxygen.

Historically, poor knowledge of pathogenic 'diversity' in aquatic environments has led to frequent emergence of previously unknown diseases, usually causing severe economic losses, negative impacts on food security (especially for small scale farmers) and disrupting the wider value chain. Some diseases are relatively restricted to regional level or can be controlled relatively quickly, while others have the potential for rapid spread to distant regions with trade and impose long-term economic and social effects. Prevalence of disease is considered one of the major reasons of poor fish production in aquaculture and threatens food security on a global scale. It is estimated that fish diseases cost the aquaculture sector US$ 6 billion annually in loss worldwide, thus disease is a major limiting factor for the expansion of the aquaculture industry (Chapter 12).

II. Fish Diseases and Parasites in Namibia

Fishes are subject to a wide range of diseases and parasites. Some of these are fairly well known in Namibia and have had a negative impact on the aquaculture industry or have caused losses to fish keepers and ornamental fish traders (Chapter 12). Some of the most common diseases and parasites found in Namibia are included in Figure 1:

A. Typical Water Mould Infection (Saprolegniosis)

Water moulds are one of the most common infections of freshwater fishes and have traditionally been referred to as aquatic fungi. Water moulds are classical opportunists that usually feed on dead organic matter. Skin wounds caused by mechanical trauma such as handling, crowding and heavy feeding rates provide a portal of entry for water mould. Mucus secreted by the skin of a fish is an important means of protecting against water moulds. Outbreaks often occur in the cooler months of the year and therefore it is often called 'winter kill'. In Hardap Dam, this disease is prevalent at the end of winter and the beginning of summer when the variation in minimum and maximum water temperatures is greatest. The most common lesions are caused by the oomycete *Saprolegnia parasitica,* which is why the disease is called Saprolegniosis. The typical water mould infection appears as a white, cottony, fur-like growth on the skin or the gills.

B. Columnaris Infection (Myxobacterial Disease, Fin Rot, Cotton Wool Disease)

Columnaris is a common bacterial disease that usually affects the skin or gills of freshwater fishes. *Flavobacterium columnare* is the most prevalent member of the group and can rapidly infect fish populations and cause large mortalities. Flavobacteria are naturally occurring on healthy fish in aquatic ecosystems. Risk factors for bacterial infections to fish include physical injury, low oxygen levels, organic pollutants and high nitrate levels. Surplus fish food in the water column also supports the growth of bacteria, representing a pathway for infection. Infections often present a whitish plaque on the skin and may also appear as lesions, haemorrhages and fin rot, especially the caudal fin.

C. Ich Infection (Freshwater White Spot Disease, Ichthyophthiriosis)

Protozoan parasites such as *Ichthyophthirius*, otherwise known as Ich or White Spot disease, are amongst the most common diseases in home aquaria and ponds. Virtually all freshwater fishes are susceptible to infection and scaleless fishes such as catfishes are especially vulnerable. These unicellular parasites feed in a small cavity made and occupied by the ich cell in the skin or gill epithelium of the fish. After it feeds, it breaks through the skin, falls off the host and divides by binary fission to quickly overwhelm a population. This disease is typically a relatively warm water disease (15-25 °C), but can occur at low temperatures and is especially prevalent in stressed fish after overwintering. Stress and high fish densities play major roles in the severity of outbreaks.

D. Nematode Infection (Roundworms or Threadworms)

Freshwater fishes are either intermediate or final hosts for nematodes. Nematode larval stages are often recorded in the body cavity and intestines of freshwater fishes, while the adults occur in the gut of piscivorous birds. Eggs are shed into the water via the faeces of a host, usually a bird, and develop into second-stage larvae, which subsequently infect *copepods* (zooplankton). When the copepods are eaten by fish, third-stage larvae develop. These larvae may penetrate the gut wall, encyst the musculature and develop into adults only in a definitive host when it eats an infected fish, thus completing the life cycle. Low infection rates are expected in natural environments since parasites are normally in equilibrium with their hosts, which ensures that the parasite does not kill the intermediate host and reaches its final host to complete the life cycle. Infections may debilitate fish and make them unsuitable for human consumption if not properly prepared. Most catfish in Namibian waters are heavily infected with roundworms especially between their intestines. Although unsightly and normal, the majority of the meat from infected catfishes is consumable if properly prepared.

E. Trematodes (Digenean Fluke Infection, Metacercarial Infection, Black Spot)

Digeneans or parasitic flatworms (known as flukes) occur either as adults within the internal organs or as larvae in the form of white- or black-spot cysts in the skin. Adult flatworms usually produce eggs that pass from the gut of the final host (fish, bird or mammal). The eggs hatch and infect a mollusc (usually a snail). In the snail, cercaria develop from the eggs and are released to infect fish, then to be known as metacercaria which create cysts in muscle tissue. Metacercaria become adults and the life cycle is completed when the fish is eaten by the final host.

Freshwater aquarium fish are commonly infected if collected from the wild. These infections usually do not progress further in the aquaria, but there is the potential for fish to become infected from parasites released by the snail intermediate host, especially in aquaria that use snails to control algae. Trematodes can infect humans if fish are not cooked well or heavily salted when dried.

F. Cestode Infection (Tapeworm Infection)

Tapeworms have a complex life cycle that requires 1 or 2 intermediate hosts. Larval cestodes can be some of the most damaging parasites to freshwater fishes and especially decrease carcass value if they are present as dense cysts in the muscle. The Asian *Bothriocephalus acheilognathi* was introduced with imported Grass Carp and has infected many fish species in southern Africa. The intermediate hosts of this parasite are copepods which are eaten by zooplankton-feeding fishes. Birds eating infected fish pose a serious risk at spreading these parasites and can be especially destructive to aquaculture facilities.

G. Leech Infestation

Leeches are free-ranging external parasites that attach onto fishes and suck blood from their host. They have a direct life cycle not requiring intermediate hosts and in general resemble large digenean trematodes. Leeches have a complete digestive tract with a mouth in the anterior sucker and an anus in the posterior sucker; they are segmented like other annelids. Leeches can cause large ulcers in the mouth or on the skin and heavily-infested fish often have chronic anaemia.

H. Copepod Infestation (Anchor Worm)

Anchor worm is a general term for highly modified copepods that possess anchor-like processes for securing themselves to their hosts as external parasites. Females penetrate under the scales of fish creating an open wound. This produces an acute haemorrhage and ulcers at the area where the parasite attaches with fatality sometimes occurring as a result of blood loss and secondary bacterial or water mould infections. The most important species recorded in Namibia are *Lernaea barbicola* and *L. cyprinacea;* the latter was introduced to Africa from Asia via the introduction of

Common Carp. Anchor worms are not generally host specific and are capable of infecting members of native Cichlidae, Cyprinidae, Clariidae, Characidae and silurid catfish as well as non-native Centrarchidae. Generally, they are uncommon in freshwater aquarium fish that undergo mandatory inspection prior to being shipped or translocated.

I. Branchiuran Infestation (*Argulus* Infestation, Fish Louse)

Fish lice are free-living crustacean ectoparasites that feed by inserting a pre-oral sting (stylet) into the host and sucking body fluids with the proboscis-like mouth. Fish usually show erratic swimming behaviour as a result of the irritation caused by the stylet piercing the skin. The stylet also injects enzymes to break down tissues, causing irritation. Hooks and spines on the appendages may also cause mechanical damage that can lead to haemorrhaging and anaemia. A factor in severe infestations is the concern for secondary infections occurring in wounds, allowing for the transmission of viral and bacterial infections. The most significantly important species is *Argulus japonicus*, an exotic invasive fish parasite originating from Asia with the introduction of Common Carp and Goldfish. It was previously thought that wild fish populations were not susceptible to mass mortalities caused by fish lice; however, elevated water temperatures can result in mass infestations and fish kills. Fish lice are uncommon in freshwater aquarium fish, but may occur if wild or pond-raised fish are introduced into aquaria.

III. Diseases that Threaten Namibian Fisheries

Aquatic animal health and aquatic biosecurity have received significant attention in Namibia which must meet the aquatic animal health standards of international markets (e.g., the European Union). Outbreaks of serious diseases in wild and farmed freshwater fish caused the Namibian government to recognise the vulnerability of the human population and aquatic environments to transboundary aquatic animal diseases; this has spurred a strong interest in aquatic animal health and improved aquatic biosecurity at international, national and local levels.

There are many native diseases of freshwater fishes in Namibia that are regarded as naturally occurring and originating from African aquatic systems. These are less relevant to transboundary control measures beyond a requirement that fish should be free from visible parasites and lesions. However, some of the most highly contagious pathogens have accompanied the introduction of non-native fishes into Namibia. These relatively "new" *exotic* diseases have the potential for causing irreversible damage to both wild and farmed fish populations.

Some of these are regarded as serious diseases that can have negative implications for the country as a whole, i.e., they are notifiable under the World Organisation for Animal Health (OIE). Namibia is a member country of the OIE and required to recognise disease and pest-free areas for trade purposes in the context of the OIE's Application of Sanitary and Phytosanitary Measures. By acquiring and maintaining its official status, Namibia demonstrates transparency and promotes animal health and public health worldwide, fundamental to gaining the trust of its trade partners, neighbouring countries and the international community. Notifiable diseases are considered in detail within the OIE Manual of Diagnostic Tests for Aquatic Animals that provides a uniform approach for health certification in connection with trade in aquatic animals and products.

Figure 1. (A) Ich, or White Spot disease, is one of the most common diseases in home aquaria and ponds and is caused by protozoan parasites such as ***Ichthyophthirius****;* (B) Skin wounds caused by mechanical trauma such as handling, crowding and heavy feeding rates provide a portal of entry for water mould which in severe cases can lead to infection with ulcer-forming bacteria which can rapidly infect fish populations and cause large mortalities; (C) Myxosporean parasites in gills; (D) Nematode larval stages are often recorded in the body cavity and intestines of Tigerfish; (E) Digeneans or parasitic flatworms (known as flukes) occur either as adults within the internal organs or as larvae in the form of black-spot cysts in the skin; (F) Anchor worm (*Lernaea cyprinacea*) under a microscope, showing it's four appendages (anchors) and slim body[5]; (G) Parasitic crustaceans are ectoparasites that feed by inserting a pre-oral sting (stylet) into the host (e.g., mouth or gills) and sucking body fluids with the proboscis-like mouth (Photos: (A) G. Merron; (B-G) F. Jacobs, K. McHugh and L.P. van der Westhuizen).

Several fish diseases in Namibia are declared to be serious under the National List of Diseases of Aquatic Organisms (Namibia Aquaculture Act 2002) and must be reported to the OIE in the event of an outbreak. Most of these declarations are aimed toward the regulation of trade in ornamental fish and the aquaculture industry. Although all the listed diseases have the potential to cause severe economic loss, only Epizootic Ulcerative Syndrome (EUS) is of known concern to date to wild and cultured freshwater fish populations in Namibia.

EUS is a fish disease caused by the fungus *Aphanomyces invadans* from the group commonly known as water moulds. EUS has a complex infectious aetiology, and is clinically characterised by the presence of invasive *Aphanomyces* infections that result in the death of soft tissue (necrotising ulcerative lesions), typically leading to severe inflammation (granulomatous response) and mortality. Tissue necrosis in early lesions causes an intense inflammatory reaction characterised by local reddening and swelling of the skin, hence it is also known as Red Spot Disease (RSD). EUS spreads via zoospores in aquatic environments and secondary zoospores enter the skin of fish following mechanical or environmental damage. Once the zoospores germinate, hyphae invade the skin and musculature, causing infection. The disease is most commonly seen when there is low temperature, heavy rainfall or a drastic change in environmental conditions. Notification to the OIE is mandatory once an outbreak occurs; eradication of diseased fish, quarantine and sterilisation of a facility is required. Once the disease is eradicated, good husbandry, surveillance and biosecurity measures are necessary to prevent recurrence. This is obviously impossible in the wild and therefore quarantine and health certificates are necessary for the movement of all live fish to prevent the spreading of the disease. Fish infected with EUS do not normally pose health hazards to humans, although the ulcerations and tissue decay could support secondary, more threatening pathogens. It is therefore recommended not to eat contaminated fish unless it is thoroughly cooked.

Globally, the first reported outbreaks of EUS were from southeast Asia and Australia in the early 1970s when it caused high mortalities of wild and cultured fish. The first report of EUS in sub-Saharan Africa was from the Upper Zambezi and Chobe Rivers in Namibia and Botswana in November 2006, followed by Zambia in 2007 (Fig. 2), but it may have been present earlier. Subsequently, 24 fish species from these regions were shown to be susceptible to EUS, including Sharptooth Catfish, Blunttooth Catfish, African Pike, Nembwe, Upper Zambezi Labeo and several of the smaller barbs (*Enteromius* spp., Fig. 3). It was, however, not until May 2007 that the disease could be positively isolated and identified. All fish farms in the Zambezi Region used water directly from the Zambezi River and farmed fish showed signs of EUS. The most severe outbreak was during 2007 in Lake Liambezi after it filled and water quality deteriorated. Another EUS outbreak was detected and confirmed by histopathology for the first time in 2008 at Mpungu, Karovo and Shipapo wa Mbambangandu Fish Farms situated next to the Kavango River (Fig. 2). No serious outbreak was recorded for the Kwando-Linyanti or Zambezi-Chobe systems during this period, but the Upper Zambezi-Chobe demonstrated a persistent, but low occurrence of infected fish.[1,2]

The presence of EUS was again confirmed during a fish biodiversity assessment of Lake Liambezi in August 2011.[3] Two Sickle-fin and three Longbeard Barbs with circular ulcerative skin lesions were collected and the presence of EUS was confirmed histologically. The first known record of EUS in Angola was recorded in 2012 in the upper catchment of the Kavango River at Cuchi, Angola.[4] Species infected were *Labeo cylindricus* and *Barbus* cf. *thamalakanensis*.

Suspected EUS lesions on Sharptooth Catfish at Onavivi Inland Aquaculture Centre and Epalela Fish Farm in northwest Namibia were observed in June 2016. Some 64 samples were collected by the EUS emergency response team and submitted to the Microbiology Laboratory at the University of Zambia in Lusaka through funding of the FAO EUS surveillance programme. More than 50% of the collected samples tested positive for EUS in both histopathology and polymerase chain reaction (PCR). Small scale fish farmers in the Ohangwena region reported a suspected EUS outbreak in 2017. Samples again tested positive for EUS (Department of Agriculture, Forestry and Fisheries, Cape Town). In 2020, Straightfin Barbs collected in *iishana* north of Oshakati showed suspected EUS lesions and subsistence fishers reported sporadic outbreaks of suspected EUS in the backwaters along the Kavango west of Rundu (Fig. 3).

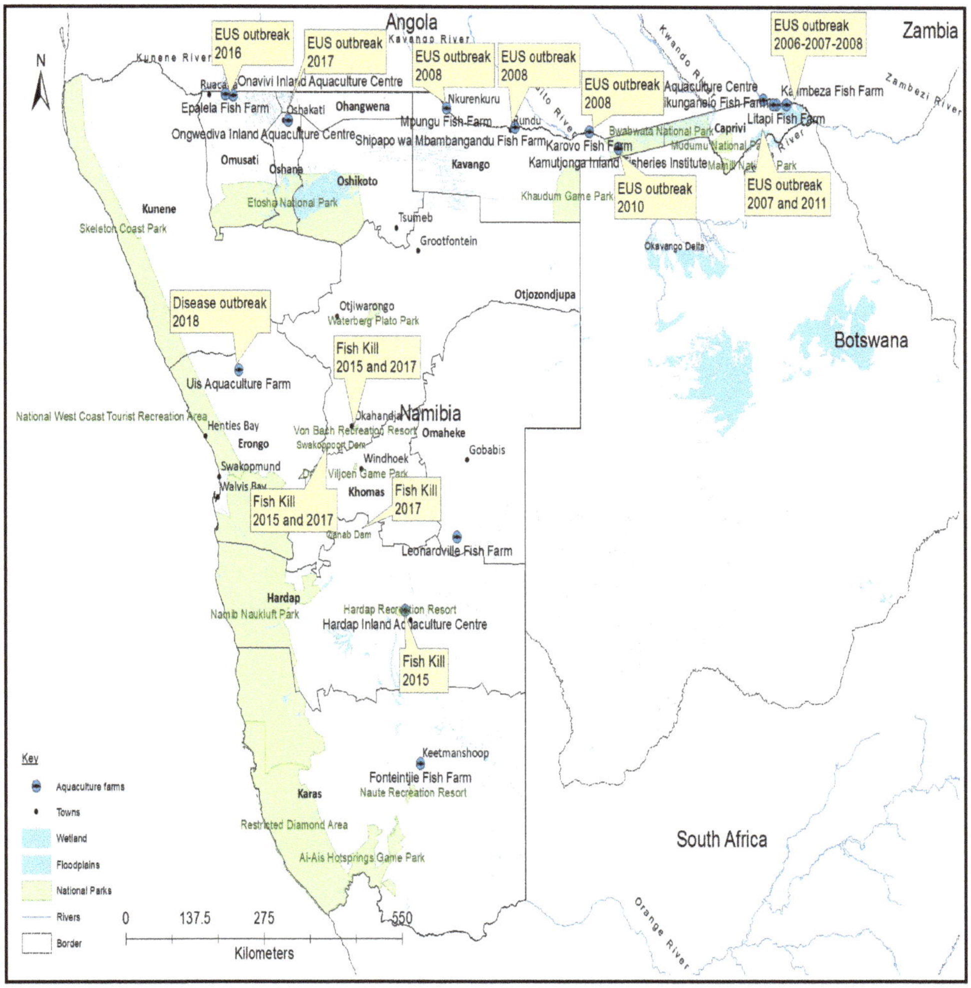

Figure 2. EUS was first reported in the Zambezi Region in November 2006. All fish farms in the Zambezi Region showed signs of EUS and presence was confirmed during the period 2006 - 2007- 2008. EUS was also detected in Lake Liambezi in 2007 and again in 2011. In the Kavango Region, EUS was detected and confirmed for the first time in 2008 at Mpungu, Karovo and Shipapo wa Mbambangandu Fish Farms. In 2016 the presence of EUS was confirmed at Onavivi Inland Aquaculture Centre and Epalela Fish Farm. In 2017, small scale fish farmers from Ohaingu reported a suspected EUS outbreak which was confirmed. An unidentified disease caused the death of approximately 400,000 fish in the Uis Aquaculture Farm in 2018 (Chapter 12). Major fish kills events were observed in Von Bach, Swakoppoort, Oanob and Hardap dam during 2015 and 2017 (map by F. Jacobs).

Figure 3. Straightfin Barb showing EUS infection that resulted in epidermal lesions and muscle exposure (Photo: B. van der Waal).

IV. MAJOR FISH KILLS IN NAMIBIA

Large-scale fish kills have become common internationally and especially in neighbouring South Africa. Although such large-scale mortalities are rare in Namibia, the frequency of occurrence is increasing. Large-scale mortalities can be as a result of natural phenomena and life cycle related activities (spawning events of salmon), human-related influences such as pollution, and terrestrial and aquatic modifications. Fish kill events in Namibia have led to the decline of recreational and commercial fish species, which may have a negative effect on the regulation of food-web dynamics and nutrient balances especially in state dams (Chapter 6).

In general, a minimum number of 25 dead fish over a 48-hour period is regarded as representative of a fish kill event as it is statistically unlikely that numbers higher than this will be the result of natural causes (e.g., old age). Fish kills are complex as fishes are wholly dependent on the characteristics of the water medium in which they occur. These characteristics may include physico-chemical parameters, biotic interactions, rainfall, discharge, habitat availability, temperature, etc., and are influenced by both natural and human-induced activities. Furthermore, potential environmental stressors on fish may act synergistically. Fish kills usually happen fast and are very visible when they occur, often causing considerable public outcry from anglers, landowners, government departments and the media. Kills evoke an emotional response in riparian communities that subsist on the stocks.

Namibia is fortunate in that most aquatic ecosystems are relatively far from major industrial activities or massive townships which can have detrimental effects on water quality and fisheries. However, prolonged droughts that are often experienced in Namibia can have devastating effects, especially in confined or closed systems like the state dams (Chapter 6). Namibia experienced major fish kill events in 2015 and 2017 in Hardap, Oanob, Von Bach and Swakoppoort Dams after heavy rains following prolonged drought periods (Figs. 2 & 4). Although some dams have experienced fish kills in the past, the stressors and their interactions are still poorly understood.

Establishing the primary cause of fish kill events in Namibia is a difficult and expensive process which most often produces vague explanations (Chapter 12). There is a common misconception in Namibia that fish kills result from poison or pollution (e.g., persistent organic pollutants and heavy metals) occurring in the catchment areas of state dams. Although 'pollution' (recognising pollution comes in many forms) can contribute to the severity of an event, it is only one of many stressors that ultimately result in a fish kill. Fish kills in Namibia usually follow periods of good rainfall, whereby reservoirs receive a strong inflow of freshwater after periods of severe drought.

Figure 4. Namibia experienced fish kill events in 2015 and 2017 in Hardap, Oanob, Von Bach and Swakoppoort Dams after heavy rains following prolonged drought periods (Photo: C. Hay).

The inflow of freshwater is usually accompanied by tonnes of dead organic matter that are deposited in the inlets of Namibia's reservoirs. There are also numerous semi- and intensive-livestock production facilities, such as feedlots, that contribute to the nutrient load in the catchment areas. High water temperatures and salinities, concentrated fish populations (i.e., high respiration levels), the aerobic decomposition of organic matter and re-suspension of anoxic sediments are the primary factors that lead to low oxygen concentrations and large mortalities from suffocation (asphyxiation). This is especially evident during the first few days after strong inflows. Surviving fish may often gulp for air at the water's surface during this time and some may recover during the day as oxygen levels rise due to wave action on the water surface and photosynthesis.

Sharptooth Catfish, Common Carp and Mozambique Tilapia dominate state reservoir fisheries; the latter two are the most effected species during fish kills in Namibian impoundments. These species usually die in the thousands and can be observed in *rigor mortis* (the stiffness that sets in shortly after death) with their mouths open, gills flared and spines arched.

In addition, the reservoirs often experience a second fish kill event following a few days after the first event. Fish in other parts of the reservoirs usually can be observed gulping air at the surface and exhibiting lethargic behaviour, signs of low dissolved oxygen levels. Some reservoirs experience a build-up of toxic hydrogen sulphide, metals and nitrites during drought years in bottom waters. This, together with extreme ambient air temperatures with very little wind to facilitate the movement and mixing of water, creates thermal *stratification* of the water column. The freshwater inflow from heavy rains forces the water column to mix, a phenomenon known as *turnover*. Fish are then exposed to both hypoxic conditions and high ambient temperatures, either of which would result in mortality. The turnover is usually associated with eutrophic conditions and large algae/phytoplankton blooms follow during subsequent days. In turn, the dramatic fluctuations of day- and night-time dissolved oxygen concentrations further deteriorate water quality to cause mass mortalities (Fig. 5).

Extremely cold temperature in southern Namibia, especially in the vicinity of Hardap Dam, has also been responsible for large die offs of Mozambique Tilapia that seems to be the most vulnerable species.

Figure 5. Large phytoplankton blooms during the day coupled with high temperatures and oxygen demand at night can result in mass mortalities. Common Carp (left) and Sharptooth Catfish (right) gulping air due to low oxygen concentrations in the water (Photos: E. Hattingh).

V. The Way Forward

The poor status of knowledge on the occurrence of freshwater fish diseases and potential risk factors within the various inland water bodies and aquaculture systems in Namibia is a serious inhibitory factor towards formulating appropriate disease control policies. However, Namibia developed a EUS Standard Operating Procedure (SOP) in 2015, to obtain information relative to (a) the health status of wild fish stocks and (b) the suitability of conditions under which fish are maintained at aquaculture or production facilities. In addition, Namibia is in the process of developing a strategic disease control programme by evaluating the occurrence of all freshwater fish diseases, mitigation and treatment methods, the risks of their occurrence and their potential impact on productivity of aquatic systems.

To date, the only dependable approach for diagnosis of fish diseases lies in the specific identification of the pathogens using laboratory methods. These methods, which are suitable for the diagnosis of isolated cases of disease, form part of Namibia's national aquatic animal health surveillance/control programmes. These programmes are incorporated in the national protocol for inland fisheries research as part of the bi-annual disease report to the OIE. Highly contagious pathogenic fish diseases are the highest priority in the food, aquarium, wild and aquaculture fisheries sectors. However, successful fish health management begins with prevention of disease rather than treatment. Prevention of fish disease is accomplished through good water quality management, nutrition and sanitation. Without this foundation, it is impossible to prevent outbreaks of opportunistic diseases.

Presently, sporadic outbreaks of EUS continue to plague Namibia's wild and farmed fish populations. Under ideal conditions, only native fishes should be used in the aquaculture fisheries sectors and importation of all potential carriers of new pathogens should be avoided. Further to this, the continuous development of the aquaculture sectors increases the risk of spread and introduction of new aquatic pathogens, pests and plants, posing a high threat to aquatic production, aquatic environmental health, food safety and food security, all within the naturally stressed environment that Namibia offers.

Fisheries support millions of people globally and are an important source of protein and income to many communities around the world. Diseases and parasites are threats to the sustainability of Namibian inland fisheries. Because of this, it is important to understand the complexity, morphology, systematics and life histories of both the indigenous forms of diseases as well as those that may be introduced, either by the introduction of exotic fishes or water transfer schemes. Terms like 'aquatic epidemiologist' and 'aquatic parasitologist' are gaining international acceptance and as a result may enable collaboration and training of African-continent aquatic scientists in novel technologies to map the biodiversity of these interesting and ecologically important *bioindicators*.

In the event of suspected disease outbreaks on freshwater fishes in Namibia please contact:

Dr. Francois J. Jacobs
Chief Fisheries Biologist
Kamutjonga Inland Fisheries Institute
Ministry of Fisheries and Marine Resources
Tel. +264 66259931; Fax. +264 66259922
Email. Francois.Jacobs@mfmr.gov.na

Or

Dr. Heidi Skrypzeck
Chief Fisheries Biologist
NatMIRC/Mariculture
Ministry of Fisheries and Marine Resources
Tel. +264 644101254; Fax. +264 64404385
Email. Heidi.Skrypzeck@mfmr.gov.na

VI. References

[1]Huchzermeyer, K,D. and B.C.W. van der Waal. 2012. Epizootic ulcerative syndrome: exotic fish disease threatens Africa's aquatic ecosystems. J. S. Afr.Vet. Assoc. 83(1): pp.39-46.

[2]Kibria, M.,G, E. Klingelhoeffer, A. Hitula, B. Currie, B.C.W. van der Waal and J. Endjala. 2009. DOA/MFMR, EUS: An on-site survey report: 43pp.

[3]McHugh, K.J., K.W. Christison, O.L.F. Weyl and N.J. Smit. 2014. Histological confirmation of epizootic ulcerative syndrome in two cyprinid species from Lake Liambezi, Zambezi Region, Namibia. Afr. Zool. 49: 311-316.

[4]Bills, R., P. Skelton and F. Almeida. 2012. A survey of the fishes of Upper Okavango River system in Angola. SAIAB Invest. Rept. No. 73: 62 pp.

[5]https://pondinformer.com/anchor-worm-treatment-guide/.

CHAPTER 14

CONSERVING NAMIBIA'S INLAND FISHES AND FISHERIES

by

Clinton J. Hay, Charles H. Hocutt, Francois J. Jacobs, Glenn S. Merron, Ben C.W. van der Waal and Ben J. van Zyl

I. Overview

INLAND FISHES AND FISHERIES OF NAMIBIA is the latest comprehensive review at the national level of this important renewable resource. Chapter 14 reiterates and expands upon some of the key points mentioned in the previous chapters.

Namibia's northern perennial rivers are some of the most important riverine wetlands in Africa, providing valuable habitat for a variety of fishes, mammals, reptiles, birds, amphibians and insects. The balance between the growing developmental requirements of Namibia's neighbouring countries (Angola, Botswana, Zambia and South Africa) and the desire to maintain water quality and quantity with self-sustaining fish populations contributing to human food security is a challenge. The ultimate aim is to improve the quality of life of all the people that utilise the aquatic resources by integrating conservation and development in a sustainable way. 'Conservation' of fishes does not mean preservation, but rather the wise utilisation of this resource for the benefit of all.

Namibia's inland fishes face many natural stresses as well as numerous man-related environmental perturbations that include water diversions for domestic supply, irrigation and hydropower; pollution from agricultural run-off and domestic sources, synergistically increasing toxicity levels; mining, especially groundwater abstraction; changes in river morphology; illegal and overfishing; and introduction of unwanted alien species and diseases. Potential developmental and environmental impacts to native fishes and their inland ecosystems include, but are not limited to the:

A. Eastern National Water Carrier (ENWC): Kavango River,
B. Agricultural Diversions: Orange, Kavango and Upper Zambezi Rivers,
C. Popa Falls Hydropower Project: Kavango River,
D. Oil and gas exploration and production: Kavango basin,
E. Baynes Hydropower Project: Kunene River,

F. Kunene Transboundary Water Supply Project (KTWSP),
G. Xangongo Water Supply Project, Angola: Kunene River,
H. Legal and illegal timbering: Kavango and Zambezi Regions, and
I. Road construction and maintenance, including bridges and culverts that impede fish passage nationwide, especially northern rivers and impermanent rivulets of the *iishana* region.

Any alterations to the natural environment caused by the above have the potential to be exacerbated by their combined effects. In turn, indigenous human communities' resident along these waterways and at least partly dependent upon fisheries for their wellbeing will also be adversely affected.

The increased commercialisation of fishery resources; illegal use of monofilament gill nets, especially fine-meshed nets and mosquito nets/seines (as opposed to traditional gear); and over-harvest, threaten Namibia's fish heritage. This issue, unfortunately, is not a simple 'cause and effect' relationship given the growing population in rural northern Namibia, the risk of 'hidden hunger'[1] and need for food security, nor is the solution bound to a single government ministry. This is a matrix of interconnected concerns requiring cross-ministerial attention.

At the national level, the mandate of fisheries management in Namibia resides within the Ministry of Fisheries & Marine Resources (MFMR), while the environmental assessment process lies within the Ministry of the Environment, Forestry and Tourism (MEFT). The Ministry of Agriculture, Water and Land Reform (MAWLR) has the principle responsibility for managing water quality and quantity, but also the potentially intra-agency conflicting role of promoting agricultural development, a major source of carbon emissions, land transformation and the introduction of organic and inorganic pollutants into the waterways. The actions of a number of other ministries also have the probability to impact inland waters, their environments and the general human populace surrounding them, e.g., Ministry of Mines and Energy, Ministry of Works and Transport and Ministry of Urban and Rural Development. As referenced above for MAWLR, the missions of these ministries may well be in direct conflict with one another and decisions may be made without adequate scientific input. It is important to note, however, that such interagency conflicts are not restricted to developing countries.[2] A cohesive plan of action is needed to ensure the sustainability of Namibia's aquatic resources, especially in the north where over 50% of the populace lives at an extreme poverty level and are highly dependent upon renewable resources.

To reach the goals of responsible management, utilisation and conservation of Namibia's inland fishery resources, it might be appropriate to consider a cooperative cross-ministerial approach with a common agenda. This will require the involvement and input from traditional, regional and central authorities. Potential benefits will be sustainable catches contributing to food and nutritional security for the riparian communities. Enhancing a sense of value of the resources will result in the protection of Namibia's biodiversity, cultural heritage and social capital. In this context, sport fishing should not be considered as a concession to privileged citizens as compared to subsistence fishers; measures for maintaining both subsistence and recreational fisheries are linked rather than being mutually exclusive, i.e., healthy ecosystems benefit all stakeholders.

All the perennial rivers of Namibia and the vastly important ephemeral Cuvelai system are shared resources with Angola, Zambia and South Africa, while the common border with Botswana harbours some of the most environmentally sensitive waterways on the continent. A number of transboundary water commissions are in place, but their overall impact on co-management of aquatic resources, including fisheries, has been limited although varying within each agreement. Inland fisheries management in many countries resides in the same ministry as the marine fisheries sector (e.g., Namibia) or in ministries of agriculture or water, thus leading to inland fisheries being an under-valued sector. This transfers to transboundary commissions where agriculture and water quality and quantity are the driving forces of these agreements.

Conflicts of interest can magnify between nations with different economic and social goals. Potential development of water resources in Angola or Zambia could significantly decrease the future water supply to northern Namibia and Botswana, the effects of which could be dramatic. By way of example, there is evidence from sediment and pollen samples that local shifts from wetter to drier conditions in the Okavango Panhandle ecosystem 7,000 to 4,000 years ago were related more to the hydrological cycle and flood pulse rather than to regional climate fluctuations.[3] This fact when interpreted in a modern context means that local management and conservation of inland aquatic resources are apt to be equally or greater in importance than the broader issue of 'global climate change'.

The practice of 'integrated water resources management' on a regional scale is an admirable goal, but is it realistic in a Namibian context? Basin-wide management is a reality throughout Europe and North America where mutual benefits are recognised and the countries or states involved have the resources to support such activities. Developing countries with a significant portion of their population partially or wholly dependent upon inland subsistence fisheries for food security have completely different priorities. What significance is an endangered species to a populace when all fishes and life history stages are considered consumable items to feed a hungry child or family? How do you manage an open-access fishery when you are restrained by manpower, training and financial resources along an international border with different management and enforcement strategies? Should fishers using less efficient traditional gears be rewarded vs. those employing modern gear...and if so, how are the gears distinguished? How do you educate rural communities that the very biosphere they have been historically dependent upon for food security is fragile and already past a threshold of sustainability, thus requiring their dedicated, devout care?

The 'Tragedy of the Commons' is an economic paradigm that occurs when individuals neglect the well-being of society in the pursuit of personal gain to exploit a common resource.[4] Classic examples are open-access fisheries and deforestation where a few benefit rather than the resource being distributed equally for the 'greater good'. Overharvest practices have the potential to drive a renewable resource toward exhaustion and eventually to extinction as a harvestable resource. Not all open-access property regimes cause resource degradation, but where it occurs it is basically the consequence of the state lacking the political will to monitor and enforce official policies. As observed in the open-access fishery of the Kafue Flats in Zambia, the problem appears to be the paradox of a state that is simultaneously *present* and *absent*: present in actions that dismantle local fishery traditions, but absent when it comes to the ability to enforce its own formal rules and laws that might protect the resources.[5] The role of Traditional Authorities has eroded throughout Africa; the top-down approach by governments in fisheries management (*present*) has either failed or been compromised by lack of resources or willingness (*absent*). Riverine and lacustrine communities have largely lost ownership of the natural resources on which they depend and with it the incentive to protect them.

While all of Namibia's inland waters and fisheries are at risk, the Orange/Fish River is already extremely compromised by water extraction, diversion and loading of organic and inorganic pollutants. This is in spite of each individual country in the Orange-Senqu River basin (Botswana, Lesotho, Namibia, South Africa) having legislation aimed toward water quality and fisheries management in addition to there being a transboundary cooperative agreement for the catchment. As a measure of the uniqueness of fish biodiversity in the lower Orange-Fish River basin, six of 12 indigenous fishes (50%) are endemic. However, nine non-native species (43%) and two hybrid cyprinids contribute to the total known number of fishes, indicating a significant level of alteration to the natural biodiversity of the sub-continent's oldest extant basin.

II. Mitigation Steps for Namibia's Sustainable Fisheries

An objective of ***INLAND FISHES AND FISHERIES OF NAMIBIA*** is to provide a basis for informative decisions on the long-term management of the Republic's inland fish stocks. It is considered that fisheries management in the 21st century will increasingly focus on conserving fish biodiversity and the organisms and environments upon which they depend. Having said this, it is important to remember that fisheries management is equally about managing people as it is about managing fish.[6]

Identified below are actions deemed important to the successful long-term management of Namibia's inland water resources and fisheries. They are not prioritised since all are considered as components to a holistic approach. It is important to note that most of these are ongoing activities that can be maintained or expanded.

A. Leadership

The management of inland waters and their resources is especially fragmented. It may be prudent, therefore, for MFMR to initiate and actively lead (the operative phrase) a cross-ministerial national agenda to protect and manage Namibia's inland waters and wetlands for their sustainable utilisation.

Similar leadership is required at the transboundary level. While numerous cross-border agreements currently exist, successful outcomes to serve Namibian interests will be premised upon government representatives acting in leadership roles rather than merely participating.

B. Legislation

The Namibian constitution is praised globally for addressing safeguards to its natural heritage. Adequate laws are in place for the most part, but there is a major gap concerning threatened and endangered species with a complete lack of action plans for species that are the most imperilled or unique (e.g., Orange River yellowfishes, Otjikoto Tilapia, Cave Catfish, Caprivi Killifish, Kunene Sand Catlet, etc) or targeted in the subsistence and commercial fisheries (e.g., various *Clarias* catfishes and bream). Gear, seasonal and catch restrictions need to be reviewed. A priority should be the efficient enforcement of current laws to maintain water quality; curb overfishing and illegal harvest in conjunction with traditional authorities and local fisher communities; and ban the introduction of non-native species.

C. Citizen Engagement

Two complimentary components of citizen participation are helping to restore many rivers back to health:

1. Fisheries Reserves: The Inland Fisheries Resources Act makes provision for fish reserves (FR) to protect, preserve and/or rehabilitate the inland aquatic environments and their fisheries. The establishment of FRs is a pro-active step to limit the increased fishing pressure in northern Namibia. Benefits of fish protected areas, as emphasised in Chapter 8, are that local communities take ownership by assisting in the planning and co-managing of their own resources. These subsistence communities recognise and desire the need for sustainable fishery management practices.[7] Data gathering by community fish monitors augments MFMR's capabilities to better understand local recreational and subsistence/consumption fishing.

2. Recreational (sport) fishing: Recreational angling in Namibia through its catch and release policies is largely a non-consumptive activity that promotes game fish as renewable resources. Non-game species that contribute significantly to the subsistence fishery are usually not targeted except during artificial lure competitions or by individual anglers, but this pressure is minimal. Sport fishing is highly regulated through both legislation and internal angling club rules. Angling clubs, lodges and individual anglers are strongly supportive of FRs, *inter alia* by direct financial assistance or in-kind and social support, e.g., the Sikunga Fisheries Reserve on the Upper Zambezi River. All angling clubs encourage intensified MFMR monitoring of fishing activities; data gathered at fishing competitions targeting game fish is analogous to the use of community-based fish monitors at FRs to supplement data collection. Anglers also assist in enforcement if called upon, which is identical to community or district watches assisting the Namibia Police (NamPol) for investigating criminal activity and poaching of wildlife and stock. Sport fishing is also an important component of ecotourism in northern Namibia and provides countless primary and secondary economic spinoffs.

D. Transboundary Collaboration

Transboundary collaboration on inland water resources within the Southern Africa Development Community (SADC) region has multiple layers of issues with fisheries being a component. Fisheries working groups should strive toward harmonised fisheries legislation; standardised research, management and enforcement protocols; and the creation of mutually respected fish reserves between neighbouring countries. The Kavango–Zambezi (KAZA) Transfrontier Conservation Area Freshwater and Fisheries Working Group is an established regional model for other cross-border fisheries management and sustainable utilisation agreements, be they basin-wide or bilateral.

E. Aquaculture

Aquaculture was specifically addressed as a development priority in Namibia's Second National Development Plan (NDP-2) through to the current NDP-5 and in the Government's VISION 2030 document. The aim of government is to create an "enabling environment" for aquaculture development. This requires specific policy, legislative and institutional interventions. Even though there are numerous limitations to successful commercial aquaculture ventures in Namibia, such investments are encouraged by the private sector. The overall efficiency of plant conversion to edible chicken, swine and beef protein globally is estimated at 20, 10 and 4%, respectively,[8] while efficiency of commercial aquaculture ventures equals or exceeds rates reported for chickens. Inland fishes raised strictly for food security on controlled diets have even a higher conversion efficiency and nutrient yield as they are consumed (nearly) whole at earlier life stages. Thus, MFMR's stated objective to address food security through selective fish breeding programmes will complement private development schemes. The potential of using *iishana* for short-term intensive fish culture for the benefit of local communities requires further examination. It is emphasised that aquaculture ventures must be properly planned and executed since the sector is a primary conduit for the introduction of non-native species, organic pollution and other environmental impacts in spite of the various benefits that are derived.[9]

F. Water Quality and Biomonitoring

Any alteration of the natural state of waterbodies can influence the interaction of fish and other biota within the aquatic regime. An adequate database is required to understand ecosystem shifts over time, be they natural or anthropogenic in origin such as pollution, changes in hydrology, climate change, overfishing, etc. Biological monitoring protocols[10] are an extension of biodiversity surveys, which are aimed toward natural and cultural heritage programmes. Since fish occupy every trophic level of the freshwater ecosystem above primary producers and are assimilators of ecosystem perturbations, they are excellent bioindicators of system 'health'.[11] While only a few studies are available on the functioning of Namibia's inland waters, these serve as conceptual baselines upon which to build surveillance programmes and early warning systems. Indeed, one of the inherent positive outcomes of this book is to serve as a baseline for the future. What is additionally required is the planning of an operational protocol for the routine surveillance and long-term monitoring of Namibia's aquatic ecosystems to facilitate sound decision making for sustainable management.

G. Capacity Building

Formal training and research in fisheries are provided at the University of Namibia (UNAM), while the Namibia University of Science and Technology (NUST) offers a course in Natural Resources. Lacking for the most part is an operational curriculum[12] in the K-12 grade schools focused upon building early awareness that links the 'health' of waterways to natural resource sustainability to food security to human health and nutrition. In turn, students must learn that (1) food security and human nutrition are components of national policy and (2) the enforcement, i.e., the honouring, of laws and traditional cultural mores are management and conservation tools.

H. Partnerships

The Namibia Chamber of Environment (NCE) maintains an active list of Non-Government Organisations (NGOs) involved in environmental activities, including several consulting firms. The Namibia Nature Foundation (NNF) is the most notable member in terms of co-sponsoring the successful community-based management (CBM) schemes and fish reserves (FRs) with MFMR and UNAM. As noted above, recreational angling clubs, lodges, other businesses and individuals are invaluable partners as well. Not to be overlooked, the National Museum in Windhoek is the single most important repository in the Republic for voucher specimens of taxonomic and biomonitoring significance.

I. Information Dissemination

A fundamental component to all of the above is a dedicated communications network to both inform and bridge the gap between stakeholders. Newsletters, talks, radio broadcasts, television programmes and formal meetings are forms of exchange to consider. The overall objective is to improve dialogue, coordination and cooperation between stakeholders. Information dissemination can be augmented by using fisheries extension officers through cooperative arrangements between MFMR and UNAM.

III. References

[1]Lowe, N.M. 2021. The global challenge of hidden hunger: Perspectives from the field. Proc. Nutri. Soc. 80(3): 283-289.

[2]Hubert, W.A. and M.C. Quist (Eds.). 2010. Inland Fisheries Management in North America (3rd Edition). Am. Fish. Soc., Bethesda, MD: 736 pp.

[3]Nash, D.J., M.E. Meadows and V.L. Gulliver. 2006. Holocene environmental change in the Okavango Panhandle, northwest Botswana. Quaternary Sci. Rev. 25: 1302–1322.

[4]Hardin, G. 1968. The tragedy of the commons. Science 162: 1243–1248.

[5]Haller, T. and S. Merten. 2008. "We are Zambians—Don't Tell Us How to Fish!" Institutional change, power relations and conflicts in the Kafue Flats fisheries in Zambia. Hum. Ecol.: DOI 10.1007/s10745-008-9191-4.

[6]Hilborn, R. 2007. Managing fisheries is managing people: What has been learned? Fish & Fisheries 8 (4):285-296.

[7]Tiyeho, D.C. 2020. Assessment of fishers' catches on the Kavango River floodplains, Namibia. M.Sc. Thesis, Univ. Namibia, Windhoek: 82 pp.

[8]Boyd, C.E. 2018. About half of global fish and shrimp production is based on manufactured feeds. https://www.globalseafood.org/advocate.

[9]Black, K.D. (Ed.). 2001. Environmental impacts of aquaculture. Blackwell, London: 320pp.

[10]Vadas, R.L. Jr, and 26 coauthors.in press.Assemblage-based biomonitoring of aquatic ecosystem health via multimetric indices: A critical review and suggestions for improving their applicability. Wat. Biol. Secur.: DOI: 10.1016/j.watbs.2022.100054.

[11]Hocutt, C.H. and J.R. Stauffer, Jr. (Eds.). 1980. Biological monitoring of fish. Lexington Books, D.C.Heath Co., Lexington, MA.: 416 pp.

[12]Gawazah, L. 2021. The application, relevance, differences and similarities between the official and operational curricula of learning in the Namibian educational context. https:// www.researchgate.net/ publication/352780880: 5 pp.

GLOSSARY and ABBREVIATIONS

Adipose fin: Medial fleshy fin without rays located behind the rayed dorsal fin when present

Alien: Any species deliberately or inadvertently introduced to an area outside its natural range by man, i.e., exotic

Anal fin: Medial ventral fin in the midline behind the anus to stabilise fish and prevent rolling

Anterior: Near or towards the front

Archipelago: An expanse of water (river, lake or sea) with many scattered islands

Aufwuchs: German term coined ca. 1920 referring to algae, microbes, small invertebrates, etc. living on aquatic plant stems, root balls, rocks and sunken logs, i.e., periphyton

Avivore: Specialised predator of birds

Backwater: A part of a river with little or no current

Bar: Typically for fish, a few to several vertical bands of dark colour along the sides interspersed with bands of lighter pigmentation

Barbel: Fleshy filament growing from around the mouth of a fish used as an organ of touch and taste

Bashing: A fishing technique whereby the water surface is hit with paddles, etc. to drive fish into gillnets set near vegetation

Bearer: Reproductive strategy where the eggs and/or young are offered protection in a body cavity, usually the mouth (external bearer) or oviduct (internal bearer) of the female

Benthic: Living near or on the bottom

Benthopelagic: Living and feeding near the bottom as well as in midwater or near the surface

Binomial: Literally translated as "two names", the system of modern classification using both the generic and specific names to identify a particular species, e.g., *Homo sapiens*

Bioindicator: Two definitions are offered: (1) A living organism or community of organisms that can imply the 'health' of the ecosystem they are living in; (2) A species used in laboratory trials to determine their tolerance to environmental changes

Biological monitoring: The use of biological data, e.g., fish and/or benthic invertebrate communities, to assess the *health* of an aquatic ecosystem; laboratory studies may be conducted to determine the effect of a particular stress

Biparental: With reference to parental care, both parents participate

Bivalve: Aquatic mollusc with hinged shell

Body depth: The greatest straight vertical depth of a fish excluding the fins, normally just anterior to the dorsal fin

Bower: An excavated saucer-shaped depression on sandy substrate where mating occurs and eggs are fertilised, but does not provide protection nor incubate the eggs, unlike a nest; usually associated with mouthbrooders

Caudal fin: Medial posterior (tail) fin used by fish for propulsion

Caudal peduncle: Section of fish's body from the anal fin to the base of the caudal fin with scale counts taken around its circumference at its narrowest portion

Cenote lake: A sinkhole formed after the collapse of the limestone surface

Cf.: Latin, *conformis*, used in taxonomy meaning like or similar to

Classification: The study and ranking of organisms based upon their evolutionary relationships and similarity to other organisms

Common name: Colloquial or regional name for a fish, e.g., Nembwe

Compressed: Body is compacted from side to side

Conservancy: An area designated for the management and sustainable use of its resources by the local populace

Copepod: Small crustaceans, either free-floating zooplankton, benthic, free-swimming or parasitic on fishes

Countershading: Fishes being darker in colour dorsally, lighter ventrally, providing camouflage when viewed (e.g., by a predator) from above or below, respectively

Crepuscular: Active at dawn and dusk

Crustacean: Mainly aquatic arthropods including copepods, crabs, lobsters, shrimps and barnacles

Ctenoid: Fish scale with many tooth-like protrusions (*ctenii*) on the edges

Cycloid: Fish scale with smooth edges

Depauperate: Lacking in numbers of species

Daphnia: Small (0.2-5 mm) planktonic crustaceans from the order Cladocera commonly known as water fleas; prey of many fish species, especially juveniles

Depressed: Body is compacted from dorsal to ventral, i.e., flattened

Detritivore: An organism that predominantly consumes detritus

Detritus: Organic matter or waste that settles to the bottom of aquatic habitats and is decomposed by microorganisms

Diagnostic characters: Anatomical features that distinguish between taxonomic categories be it at the domain, kingdom, phylum, class, order, family, genus or species level, e.g., scale presence/absence and fin spine/ray counts in fish

Diatoms: Large group of microalgae found in aquatic environments and soils

Diploid: Having paired chromosomes, i.e., one from each parent

Disease: An abnormal condition affecting the normal functioning of an organism

Disjunct: Populations physically separated from one another, i.e., not a continuous distribution

Diurnal: Active during the day

Diversity: The number of species present

Domain: The highest taxonomic category in the Linnaean Hierarchy; most biologists agree that there are three domains (Bacteria, Archaea, and Eukarya) with fishes belonging in the Eukarya

Dorsal: Upper side of fish, ergo the dorsal view would be looking down from the top

Dorsal fin: Medial fin located dorsally on the fish's body providing stability

Ecosystem: The complex interactions between biological communities within their physical and chemical environment

Efundja: Large flood of the Cuvelai basin and associated *iishana* draining southwards from Angola towards Etosha Pan with migrating fish an important feature

Eggspots: Conspicuous colour markings on the anal fins of certain cichlids that look like eggs that are related to mating behaviour in cichlids

Elbow of capture: A sharp turn in the course of a river often at a 90° angle where a captured stream has been diverted into the captor, a process termed *stream capture*

Electric organ: Modified muscle or nerve tissue in 'electric' fish to create an electric field used for defence against predators, detecting prey, communication, navigation and courtship, found only in snoutfishes in Namibia

Electrogenic: Fishes able to produce electricity

Endangered: Species with such small numbers and restricted to precise habitat conditions they are at risk of extinction (from factors such as habitat alteration, over exploitation, introduced species, etc.). Many are listed by CITES

Endemic: Native species with restricted distribution, be it in reference to a single fontein, river basin or biogeographic province

Endorheic: Catchment with no outflow

Ephemeral: With reference to waterbodies, this being seasonally dry

Erowo: Traditional hook and line, Kavango region

Estuary: A partially enclosed coastal waterbody where there is a permanent or periodic mixing

Etymology: Word origin and history of its meaning

Eurytopic: An organism capable of living under a wide range of environmental conditions

Exotic: Any species deliberately or inadvertently introduced to an area outside its natural range by man, i.e., alien or non-native species

Eye diameter: The greatest distance across the cornea

Eyespots: Eye-like markings, *ocelli*, a form of mimicry

Falcate: Sickle shaped

Family: Taxonomic group of one or more genera or tribes sharing common features and characteristics

Fecundity: Total number of eggs that are produced, usually dependent upon age and size of the female

Fertility: Percentage of viable offspring vs. total number of eggs (fecundity)

Filamentous: Thin in diameter and threadlike

Fish guard/monitor: A person appointed by the conservancy under the advice of the community and traditional authority to assist the MFMR's Inland Fisheries Inspectors to patrol, monitor and manage local fishery reserves as a component of CBM

Fisheries reserves: Proclaimed protected areas with fishing restrictions

Fishers: People who catch fish

Fishery scientist: Persons employing field, laboratory or information technologies to study the biology, life history and population dynamics of fish; their interaction in the ecosystem; and their management

Fjord: A deep, long and narrow inlet bordered by steep slopes or cliffs that was once occupied by glaciers

Flood pulse concept: Flooding is the major "driver" of lowland or floodplain rivers with the ecological character of the associated biota reflected by the frequency, magnitude and rates of rise and fall of flooding

Floodplain: Low-lying area covered by terrestrial grasses and shrubs that is seasonally flooded when the river laterally overflows its banks

Flotsam: Debris floating in the water

Fontein: A regional term for freshwater spring

Fork length (FL): Fish length from the tip of the snout to the fork of the caudal fin

Formally described: Published in the scientific literature

Frenum: A restraining membrane between the snout and upper jaw, preventing the latter from being protrusible

Fry: Informal term for young fishes, which may include the embryo and juvenile stages

Fusiform: Body shape widest in the middle and tapering both forward and posteriorly

Gastropod: Molluscs in the Class Gastropoda that include whelks, snails and slugs

Genus (plural *genera*): In taxonomy, 'genus' is the first part of a scientific name, always capitalised and either underlined or italicised, e.g., *Synodontis*, with all genera sharing similar features comprising a family

Ghieliemientjies: Afrikaans term, a regional name for minnows, especially the barbs

Gills: Respiratory organs in fishes and other aquatic organisms for respiration that extract dissolved oxygen from the water and excrete carbon dioxide into the water

Gill cover: Opercle, or bony plate covered by skin over the gills for protection

Gill filaments: Structures of the red fleshy gills and comprised of finer lamellae to increase the surface area for exchanging gas, water and ions from the water into (e.g., oxygen) and out (e.g., carbon dioxide, ammonia) of the fish

Gill rakers: Cartilaginous or bony protuberances on the gill arches that are used in suspension feeding of microprey and protecting the gill filaments

Gonads: Reproductive organs that produce gametes, being testes in males and ovaries in females

Gondwana: An ancient supercontinent 600 MYA comprised of South America, Africa, Arabian Peninsula, Madagascar, India, Australia and Antarctica that rifted apart in stages

Gonopodium: The modified anal fin of male livebearers permitting the depositing of sperm into the vent of females (non-native Poeciliidae in Namibia), i.e., it is an intromittent organ

Gravid: Female fish carrying eggs or young

Guarder: Reproductive strategy where eggs are laid in a nest and guarded by either parent

Habitat: The place where a fish lives, e.g., mainstream, rocky rapids, backwater

Haplochromine: Tribe of cichlids (Haplochromini) that includes many closely related genera that are mostly endemic to eastern, southern and northern Africa

Head length: Distance from the anterior most part of the head to the posterior end of the bony margin of the operculum

Health: An ecological term referring to the well being of an individual, species, community, ecosystem or environment

Herbivore: Species that feed on plants

Herptile: For Africa, generic term for reptiles (crocodiles, turtles, tortoises, snakes, lizards) and amphibians (frogs, toads)

Hominin: Members of the taxonomic tribe Hominini that includes humans, chimpanzees and bonobos, but excludes gorillas

Humeral process: A bony outgrowth of the pectoral girdle, the size and shape of which assists in species identification, especially catfish

Hypogean: Subterranean

Hypural plate: A bony plate located just anterior to the caudal fin to which its rays are anchored

Ichthyology: Scientific study of fishes, including their systematics, anatomy, ecology, life history development, and distribution in time and space

Identification: The recognition of individuals based upon diagnostic characteristics that may be morphological, meristic, colour, behavioural, physiological or genetical features

Indigenous: A native species, but (as compared to an endemic), not necessarily confined to a particular fontein, river or biogeographic province

Inferior: Relative to mouth location, more ventral than subterminal

Insectivore: Feeding principally on insects

Introduced: Any species deliberately or inadvertently introduced to an area outside its natural range by man, i.e., exotic, alien, non-native

Invasive: Two contrasting definitions are offered: (1) A species deliberately or inadvertently introduced to an area outside its natural range that negatively impacts the new environment, either ecologically or economically. (2) A species that is naturally expanding its range into new territory, a pioneer

Invertebrate: Animals without a backbone, e.g., sponges, jellyfish, crustaceans, molluscs, annelids, insects, etc., comprising about 95% of all animal species

Juvenile: Any fish which is not yet sexually mature

K-selection: A reproductive strategy where K refers to the carrying capacity of the environment. Such organisms reach sexual maturity slowly, have low fecundity, are long lived and invest heavily in the success of their offspring. Their success is related to ecosystem stability as compared to *r*-strategists

ka: Thousands years ago in geological time

Kingdom: The second highest taxonomic classification rank of organisms with four kingdoms recognised in the domain Eukarya: Protista (protozoa), Animalia (animals), Plantae (plants) and Fungi (mushrooms, yeasts, moulds)

Knickpoint: A sharp change in the slope of a river or channel, such as a cataract, waterfall or lake

Labyrinth: A modified structure of the gill chamber analogous to a lung that permits respiration, i.e., an exchange of atmospheric oxygen into the circulatory system while carbon dioxide is exhaled

Lacustrine: Lake-like environment

Larva (plural *larvae*): Stage of development in fishes between an embryo and juvenile. They are often not fully formed and look different to juvenile and adults of the same species

Lateral: Side of the body

Lateral line: A sensory mechanism for fish to detect water current, pressure and movement; typically a *complete* line of pored scales on each side of the body from behind the opercle to the hypural plate, but may be *incomplete*, split into two or absent

Lek: An aggregation of males engaging in competitive displays, often associated with mating and spawning

Lentic: Lake-like environment

Linnaean hierarchy: A system for the orderly classification of plants and animals derived in 1758 by Carl Linnaeus, the Father of Taxonomy, modified and enhanced since that time

Littoral: Areas of lakes and rivers close to shore with fully- to partially-submerged aquatic vegetation

Lotic: River-like environment, i.e., running water

Lunate: Crescent shaped

Ma: Dated to millions of years ago before present, or mega annum, e.g., 'it dated to 4 Ma'
Mandibular: Related to the lower jaw
Matronym: A scientific name created to honour a female
Maxillary: Related to the upper jaw
Mayfly: Aquatic insects (order Ephemeroptera) that are prey for some fish species
Mekgacha: Colloquial name in the Kalahari region for extensive networks of dry 'fossil' river valleys or channels
Meristics: Actual counts of scales and fin rays/spines
Midge: Small two-winged flies found near water and floodplains that are prey for some fish species
Migration: Many fishes migrate daily, seasonally, annually or longer over varying distances mainly to reproduce and find appropriate food sources
Mimic: Copy or imitate in behaviour or colouration
Monogamous: A relationship where an individual has only one partner
Montreux Record: Ramsar wetlands of international importance where changes in the ecological character have occurred, are occurring or likely to occur as a consequence of man's activities
Morphology: Refers to the relationships between form (size, shape) and structure of living organisms
Morphometrics: Exact measurements of body parts and/or their ratios
Mouth: Used for feeding and respiration in fish as well as nest building and protection of offspring by various species
Mouthbrood: Care for eggs and fry in the mouth, as in some cichlids
Muho: Traditional fish spears, Kavango region
Mukolo (plural *mikolo*): Wooden canoe, Kwando-Linyanti River region
My: Duration in millions of years ago, e.g., 'it lasted for 4 My'
Naked: Without scales, notably in the catfishes
Nare(s): Nostrils
Nest: An area constructed for the laying and fertilisation of eggs, offering an area for protection or incubation of eggs and/or larvae
Ngumba: Traditional bow and arrow, Kavango region
Niche: Ecological position of a species within an ecosystem that describes its role and conditions necessary for its survival
Nocturnal: Active at night
Nomenclature: The practice of giving scientific names to particular taxonomic entities
Non-guarder: Reproductive strategy where eggs are released into the water column or nest, or hidden, but no protection is offered by the parents
Nostrils: External openings of the nasal cavity; nares
Nuptial: Related to breeding in fishes, usually referring to (a) colouration where rapid short-term colour changes are associated with social signalling in courtship and mating or (b) physical changes such as the development of tubercles for stimulation of females and nest defence
Ocelli: Eyespots
Ofima: Sandy wooded ridges between iishana in the Cuvelai basin used for settlement and farming
Omnivore: Species that feed on both plants and animals
Omuramba (plural: ***omiramba***): a Herero term for ancient grassed riverbeds found in northeastern

Namibia which channel runoff and occasionally provide standing pools of water; Mekgacha

Opercle: Gill cover

Operculum: Main bone in the opercle

Orbit: Bone eye socket

Origin of fin: Anterior base of a fin

Oshana (plural ***iishana***)***:*** Shallow seasonally-inundated grassed channels in the Cuvelai-Etosha basin carrying water from local rains or floods

Otolith(s): A hard calcium carbonate structure, commonly known as "ear stone", used for the detection and processing of sound in fish. May be used in aging studies in certain species if the calcium structure is laid down annually

Ovary: Female reproductive gland

Oviparous: Egg laying

Ovoviviparous: Producing young from eggs hatched in the female's body, but without nourishment being received from the parent, e.g., mouthbrooders

Oxbow lagoon: Typically, a bow-shaped backwater cut off from the main channel by the river changing its course

Palaeo: Old or ancient

Pan: A natural shallow endorheic depression formed by the seasonal evaporation of rainwater leaving a bed of salt and other minerals

Papillose: A skin surface in fish with raised papillae or pimples most often found on taste buds, barbels and urogenitalia of certain species

Papyrus: A type of wetland sedge

Patronym: A scientific name created to honour a male

Pectoral fin(s): Anterior fins on each side of the body associated with the pectoral (shoulder) girdle that generally control directional movement (side-to-side and up and down), but also used for jumping, gliding and crawling in some species

Pectoral spine: Spine associated with the pectoral fin

Pelagic: Area in the water column that is neither close to the bottom nor near the surface, so midwater

Pelvic fin(s): Posterior fins on each side of the body associated with the pelvic girdle (hip) to help balance the fish during locomotion, prevent it from rolling and braking. They are modified in some species for (a) walking across the bottom and (b) serving as an intromittent organ for reproduction in sharks, skates and rays. They may be absent, e.g., Mastacembelidae

Perennial: With reference to rivers, flowing year round vs. seasonal flow regimes of ephemeral rivers

Periglacial: An area adjacent to a glacier or landform repeatedly exposed to freezing and thawing

Periphyton: Community of organisms attached to submerged substrates, i.e., aufwuchs

Pharyngeal: Section of the mouth and nasal cavity that is modified in fishes as the gill arches

Phylogenetics: The study of evolutionary relationships within and between taxonomic ranks

Phylum (plural ***phyla***)**:** The highest taxonomic category beneath kingdom in the Linnean heirarchy

Phytoplankton: Microscopic plants that drift or float freely in the water column, carry out photosynthesis and are the basis of primary production in aquatic ecosystems

Pioneer: The first inhabitants to successfully colonise, live and survive in new or disturbed environments

Piscivore: Feeds on fishes

Plankton: Life stages of plants (*phytoplankton*). protozoa, jellyfish, crustaceans (*zooplankton*) and fish (*ichthyoplankton*) that generally drift with water and wind currents. They are at the base of the aquatic food chain as primary producers and consumers

Playa: Pan, a dry flat area at the bottom of a desert basin, but sometimes temporarily covered with water

Posterior: Towards the rear

Postorbital length: Distance from the posterior-most portion of the orbit to the end of the bony margin of the operculum

Potamodromous: A fish that migrates and completes its life cycle entirely in freshwater

Predator: Organisms that prey on animals

Prey: Animals that are consumed by predators

Proto: Meaning 'first' or 'dawn'

Protrusible: Part of the body that can be extended, e.g., jaws in some fishes

r*-selection:* A reproductive strategy where *r* refers to reproduction. Such organisms have a short life span, reach sexual maturity rapidly, have high fecundity and invest minimally in the success of their offspring. They are more apt to thrive in unstable environments as compared to K-strategists

Radiotelemetry: The use of radio-transmitters and receivers to track animals

Ramsar (site): A wetland designated to be of international importance under the Ramsar Convention, an intergovernmental treaty established in 1971 in Ramsar, Iran for the conservation and sustainable use of wetlands and waterfowl habitats

Ray: Skeletal supports that are found in the fins of most fishes

Redspot disease: Caused by water mould *Aphanomyces invadans* and infecting many fresh- and brackish-water fishes; also known as epizootic ulcerative syndrome (EUS)

Refuge (plural ***refugia***): An area where an organism can survive or escape

Relict: Species/populations surviving in isolation from a former, more widespread distribution

Reticulated: A net-like pattern or arrangement

Rheophilic: Preferring flowing water habitats

Richness: Refers to numbers of species present, i.e., the higher the number, the more 'rich'

Riparian: Land zones adjacent to river and lake margins

Scientific name: Two-part name (binomial nomenclature) of every recognised species usually derived from Latin or Greek and consisting of genus and species epithets. It is always italicised or underlined, e.g., *Clarias gariepinus* or Clarias gariepinus

Serial spawner: Fish species that spawns several times within a given time frame, usually in response to environmental cues such as rainfall

Serrated: Saw-like, usually referring to the saw-toothed edge of a fin spine or ray

Sexual dimorphism: Males and females of the same species exhibit different physical features

Sexual maturity: Capability to reproduce, usually considered adulthood

Shikuku: Traditional fish funnels, Kavango region

Sididi: Traditional push baskets, Kavango region

Sintunga: Traditional fish corral traps, Kavango region

Snout: From the front of the orbit to the forward-most margin of the upper jaw; snout length is the distance between these two points

Soft ray: Segmented skeletal structures covered with a translucent membrane giving fin support

sp. (plural spp.): Abbreviation for the word 'species'

Species: Historically considered as a group of closely related organisms with a common history that are/were reproductively isolated from other such species. However, it is recognised that hybridisation may occur when closely related species are brought into proximity with one another or under the conditions of ecosystem stress (See Chapter 2 for further discussion of 'species')

Spine: Smooth or serrated, unbranched, unsegmented skeletal structures giving fin support

Standard length (SL): Length of fish measured from the tip of the snout to the point just before the tail at which a crease is formed when bent forward, location of the hypural plate

Stratification: The occurrence of two distinct thermal or chemical layers in a body of water

Stream capture: Stream capture, or stream piracy, is a process where one river (captive) is diverted from its original basin into another (captor), the location often distinguished by an *elbow of capture.* The captor typically has a shorter distance to the sea and a stronger gradient, advantages for cutting into the common divide

Stripe: Horizontal band of colour (compared to a vertical bar)

Subsistence: Supporting oneself at a minimal level usually based on agriculture, hunting and gathering

Subspecies: Taxonomic category below species

Substrate: Surface that an organism lives and grows upon or obtains its food from

Subterminal: Below the end of the snout, not ventral or inferior

Sympatric: Two or more (closely-related) species living in the same geographic area

Systematics: The study of the diversity of organisms and includes taxonomy, classification, nomenclature and identification

Tail fin: Caudal fin used for propulsion

Tambi: Traditional scoop basket, Kavango region

Taxonomy: The theory and practice of examining the variation within species, genera, families and orders and classifying organisms based on shared characteristics

Terminal: At the posterior end of the body

Terrestrial: From or on the land

Territorial: Defending a nest or other space

Testis (plural ***testes***): Male reproductive organ producing gametes or sperm (milt) in fish

Tetraploid: Having two pairs of chromosomes from each parent, i.e., a total of four

Tilapiine: Tribe within the Cichlidae family

Tillite: Sedimentary rocks composed especially of glacial till or fine 'rock flour' ground by the mechanical crushing of rocks by glaciers

Toponym: A scientific name created to recognise a place or geographic location

Total length (TL): Length measured in a straight line from the tip of the snout to the tip of the tail

Traditional authority: Consisting of the Chief or Head of the traditional community and senior traditional councillors appointed or elected

Translocated: Species that have been deliberately transferred by man from one drainage to another outside the natural range

Tribe: A taxonomic rank between family and genera with all representatives sharing common characteristics

Trophic level: The level within the food chain that organisms occupy: primary producer, primary consumer, secondary consumer, tertiary consumer or apex predator

Truncate: Ending suddenly with a straight edge or square cut

Tubercle: Small rounded bump on body, fins or head of fish, usually males during breeding season

Turbidity: The presence of suspended particles in water decreasing transparency

Turnover: A wind- or current-driven process that causes mixing of the whole water column and breakdown of chemically- or thermally-stratified waters

Type: The individual specimen (*holotype*) or several representatives (*syntypes*) upon which the original species description is based upon, collected at the *type locality*

Unbranched ray: Simple segmented fin ray that can be flexible

Vent: Combined external opening of the digestive, urinary and reproductive tracts in fish

Ventral: Underside of a fish

Vertebrate: Animals with backbones (vertebrae), inclusive of fish, amphibians, reptiles (including birds) and mammals

Viviparous: Giving birth to live young that have received nourishment from the parent via a structure analogous to a placenta

Vlei: Shallow lake or waterbody that is usually seasonal

Vomerine: Referring to the predatory teeth for gripping located on the vomer, a thin flat bone on the roof of the mouth between the nostrils

Wetland: Ecosystem where water is at or near the surface for varying periods during the year, e.g., marshes, swamps and bogs

Yolk sac: Membranous sac attached to an embryo which provides nourishment (egg yolk) to developing fish

YOY: Young-of-the-year, all of a species' offspring younger than one year of age, often used interchangeably with juvenile(s)

Zoogeography: Branch of zoology concerning animal distribution in time and space

Zooplankton: Plankton comprising small animals and immature stages of larger animals, representing the primary consumers in aquatic food chains and the basis of nutrition for higher consumers

Index of Scientific and Common Names

SUBJECT INDEX

ADDENDUM

Reference is made in Chapter 3 (Page 54) of evidence in support of an ancient transfer of fishes across the Great Escarpment from an interior drainage into the Kuiseb-Gaub River, most prominent being Chubbyhead Barb (Page 96) for which there are several documented collections from the Kuiseb basin. Photographs below are of single specimens of River Sardine and an unidentified cyprinid originally housed at the Gobabeb Training & Research Centre. As referenced in the text, provenance is lacking for these individuals; however, the Centre is located adjacent to the Kuiseb River and it was confirmed that it would be unlikely that the specimens were derived from drainages elsewhere. These photographs are presented to ensure their record is not lost. Their stocks may have had access to the same capture as Chubbyhead Barb or, alternately, source of introduction.

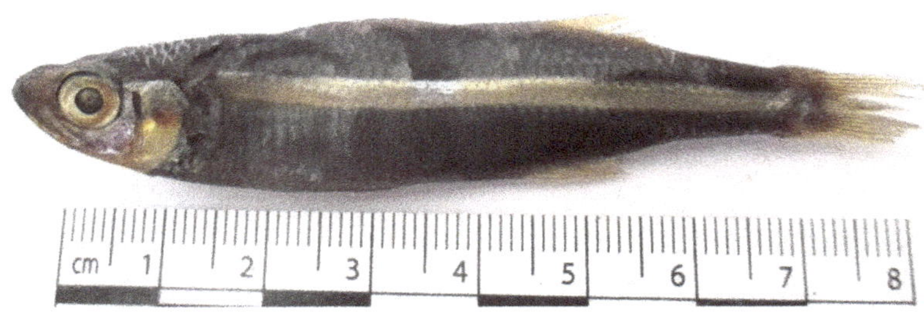

E. Marais

River Sardine

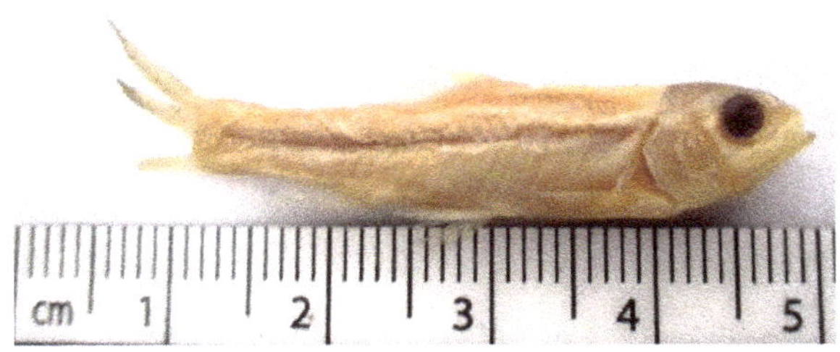

E. Marais

Unknown cyprinid

Two important collection activities were conducted in the Kunene River drainage during the publication of this work, including the adjacent Baynes Mountains. This is perhaps the most inaccessible and least hospitable area of Namibia. Ben van Zyl led expeditions in July 2022 and again in June 2023, the latter as the publisher goes to press. Van Zyl was accompanied by Jay R. Stauffer, Jr. and Burger Oelofsen (amongst many Oelofsen family members!). They collected a cryptic specimen in July 2022 presumed to be of the family Kneriidae, especially based upon a high lateral line scale count of 95.

J. Stauffer

The second survey in June 2023 confirmed the presence of several age groups of this species. Additionally, it is thought a sibling was collected in a nearby tributary. Their taxonomic status is under review.

The second survey also yielded another kneriid, tentatively assigned as *Kneria polli* based on the presence of a prominent spot at the base of the pelvic fins and the presence of the species in the upper Cunene River, Angola; however, identification requires confirmation. A photograph is compared to a line drawing from Prof. Max Poll's (1967) *Contribution à la faune ichthyologique de l'Angola*.

B. van Zyl

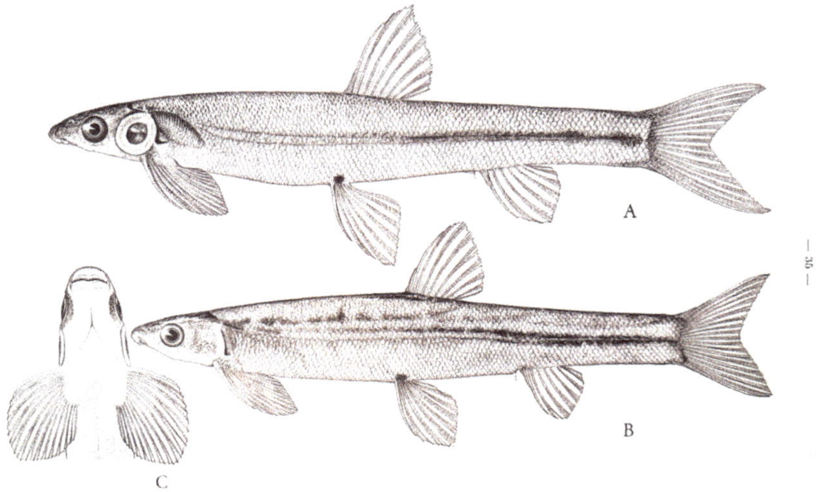

Fig. 2 — *Kneria polli* TREWAVAS, région de Chibia, sous-afflt. riv. Cunene; A: ♂, 71 mm, × 3,1; B: ♀, 65 mm, × 3,05; C: ♂, 71 mm, dessous de la tête, × 3,86.

M. Poll

These latter findings if verified will increase the number of fishes known in Namibia and endemic to the Kunene River basin.

CH Hocutt
July 2023

www.ingramcontent.com/pod-product-compliance
Lightning Source LLC
Chambersburg PA
CBHW041747220926
51847CB00013B/133
* 9 7 8 1 9 3 8 8 1 4 3 8 9 *